INTRODUCTION TO INTEGRAL CALCULUS

INTRODUCTION TO INTEGRAL CALCULUS

Systematic Studies with Engineering Applications for Beginners

Ulrich L. Rohde
Prof. Dr.-Ing. Dr. h. c. mult.
BTU Cottbus, Germany
Synergy Microwave Corporation Peterson, NJ, USA

G. C. Jain
(Retd. Scientist) Defense Research and Development Organization
Maharashtra, India

Ajay K. Poddar
Chief Scientist, Synergy Microwave Corporation,
Peterson, NJ, USA

A. K. Ghosh
Professor, Department of Aerospace Engineering
Indian Institute of Technology – Kanpur
Kanpur, India

A JOHN WILEY & SONS, INC., PUBLICATION

Published by John Wiley & Sons, Inc., Hoboken, New Jersey
Published simultaneously in Canada

Library of Congress Cataloging-in-Publication Data:

Introduction to integral Calculus : systematic studies with engineering applications for beginners /
Ulrich L. Rohde.
    p. cm.
  Includes bibliographical references and index.
ISBN 978-1-118-11776-7 (cloth)
1. Calculus, Integral–Textbooks. I. Rohde, Ulrich L.
QA308.I58 2012
515'.43–dc23

2011018422

CONTENTS

FOREWORD

"What is Calculus?" is a classic deep question. Calculus is the most powerful branch of mathematics, which revolves around calculations involving varying quantities. It provides a system of rules to calculate quantities which cannot be calculated by applying any other branch of mathematics. Schools or colleges find it difficult to motivate students to learn this subject, while those who do take the course find it very mechanical. Many a times, it has been observed that students incorrectly solve real-life problems by applying Calculus. They may not be capable to understand or admit their shortcomings in terms of basic understanding of fundamental concepts! The study of Calculus is one of the most powerful intellectual achievements of the human brain. One important goal of this manuscript is to give beginner-level students an appreciation of the beauty of Calculus. Whether taught in a traditional lecture format or in the lab with individual or group learning, Calculus needs focusing on numerical and graphical experimentation. This means that the ideas and techniques have to be presented clearly and accurately in an articulated manner.

The ideas related with the development of Calculus appear throughout mathematical history, spanning over more than 2000 years. However, the credit of its invention goes to the mathematicians of the seventeenth century (in particular, to Newton and Leibniz) and continues up to the nineteenth century, when French mathematician Augustin-Louis Cauchy (1789–1857) gave the definition of the limit, a concept which removed doubts about the soundness of Calculus, and made it free from all confusion. The history of controversy about Calculus is most illuminating as to the growth of mathematics. The soundness of Calculus was doubted by the greatest mathematicians of the eighteenth century, yet, it was not only applied freely but great developments like differential equations, differential geometry, and so on were achieved. Calculus, which is the outcome of an intellectual struggle for such a long period of time, has proved to be the most beautiful intellectual achievement of the human mind.

There are certain problems in mathematics, mechanics, physics, and many other branches of science, *which cannot be solved by ordinary methods of geometry or algebra alone*. To solve these problems, we have to use a new branch of mathematics, known as *Calculus*. It uses not only the ideas and methods from arithmetic, geometry, algebra, coordinate geometry, trigonometry, and so on, but also *the notion of limit*, which is a *new idea* which lies at the *foundation of Calculus*. Using this notion as a tool, *the derivative* of a function (which is a variable quantity) is defined as the limit of a particular kind. In general, *Differential Calculus* provides a method for calculating "*the rate of change*" of the value of the variable quantity. On the other hand, *Integral Calculus* provides methods for calculating the total effect of such changes, under the given conditions. The phrase *rate of change* mentioned above stands for the actual rate of change of a variable, and *not its average rate of change*. The phrase "rate of change" might look like a foreign language to beginners, but concepts like *rate of change, stationary point*, and *root*, and so on, have precise mathematical meaning, agreed-upon all over the world. Understanding such words helps a lot in understanding the mathematics they convey. At this stage, it must also be made clear that whereas algebra, geometry, and trigonometry are the tools which are used in the study of Calculus, they should not be confused with the subject of Calculus.

This manuscript is the result of joint efforts by Prof. Ulrich L. Rohde, Mr. G. C. Jain, Dr. Ajay K. Poddar, and myself. All of us are aware of the practical difficulties of the students face while learning Calculus. I am of the opinion that with the availability of these notes, students should be able to learn the subject easily and enjoy its beauty and power. In fact, for want of such simple and systematic work, most students are learning the subject as a set of rules and formulas, which is really unfortunate. I wish to discourage this trend.

Professor Ulrich L. Rohde, Faculty of Mechanical, Electrical and Industrial Engineering (RF and Microwave Circuit Design & Techniques) Brandenburg University of Technology, Cottbus, Germany has optimized this book by expanding it, adding useful applications, and adapting it for today's needs. Parts of the mathematical approach from the Rohde, Poddar, and Böeck textbook on wireless oscillators (*The Design of Modern Microwave Oscillators for Wireless Applications: Theory and Optimization*, John Wiley & Sons, ISBN 0-471-72342-8, 2005) were used as they combine differentiation and integration to calculate the damped and starting oscillation condition using simple differential equations. This is a good transition for more challenging tasks for scientific studies with engineering applications for beginners who find difficulties in understanding the problem-solving power of Calculus.

Mr. Jain is not a teacher by profession, but his curiosity to go to the roots of the subject to prepare the so-called *concept-oriented notes for systematic studies in Calculus* is his contribution toward creating interest among students for learning mathematics in general, and Calculus in particular. This book started with these concept-oriented notes prepared for teaching students to face real-life engineering problems. Most of the material pertaining to this manuscript on calculus was prepared by Mr. G. C. Jain in the process of teaching his kids and helping other students who needed help in learning the subject. Later on, his friends (including me) realized the beauty of his compilation and we wanted to see his useful work published.

I am also aware that Mr. Jain got his notes examined from some professors at the Department of Mathematics, Pune University, India. I know Mr. Jain right from his scientific career at Armament Research and Development Establishment (ARDE) at Pashan, Pune, India, where I was a Senior Scientist (1982–1998) and headed the Aerodynamic Group ARDE, Pune in DRDO (Defense Research and Development Organization), India. Coincidently, Dr. Ajay K. Poddar, Chief Scientist at Synergy Microwave Corp., NJ 07504, USA was also a Senior Scientist (1990–2001) in a very responsible position in the Fuze Division of ARDE and was aware of the aptitude of Mr. Jain.

Dr. Ajay K. Poddar has been the main driving force towards the realization of the conceptualized notes prepared by Mr. Jain in manuscript form and his sincere efforts made timely publication possible. Dr. Poddar has made tireless effort by extending all possible help to ensure that Mr. Jain's notes are published for the benefit of the students. His contributions include (but are not limited to) valuable inputs and suggestions throughout the preparation of this manuscript for its improvement, as well as many relevant literature acquisitions. I am sure, as a leading scientist, Dr. Poddar will have realized how important it is for the younger generation to avoid shortcomings in terms of basic understanding of the fundamental concepts of Calculus.

I have had a long time association with Mr. Jain and Dr. Poddar at ARDE, Pune. My objective has been to proofread the manuscript and highlight its salient features. However, only a personal examination of the book will convey to the reader the broad scope of its coverage and its contribution in addressing the proper way of learning Calculus. I hope this book will prove to be very useful to the students of Junior Colleges and to those in higher classes (of science and engineering streams) who might need it to get rid of confusions, if any.

My special thanks goes to Dr. Poddar, who is not only a gifted scientist but has also been a mentor. It was his suggestion to publish the manuscript in two parts (Part I: Introduction to Differential Calculus: Systematic Studies with Engineering Applications for Beginners and Part II: Introduction to Integral Calculus: Systematic Studies with Engineering Applications for Beginners) so that beginners could digest the concepts of Differential and Integral Calculus without confusion and misunderstanding. It is the purpose of this book to provide a clear understanding of the concepts needed by beginners and engineers who are interested in the application of Calculus of their field of study. This book has been designed as a supplement to all current standard textbooks on Calculus and each chapter begins with a clear statement of pertinent definitions, principles, and theorems together with illustrative and other descriptive material. Considerably more material has been included here than can be covered in most high schools and undergraduate study courses. This has been done to make the book more flexible; to provide concept-oriented notes and stimulate interest in the relevant topics. I believe that students learn best when procedural techniques are laid out as clearly and simply as possible. Consistent with the reader's needs and for completeness, there are a large number of examples for self-practice.

The authors are to be commended for their efforts in this endeavor and I am sure that both Part I and Part II will be an asset to the beginner's handbook on the bookshelf. I hope that after reading this book, the students will begin to share the enthusiasm of the authors in understanding and applying the principles of Calculus and its usefulness. With all these changes, the authors have not compromised our belief that the fundamental goal of Calculus is to help prepare beginners enter the world of mathematics, science, and engineering.

Finally, I would like to thank Susanne Steitz-Filler, Editor (Mathematics and Statistics) at John Wiley & Sons, Inc., Danielle Lacourciere, Senior Production Editor at John Wiley & Sons, Inc., and Sanchari S. at Thomosn Digital for her patience and splendid cooperation throughout the journey of this publication.

AJOY KANTI GHOSH
PROFESSOR & FACULTY INCHARGE (FLIGHT LABORATORY)
DEPARTMENT OF AEROSPACE ENGINEERING
IIT KANPUR, INDIA

PREFACE

In general, there is a perception that Calculus is an extremely difficult subject, probably because the required number of good teachers and good books are not available. We know that books cannot replace teachers, but we are of the opinion that, good books can definitely reduce dependence on teachers, and students can gain more confidence by learning most of the concepts on their own. In the process of helping students to learn Calculus, we have gone through many books on the subject, and realized that whereas a large number of good books are available at the graduate level, there is hardly any book available for introducing the subject to beginners. The reason for such a situation can be easily understood by anyone who knows the subject of Calculus and hence the practical difficulties associated with the process of learning the subject. In the market hundreds of books are available on Calculus. All these books contain a large number of important solved problems. Besides, the rules for solving the problems and the list of necessary formulae are given in the books, without discussing anything about the basic concepts involved. Of course, such books are useful for passing the examination(s), but Calculus is hardly learnt from these books. Initially, the coauthors had compiled *concept-oriented notes for systematic studies in differential and integral Calculus,* intended for beginners. These notes were used by students, in school- and undergraduate-level courses. The response and the appreciation experienced from the students and their parents encouraged us to make these notes available to the beginners. It is due to the efforts of our friends and well-wishers that our dream has now materialized in the form of two independent books: Part I for Differential Calculus and Part II for Integral Calculus. Of course there are some world class authors who have written useful books on the subject at introductory level, presuming that the reader has the necessary knowledge of prerequisites. Some such books are *What is Calculus About?* (By Professor W.W. Sawyer), *Teach Yourself Calculus* (By P. Abbott, B.A), *Calculus Made Easy* (By S.P. Thomson), and *Calculus Explained* (By W.J. Reichmann). Any person with some knowledge of Calculus will definitely appreciate the contents and the approach of the authors. However, a reader will be easily convinced that most of the beginners may not be able to get (from these books) the desired benefit, for various reasons. From this point of view, both parts (Part I and Part II) of our book would prove to be unique since it provides a comprehensive material on Calculus, for the beginners. First six chapters of Part I would help the beginner to come up to the level, so that one can easily learn *the concept of limit,* which is in the foundation of calculus. The purpose of these works is to provide *the basic (but solid) foundation of Calculus to beginners.* The books aim to show them the *enjoyment in the beauty and power of Calculus and develop the ability to select proper material needed for their studies in any technical and scientific field, involving Calculus.*

One reason for such a high dropout rate is that at beginner levels, Calculus is so poorly taught. Classes tend to be so boring that students sometimes fall asleep. Calculus textbooks get fatter and fatter every year, with more multicolor overlays, computer graphics, and photographs of eminent mathematicians (starting with Newton and Leibniz), yet they never seem easier to comprehend. We look through them in vain for simple, clear exposition, and for problems that will hook a student's interest. Recent years have seen a great hue and cry in mathematical circles over ways to improve teaching Calculus to beginner and high-school students. Endless conferences have been held, many funded by the federal government, dozens of experimental programs are here and there. Some leaders of reform argue that a traditional textbook gets weightier but lacks the step-by-step approach to generate sufficient interest to learn Calculus in beginner, high school, and undergraduate students. Students see no reason why they should master tenuous ways of differentiating and integrating by hand when a calculator or computer will do the job. Leaders of Calculus reform are not suggesting that calculators and computers should no longer be used; what they observe is that without basic understanding about the subject, solving differentiation and integration problems will be a futile exercise. Although suggestions are plentiful for ways to improve Calculus understanding among students and professionals, a general consensus is yet to emerge.

The word "Calculus" is taken from Latin and it simply means a "stone" or "pebble", which was employed by the Romans to assist *the process of counting*. By extending the meaning of the word "*Calculus*", it is now applied to wider fields (of calculation) which involve processes other than mere counting. In the context of this book (with the discussion to follow), the word "*Calculus*" is an abbreviation for *Infinitesimal Calculus* or to one of its two separate but complimentary branches—*Differential Calculus* and *Integral Calculus*. It is natural that the above terminology may not convey anything useful to the beginner(s) until they are acquainted with the processes of *differentiation* and *integration*. This book is a true textbook with examples, it should find a good place in the market and shall compare favorably to those with more complicated approaches.

The author's aim throughout has been to provide a tour of Calculus for a beginner as well as strong fundamental basics to undergraduate students on the basis of the following questions, which frequently came to our minds, and for which we wanted satisfactory and correct answers.

(i) *What is Calculus?*

(ii) *What does it calculate?*

(iii) *Why do teachers of physics and mathematics frequently advise us to learn Calculus seriously?*

(iv) *How is Calculus more important and more useful than algebra and trigonometry or any other branch of mathematics?*

(v) *Why is Calculus more difficult to absorb than algebra or trigonometry?*

(vi) *Are there any problems faced in our day-to-day life that can be solved more easily by Calculus than by arithmetic or algebra?*

(vii) *Are there any problems which cannot be solved without Calculus?*

(viii) *Why study Calculus at all?*

(ix) *Is Calculus different from other branches of mathematics?*

(x) *What type(s) of problems are handled by Calculus?*

At this stage, we can answer these questions only partly. However, as we proceed, the associated discussions will make the answers clear and complete. To answer one or all of the above questions, it was necessary to know: *How does the subject of Calculus begin?*; *How can we learn Calculus?* and *What can Calculus do for us?* The answers to these questions are hinted at in the books: *What is Calculus about?* and *Mathematician's Delight*, both by W.W. Sawyer. However, it will depend on the curiosity and the interest of the reader to study, understand, and absorb the subject. The author use *very simple and nontechnical language to convey the ideas involved*. However, if the reader is interested to learn the operations of Calculus faster, then he may feel disappointed. This is so, because the nature of Calculus and the methods of learning it are very different from those applicable in arithmetic or algebra. Besides, one must have a real interest to learn the subject, patience to read many books, and obtain proper guidance from teachers or the right books.

Calculus is a higher branch of mathematics, which enters into the process of calculating changing quantities (and certain properties), in the field of mathematics and various branches of science, including social science. It is called *Mathematics of Change*. We cannot begin to answer any question related with change unless we know: *What is that change and how it changes?* This statement takes us closer to the concept of function $y = f(x)$, wherein "y" is related to "x" through a rule "f". We say that "y" is a function of x, by which we mean that "y" depends on "x". (We say that "y" is a *dependent variable*, depending on the value of x, an *independent variable*.) From this statement, it is clear that as the value of "x" changes, there results a corresponding change in the value of "y", depending on the *nature of the function "f"* or *the formula defining "f"*.

The *immense practical power of Calculus is due to its ability to describe and predict the behavior of the changing quantities "y" and "x"*. In case of linear functions [which are of the form $y = mx + b$], an amount of change in the value of "x" causes a proportionate change in the value of "y". However, in the case of other functions (like $y = x^2 - 5$, $y = x^3$, $y = x^4 - x^3 + 3$, $y = \sin x$, $y = 3e^x + x$, etc.) which are not linear, *no such proportionality exists*. Our interest lies in studying the behavior of the dependent variable "y"$[=f(x)]$ with respect to the change in (the value of) the independent variable "x". In other words, we wish to find *the rate at which "y" changes with respect to "x"*.

We know that *every rate* is the *ratio of change that* may occur in quantities which are related to one another through a rule. It is easy to compute *the average rate at which the value of y changes when x is changed from x_1 to x_2*. It can be easily checked that (*for the nonlinear functions*) these average rate(s) are different *between different values of x*. [Thus, if $|x_2 - x_1| = |x_3 - x_2| = |x_4 - x_3| = \ldots\ldots$, (for all $x_1, x_2, x_3, x_4, \ldots$) then we have $f(x_2) - f(x_1) \neq f(x_3) - f(x_2) \neq f(x_4) - f(x_3) \neq \ldots\ldots$]. Thus, we get that the rate of change of y is different *in between different values of x.*

Our interest lies in computing *the rate of change* of "y" at every value of "x". *It is* known as *the instantaneous rate of change of "y" with respect to "x"*, and we call it the "*rate function*" *of "y" with respect to "x"*. It is also called the *derived function* of "y" with respect to "x" and denoted by the symbol $y'[=f'(x)]$. The derived function $f'(x)$ is also called the derivative of $y[=f(x)]$ with respect to x. The equation $y' = f'(x)$ tells that the *derived function $f'(x)$ is also a function* of x, derived (or obtained) from the original function $y = f(x)$. There is another (useful) symbol for the *derived function*, denoted by dy/dx. This symbol *appears like a ratio, but it must be treated as a single unit*, as we will learn later. The equation $y' = f'(x)$ gives us the *instantaneous rate of change* of y with respect to x, for every value of "x", for which $f'(x)$ is defined.

To define the *derivative formally* and to *compute it symbolically* is the subject of *Differential Calculus*. In the process of defining the derivative, various subtleties and puzzles will inevitably arise. Nevertheless, *it will not be difficult to grasp the concept (of derivatives) with our systematic approach*. The relationship between $f(x)$ and $f'(x)$ is the *main theme*. We will study what it means for $f'(x)$ to be "*the rate function*" of $f(x)$, and what each function says about the other. It is important to understand clearly *the meaning of the instantaneous rate of change of $f(x)$ with respect to x*. These matters are systematically discussed in this book. Note that we have *answered the first two questions* and now proceed to answer the *third one*.

There are certain problems in mathematics and other branches of science, which cannot be solved by ordinary methods known to us in arithmetic, geometry, and algebra alone. In Calculus, we can study the properties of a function without drawing its graph. However, it is important to be aware of the underlying presence of the curve of the given function. Recall that this is due to the introduction of coordinate geometry by Decartes and Fermat. Now, consider the curve defined by the function $y = x^3 - x^2 - x$. We know that, the slope of this curve changes from point to point. If it is desired to find its slope at $x = 2$, then Calculus alone can help us give the answer, which is 7. No other branch of mathematics would be useful.

Calculus uses not only the ideas and methods from arithmetic, geometry, algebra, coordinate geometry, trigonometry, and so on but also the *notion of limit*, which is a *new idea* that lies at the foundation of Calculus. Using the *notion of limit as a tool, the derivative of a function is defined as the limit of a particular kind.* (It will be seen later that the derivative of a function is *generally* a new function.) Thus, *Calculus provides a system of rules for calculating changing quantities which cannot be calculated otherwise.* Here it may be mentioned that the concept of limit is equally important and applicable in Integral Calculus, which will be clear when we study the concept of the definite integral in Chapter 5 of Part II. Calculus is the most beautiful and powerful achievement of the human brain. It has been developed over a period of more than 2000 years. *The idea of derivative of a function is among the most important concepts in all of mathematics and it alone distinguishes Calculus from the other branches of mathematics.*

The derivative and *an integral* have found many diverse uses. The list is very long and can be seen in any book on the subject. *Differential calculus* is a subject which can be applied to anything which *moves*, or *changes* or *has a shape*. It is useful for the study of machinery of all kinds - for electric lighting and wireless, optics and thermodynamics. It also helps us to answer questions about the *greatest* and *smallest values* a function can take. Professor W.W. Sawyer, in his famous book *Mathematician's Delight*, writes: *Once the basic ideas of differential calculus have been grasped, a whole world of problems can be tackled without great difficulty. It is a subject well worth learning.*

On the other hand, *integral calculus* considers the problem of *determining a function from the information about its rate of change.* Given a formula for the velocity of a body, as a function of time, we can use integral calculus to produce a formula that tells us how far the body has traveled from its starting point, at any instant. It provides methods for the calculation of quantities such as areas and volumes of curvilinear shapes. It is also *useful for the measurement of dimensions of mathematical curves.*

The concepts basic to Calculus can be traced, in uncrystallized form, to the time of the ancient Greeks (around 287–212 BC). However, it was only in the sixteenth and the early seventeenth centuries *that mathematicians developed refined techniques for determining tangents to curves and areas of plane regions.* These mathematicians and their ingenious techniques set the stage for *Isaac Newton* (1642–1727) and *Gottfried Leibniz* (1646–1716), who are usually credited with the "*invention*" of Calculus.

Later on, the concept of the definite integral was also developed. *Newton and Leibniz* recognized the importance of the fact that finding derivatives and finding integrals (i.e., antiderivatives) are *inverse processes, thus making possible the rule for evaluating definite integrals.* All these matters are systematically introduced in Part II of the book. (There were many difficulties in the foundation of the subject of Calculus. Some problems reflecting conflicts and doubts on the soundness of the subject are reflected in the "Historical Notes" given at the end of Chapter 9 of Part I.) During the last 150 years, Calculus has matured bit by bit. In the middle of the nineteenth century, French Mathematician *Augustin-Louis Cauchy* (1789–1857) *gave the definition of limit, which removed all doubts about the soundness of Calculus and made it free from all confusion.* It was then that Calculus had become, mathematically, much as we know it today.

To obtain the derivative of a given function (and to apply it for studying the properties of the function) is the subject of the *'differential calculus'*. On the other hand, computing a *function whose derivative is the given function is the subject of integral calculus.* [The function so obtained is called an *anti-derivative* of the given function.] In the operation of computing the antiderivative, *the concept of limit is involved indirectly.* On the other hand, in defining *the definite integral of a function, the concept of limit enters the process directly.* Thus, *the concept of limit* is involved in both, *differential* and *integral calculus.* In fact, we might define *calculus as the study of limits.* It is therefore important that we have a *deep understanding of this concept.* Although, the topic of *limit* is rather *theoretical in nature*, it has been presented and discussed in a very simple way, in the Chapters 7(a) and 7(b) of Part-I (i.e. Differential Calculus) and in Chapter 5 of Part-II (i.e. Integral Calculus). Around the year 1930, the increasing use of Calculus in engineering and sciences, created a necessary requirement to encourage students of engineering and science to learn Calculus. During those days, Calculus was considered an extremely difficult subject. Many authors came up with introductory books on Calculus, but most students could not enjoy the subject, because the basic concepts of the Calculus and its interrelations with the other subjects were probably not conveyed or understood properly. The result was that most of the students learnt *Calculus* only as a *set of rules* and *formulas.* Even today, many students (at the elementary level) "learn" Calculus in the same way. For them, it is easy to remember formulae and apply them without bothering to know: *How the formulae have come and why do they work*?

The best answer to the question *"Why study Calculus at all?"* is available in the book: *Calculus from Graphical, Numerical and Symbolic Points of View* by Arnold Ostebee and Paul Zorn. There are plenty of good practical and "educational" reasons, which emphasize that one must study Calculus:

- Because it is good for applications;
- Because higher mathematics requires it;
- Because its good mental training;
- Because other majors require it; and
- Because jobs require it.

Also, another reason to study Calculus (according to the authors) is that Calculus is among our deepest, richest, farthest-reaching, and most beautiful intellectual achievements. This manuscript differs in certain respects, from the conventional books on Calculus for the beginners.

Organization
The work is divided into two independent books: Book I—*Differential Calculus* (*Introduction to Differential Calculus: Systematic Studies with Engineering Applications for Beginners*) and Book II–*Integral Calculus* (*Introduction to Integral Calculus: Systematic Studies with Engineering Applications for Beginners*).

Part I consists of 23 chapters in which certain chapters are divided into two sub-units such as 7a and 7b, 11a and 11b, 13a and 13b, 15a and 15b, 19a and 19b. Basically, these sub-units are different from each other in one way, but they are interrelated through concepts.

Part II consists of nine chapters in which certain chapters are divided into two sub-units such as 3a and 3b, 4a and 4b, 6a and 6b, 7a and 7b, 8a and 8b, and finally 9a and 9b. The division of chapters is based on the same principle as in the case of Part I. Each chapter (or unit) in both the parts begins with an introduction, clear statements of pertinent definitions, principles and theorems. Meaning(s) of different theorems and their consequences are discussed at length, before they are proved. The solved examples serve to illustrate and amplify the theory, thus bringing into sharp focus many fine points, to make the reader comfortable.

The contents of each chapter are accompanied by all the necessary details. However, some useful information about certain chapters is furnished below. Also, illustrative and other descriptive material (along with notes and remarks) is given to help the beginner understand the ideas involved easily.

Book II (*Introduction to Integral Calculus: Systematic Studies with Engineering Applications for Beginners*):

- Chapter 1 deals with the operation of antidifferentiation (also called integration) as the inverse process of differentiation. Meanings of different terms are discussed at length. The comparison between the operations of differentiation and integration are discussed.
- Chapter(s) 2, 3a, 3b, 4a, and 4b deal with different methods for converting the given integrals to the standard form, so that the antiderivatives (or integrals) of the given functions can be easily written using the standard results.
- Chapter 5 deals with the discussion of the concept of area, leading to the concept of the definite integral and certain methods of evaluating definite integrals.
- Chapter 6a deals with the first and second fundamental theorems of Calculus and their applications in computing definite integrals.
- Chapter 6b deals with the process of defining the natural logarithmic function using Calculus.
- Chapter 7a deals with the methods of evaluating definite integrals using the second fundamental theorem of Calculus.
- Chapter 7b deals with the important properties of definite integrals established using the second fundamental theorem of Calculus and applying them to evaluate definite integrals.
- Chapter 8a deals with the computation of plane areas bounded by curves.
- Chapter 8b deals with the application of the definite integral in computing the lengths of curves, the volumes of solids of revolution, and the curved surface areas of the solids of revolution.
- Chapter 9a deals with basic concepts related to differential equations and the methods of forming them and the types of their solutions.
- Chapter 9b deals with certain methods of solving ordinary differential equations of the first order and first degree.

An important advice for using both the parts of this book:

- The CONTENTS clearly indicate how important it is to go through the prerequisites. Certain concepts [like $(-1) \cdot (-1) = 1$, and why division by zero is not permitted in mathematics, etc] which are generally accepted as rules, are discussed logically. The *concept of infinity* and its algebra are very important for learning calculus. The ideas and definitions of functions introduced in Chapter-2, and extended in Chapter-6, are very useful.
- The role of co-ordinate geometry in defining trigonometric functions and in the development of calculus should be carefully learnt.

- The theorems, in both the Parts are proved in a very simple and convincing way. The solved examples will be found very useful by the students of plus-two standard and the first year college. Difficult problems have been purposely not included in solved examples and the exercise, to maintain the interest and enthusiasm of the beginners. The readers may pickup difficult problems from other books, once they have developed interest in the subject.

- Concepts of *limit, continuity* and *derivative* are discussed at length in chapters 7(a) & 7(b), 8 and 9, respectively. The one who goes through from chapters-1 to 9 has practically learnt more than 60% of differential calculus. The readers will find that remaining chapters of differential calculus are easy to understand. Subsequently, readers should not find any difficulties in learning the concepts of integral calculus and the process of integration including the methods of computing definite integrals and their applications in fining areas and volumes, etc.

- The differential equations right from their formation and the methods of solving certain differential equations of first order and first degree will be easily learnt.

- Students of High Schools and Junior College level may *treat this book as a text book for the purpose of solving the problems and may study desired concepts from the book treating it as a reference book*. Also the students of higher classes will find this book very useful for understanding the concepts and treating the book as a reference book for this purpose. *Thus, the usefulness of this book is not limited to any particular standard. The reference books are included in the bibliography*.

I hope, above discussion will be found very useful to all those who wish to learn the basics of calculus (or wish to revise them) for their higher studies in any technical field involving calculus.

Suggestions from the readers for typos/errors/improvements will be highly appreciated.

Finally, efforts have been made to the ensure that the interest of the beginner is maintained all through. It is a fact that reading mathematics is very different from reading a novel. However, we hope that the readers will enjoy this book like a novel and learn Calculus. We are very sure that if beginners go through the first six chapters of Part I (i.e., prerequisites), then they may not only learn Calculus, but will start loving mathematics.

DR. -ING. AJAY KUMAR PODDAR
CHIEF SCIENTIST
SYNERGY MICROWAVE CORPORATION
NJ 07504, USA
FORMER SENIOR SCIENTIST (DRDO, INDIA)

Spring 2011

BIOGRAPHIES

Ulrich L. Rohde holds a Ph.D. in Electrical Engineering (1978) and a Sc.D. (Hon., 1979) in Radio Communications, a Dr.-Ing (2004), a Dr.-Ing Habil (2011), and several honorary doctorates. He is President of Communications Consulting Corporation; Chairman of Synergy Microwave Corp., Paterson, NJ; and a partner of Rohde & Schwarz, Munich, Germany. Previously, he was the President of Compact Software, Inc., and Business Area Director for Radio Systems of RCA, Government Systems Division, NJ. Dr. Rohde holds several dozen patents and has published more than 200 scientific papers in professional journals, has authored and coauthored 10 technical books. Dr. Rohde is a Fellow Member of the IEEE, Invited Panel Member for the FCC's Spectrum Policy Task Force on Issues Related to the Commission's Spectrum Policies, ETA KAPPA NU Honor Society, Executive Association of the Graduate School of Business-Columbia University, New York, the Armed Forces Communications & Electronics Association, fellow of the Radio Club of America, and former Chairman of the Electrical and Computer Engineering Advisory Board at New Jersey Institute of Technology. He is elected to the "First Microwave & RF Legends" (Global Voting from professionals and academician from universities and industries: Year 2006). Recently Prof. Rohde received the prestigious "Golden Badge of Honor" and university's highest Honorary Senator Award in Munich, Germany.

G.C. Jain graduated in science (Major—Advance Mathematics) from St. Aloysius College, Jabalpur in 1962. Mr. Jain has started his career as a Technical Supervisor (1963–1970), worked for more than 38 years as a Scientist in Defense Research & Development Organization (DRDO). He has been involved in many state-of-the-art scientific projects and also responsible for stabilizing MMG group in ARDE, Pune. Apart from scientific activities, Mr. Jain spends most of his time as a volunteer educator to teach children from middle and high school.

Ajay K. Poddar graduated from IIT Delhi, Doctorate (Dr.-Ing.) from TU-Berlin (Technical University Berlin) Germany. Dr. Poddar is a *Chief Scientist*, responsible for design and development of state-of-the-art technology (oscillator, synthesizer, mixer, amplifier, filters, antenna, and MEMS based RF & MW components) at Synergy Microwave Corporation, NJ. Previously, he worked as a Senior Scientist and was involved in many state-of-the-art scientific projects in DRDO, India. Dr. Poddar holds more than dozen US, European, Japanese, Russian, Chinese patents, and has published more than 170 scientific papers in international conferences and professional journals, contributed as a coauthor of three technical books. He is a recipient of several scientific achievement awards, including RF & MW state-of-the-art product awards for the year 2004, 2006, 2008, 2009, and 2010. Dr. Poddar is a senior member of professional societies IEEE (USA), *AMIE (India)*, and *IE (India)* and involved in technical and academic review committee, including the Academic Advisory Board member Don Bosco Institute of Technology, Bombay, India (2009–to date). Apart from academic and scientific activities,

Dr. Poddar is involved in several voluntary service organizations for the greater cause and broader perspective of the society.

A.K. Ghosh graduated and doctorate from IIT Kanpur. Currently, he is a Professor & Faculty Incharge (Flight Laboratory) Accountable Manager (DGCA), Aerospace Engineering, IIT Kanpur, India (one of the most prestigious institutes in the world). Dr. Ghosh has published more than 120 scientific papers in international conferences and professional journals; recipient of DRDO Technology Award, 1993, young scientist award, Best Paper Award—In-house Journal "Shastra Shakti" ARDE, Pune. Dr. Ghosh has supervised more than 30 Ph.D. students and actively involved in several professional societies and board member of scientific review committee in India and abroad. Previously, he worked as a Senior Scientist and Headed Aerodynamic Group ARDE, Pune in DRDO, India.

INTRODUCTION

In less than 15 min, let us realize that calculus is capable of computing many quantities accurately, which cannot be calculated using any other branch of mathematics.

To be able to appreciate this fact, we consider a "nonvertical line" that makes an angle "θ" with the positive direction of x-axis, and that $\theta \neq 0$. We say that the given line is "inclined" at an angle "θ" (or that the inclination of the given line is "θ").

The important idea of our interest is the "slope of the given line," which is expressed by the trigonometric ratio "tan θ." Technically the slope of the line tells us that if we travel by "one unit," in the positive direction along the x-axis, then the number of units by which the height of the line rises (or falls) is the measure of its slope.

Also, it is important to remember that the "slope of a line" is a constant for that line. On the other hand "the slope of any curve" changes from point to point and it is defined in terms of the slope of the "tangent line" existing there. To find the slope of a curve $y = f(x)$ at any value of x, the "differential calculus" is the only branch of Mathematics, which can be used even if we are unable to imagine the shape of the curve.

At this stage, it is very important to remember (in advance) and understand clearly that whereas, the subject of Calculus demands the knowledge of algebra, geometry, coordinate geometry and trigonometry, and so on (as a prerequisite), but they do know from the subject of Calculus. Hence, calculus should not be confused as a combination of these branches.

Calculus is a different subject. The backbone of Calculus is the "concept of limit," which is introduced and discussed at length in Part I of the book. The first eight chapters in Part I simply offer the necessary material, under the head: What must you know to learn Calculus? We learn the concept of "derivative" in Chapter 9. In fact, it is the technical term for the "slope."

The ideas developed in Part I are used to define an inverse operation of computing antiderivative. (In a sense, this operation is opposite to that of computing the derivative of a given function.)

Most of the developments in the field of various sciences and technologies are due to the ideas developed in computing derivatives and antiderivatives (also called integrals). The matters related with integrals are discussed in "Integral Calculus."

The two branches are in fact complimentary, since the process of integral calculus is regarded as the inverse process of the differential calculus. As an application of integral calculus, the area under a curve $y = f(x)$ from $x = a$ to $x = b$, and the x-axis can be computed only by applying the integral calculus. No other branch of mathematics is helpful in computing such areas with curved boundaries.

Prof. Ulrich L. Rohde

ACKNOWLEDGMENT

There have been numerous contributions by many people to this work, which took much longer than expected. As always, Wiley has been a joy to work with through the leadership, patience and understanding of Susanne Steitz-Filler.

It is a pleasure to acknowledge our indebtedness to Professor Hemant Bhate (Department of Mathematics) and Dr. Sukratu Barve (Center for Modeling and Simulation), University of Pune, India, who read the manuscript and gave valuable suggestions for improvements.

We wish to express our heartfelt gratitude to the Shri K.N. Pandey, Dr. P. K. Roy, Shri Kapil Deo, Shri D.K. Joshi, Shri S.C. Rana, Shri J. Nagarajan, Shri A. V. Rao, Shri Jitendra C. Yadhav, and Dr. M. B. Talwar for their logistic support throughout the preparation of the manuscripts. We are thankful to Mrs. Yogita Jain, Dr. (Mrs.) Shilpa Jain, Mrs. Shubhra Jain, Ms. Anisha Apte, Ms. Rucha Lakhe, Ms. Radha Borawake, Mr. Parvez Daruwalla, Mr. Vaibhav Jain, Mrs. Shipra Jain, and Mr. Atul Jain, for their support towards sequencing the material, proof reading the manuscripts and rectifying the same, from time to time.

We also express our thanks to Mr. P. N. Murali, , Mr. Nishant Singhai, Mr. Nikhil Nanawaty and Mr. A.G. Nagul, who have helped in typing and checking it for typographical errors from time to time.

We are indebted to Dr. (Ms.) Meta Rohde, Mrs. Sandhya Jain, Mrs. Kavita Poddar and Mrs. Swapna Ghosh for their encouragement, appreciation, support and understanding during the preparation of the manuscripts. We would also like to thank Tiya, Pratham, Harsh, Devika, Aditi and Amrita for their compassion and understanding. Finally, we would like to thank our reviewers for reviewing the manuscripts and expressing their valuable feedback, comments and suggestions.

1 Antiderivative(s) [or Indefinite Integral(s)]

1.1 INTRODUCTION

In mathematics, we are familiar with *many pairs of inverse operations*: addition and subtraction, multiplication and division, raising to powers and extracting roots, taking logarithms and finding antilogarithms, and so on. In this chapter, we discuss the inverse operation of *differentiation*, which we call *antidifferentiation*.

Definition (1): A function $\phi(x)$ is called *an antiderivative* of the given function $f(x)$ on the interval $[a, b]$, if at all points of the interval $[a, b]$,

$$\phi'(x) = f(x)^{(1)}$$

Of course, it is logical to use the terms differentiation and antidifferentiation to mean the operations, which must be inverse of each other. However, *the term integration* is frequently used to stand for the process of antidifferentiation, and the term *an integral* (or an indefinite integral) is generally used to mean *an antiderivative* of a function.

The reason behind using the terminology "an integral" (or an indefinite integral) will be clear only after we have studied the concept of "the definite integral" in Chapter 5. The relation between "the definite integral" and "an antiderivative" or an indefinite integral of a function is established through *first and second fundamental theorems of Calculus*, discussed in Chapter 6a.

For the time being, we agree to use these terms freely, with an understanding that the terms: "an antiderivative" and "an indefinite integral" have the same meaning for all practical purposes and that the logic behind using these terms will be clear later on. If a function *f* is differentiable in an interval *I*, [i.e., if its derivative *f'* exists at each point in *I*] then a natural question arises: *Given f'(x) which exists at each point of I, can we determine the function f(x)?* In this chapter, we shall consider this reverse problem, and study some *methods of finding f(x) from f'(x)*.

Note: We know that the derivative of a function $f(x)$, if it exits, *is a unique function*. Let $f'(x) = g(x)$ and that $f(x)$ and $g(x)$ [where $g(x) = f'(x)$] both exist for each $x \in I$, then we say that *an antiderivative (or an integral) of the function g(x) is f(x)*.[2]

1-Anti-differentiation (or integration) as the inverse process of differentiation.

[1] Note that if x is an end point of the interval $[a, b]$, then $\phi'(x)$ will stand for the one-sided derivative at x.

[2] Shortly, it will be shown that an integral of the function $g(x)[= f'(x)]$ can be expressed in the form $f(x) + c$, where c is any constant. Thus, any two integrals of $g(x)$ can differ only by some constant. We say that an integral (or an antiderivative) of a function is "unique up to a constant."

Introduction to Integral Calculus: Systematic Studies with Engineering Applications for Beginners, First Edition. Ulrich L. Rohde, G. C. Jain, Ajay K. Poddar, and A. K. Ghosh.
© 2012 John Wiley & Sons, Inc. Published 2012 by John Wiley & Sons, Inc.

To understand the concept of an antiderivative (or an indefinite integral) more clearly, consider the following example.

Example: Find an antiderivative of the function $f(x) = x^3$.

Solution: From the definition of the derivative of a function, and its relation with the given function, it is natural to guess that an integral of x^3 must have the term x^4. Therefore, we consider the derivative of x^4. Thus, we have

$$\frac{d}{dx} x^4 = 4x^3.$$

Now, from the definition of antiderivative (or indefinite integral) we can write that antiderivative of $4x^3$ is x^4. Therefore, antiderivative of x^3 must be $x^4/4$. In other words, the function $\phi(x) = x^4/4$ is an antiderivative of x^3.

1.1.1   The Constant of Integration

When a function $\phi(x)$ *containing a constant term* is differentiated, *the constant term does not appear in the derivative*, since its derivative is zero. For instance, we have,

$$\frac{d}{dx}\left(x^4 + 6\right) = 4x^3 + 0 = 4x^3;$$

$$\frac{d}{dx} x^4 = 4x^3; \text{ and}$$

$$\frac{d}{dx}\left(x^4 - 5\right) = 4x^3 - 0 = 4x^3.$$

Thus, by the definition of antiderivative, we can say that the functions $x^4 + 6$, x^4, $x^4 - 5$, and in general, $x^4 + c$ (where $c \in \mathbf{R}$), all are antiderivatives of $4x^3$.

Remark: From the above examples, it follows that a given function $f(x)$ can have *infinite number of antiderivatives*. Suppose the antiderivative of $f(x)$ is $\phi(x)$, then not only $\phi(x)$ but also functions like $\phi(x) + 3$, $\phi(x) - 2$, and so on all are called antiderivatives of $f(x)$. Since, the constant term involved with an antiderivative can be any real number, an antiderivative is called an indefinite integral, the indefiniteness being due to the constant term.

In the process of antidifferentiation, *we cannot determine the constant term*, associated with the (original) function $\phi(x)$. Hence, from this point of view, an antiderivative $\phi(x)$ of the given function $f(x)$ will always be *incomplete up to a constant*. Therefore, to get a complete antiderivative of a function, an *arbitrary constant* (which may be denoted by "*c*" or "*k*" or any other symbol) must be added to the result. This *arbitrary constant* represents the *undetermined constant term* of the function, and is called the *constant of integration*.

1.1.2   The Symbol for Integration (or Antidifferentiation)

The symbol chosen for expressing the operation of integration is "$\int$"; it is the old fashioned elongated "S", and it is selected as being the first letter of the word "Sum", which is another aspect of integration, as will be seen later.[3]

---

[3] The symbol $\int$ is also looked upon as a modification of the summation sign $\sum$.

Thus, if an integral of a function $f(x)$ is $\phi(x)$, we write

$$\int f(x)\mathrm{d}x = \phi(x) + c, \text{ where } c \text{ is } \textit{the constant of integration.}$$

Remark: The differential "$\mathrm{d}x$" [written by the side of the function $f(x)$ to be integrated] *separately* does not have a meaning. However, "$\mathrm{d}x$" indicates *the independent variable "x"*, with respect to which *the original differentiation was made.* It also suggests that the reverse process of integration has to be performed with respect to x.

Note: The concept of differentials "$\mathrm{d}y$" and "$\mathrm{d}x$" is discussed at length, in Chapter 16. There, we have discussed how the derivative of a function $y = f(x)$ can be looked upon as the ratio $\mathrm{d}y/\mathrm{d}x$ of differentials. Besides, it is also explained that the equation $\mathrm{d}y/\mathrm{d}x = f'(x)$ can be expressed in the form

$$\mathrm{d}y = f'(x)\mathrm{d}x,$$

which defines the differential of the dependent variable [i.e., the differential of the function $y = f(x)$].

Accordingly, $\int f(x)\mathrm{d}x$ stands to mean that $f(x)$ is to be integrated with respect to x. In other words, we have to find (or identify) a function $\phi(x)$ such that $\phi'(x) = f(x)$. Once this is done, we can write

$$\int f(x)\mathrm{d}x = \phi(x) + c, \ (c \in \mathbf{R}).$$

Now, we are in a position to clarify *the distinction between an antiderivative and an indefinite integral.*

Definition: If the function $\phi(x)$ is an antiderivative of $f(x)$, then the expression $\phi(x) + c$ is called *the indefinite integral* of $f(x)$ and it is denoted by the symbol $\int f(x)\mathrm{d}x$.

Thus, by definition,

$$\int f(x)\mathrm{d}x = \phi(x) + c, (c \in \mathbf{R}), \textit{ provided } \phi'(x) = f(x).$$

Remark: Note that the function in the form $\phi(x) + c$ exhausts all the antiderivatives of the function $f(x)$. On the other hand, the function $\phi(x)$ with a constant [for instance, $\phi(x) + 3$, or $\phi(x) - 7$, or $\phi(x) + 0$, etc.] is called an antiderivative or an indefinite integral (or simply, an integral) of $f(x)$.

1.1.3   Geometrical Interpretation of the Indefinite Integral

From the geometrical point of view, the indefinite integral of a function is a collection (or family) of curves, each of which is obtained by translating any one curve [representing $\phi(x) + c$] parallel to itself, upwards or downwards along the y-axis. A natural question arises: *Do antiderivatives exist for every function f(x)?* The answer is *NO.*

Let us note, however, *without proof, that if a function f(x) is continuous on an interval* $[a, b]$, *then the function has an antiderivative.*

Now, let us integrate the function $y = \int f(x) = 2x$. We have,

$$\int f(x)\mathrm{d}x = \int 2x \, \mathrm{d}x = x^2 + c \tag{1}$$

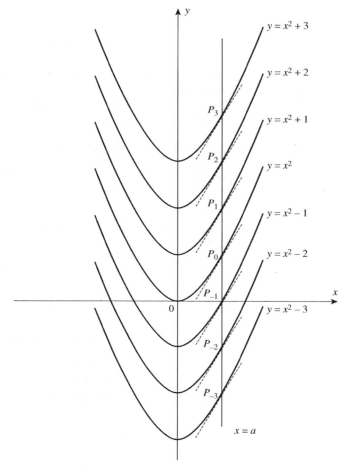

FIGURE 1.1   Shows some curves of $f(x) = 2x$ family.

For different values of c, *we get different antiderivatives* of $f(x)$. But, these antiderivatives (or indefinite integrals) are very similar geometrically. By assigning different values to c, we get different members of the family. *All these members considered together constitute the indefinite integral of $f(x) = 2x$.* In this case, each antiderivative represents a parabola with its axis along the y-axis.[4]

Note that for each positive value of c, there is a parabola of the family which has its vertex on the positive side of the y-axis, and for each negative value of c, there is a parabola which has its vertex on the negative side of the y-axis.

Let us consider the intersection of all these parabolas by a line $x = a$. In Figure 1.1, we have taken $a > 0$ (the same is true for $a < 0$). If the line $x = a$ intersects the parabolas $y = x^2$,

---

[4] For $c = 0$, we obtain $y = x^2$, a parabola with its vertex on the origin. The curve $y = x^2 + 1$ for $c = 1$, is obtained by shifting the parabola $y = x^2$ one unit along y-axis in positive direction. Similarly, for $c = -1$, the curve $y = x^2 - 1$ is obtained by shifting the parabola $y = x^2$ one unit along y-axis in the negative direction. Similarly, all other curves can be obtained.

$y = x^2 + 1$, $y = x^2 + 2$, $y = x^2 - 1$, $y = x^2 - 2$, at P_0, P_1, P_2, P_{-1}, P_{-2}, and so on, respectively, then dy/dx (i.e., the slope of each curve) at $x = a$ is $2a$. This indicates that *the tangents to the curves $\phi(x) = x^2 + c$ at $x = a$ are parallel. This is the geometrical interpretation of the indefinite integral.*

Now, suppose we want to find the curve that passes through the point $(3, 6)$. These values of x and y can be substituted in the equation of the curve. Thus, on substitution in the equation $y = x^2 + c$,

We get, $6 = 3^2 + c$

$$\therefore \quad c = -3$$

Thus, $y = x^2 - 3$ is the equation of the particular curve which passes through the point $(3, 6)$. Similarly, we can find the equation of any curve which passes through any given point (a, b). In the relation,

$$\int f(x)dx = \phi(x) + c, (c \in \mathbf{R}).$$

- The function $f(x)$ is called *the integrand.*
- The expression under the integral sign, that is, "$f(x)dx$" is called the element of integration.

Remark: By the definition of an integral, we have,

$$f(x) = [\phi(x) + c]' = \phi'(x).$$

Thus, we can write,

$$\int f(x)dx = \int \phi'(x)dx$$

$$= \int d[\phi(x)]$$

Observe that the last expression $\int d[\phi(x)]$ does not have "dx" attached to it (Why?). Recall that $d[\phi(x)]$ stands for *the differential of the function $\phi(x)$*, which is denoted by $\phi'(x)dx$, as discussed in Chapter 16 of Part I. Thus, we write,

$$\int f(x)dx = \int \phi'(x)dx = \int d[\phi(x)] = \phi(x) + c. \tag{2}$$

Equation(2) tells us that when we *integrate $f(x)$* [or *antidifferentiate the differential of a function $\phi(x)$*] we obtain the function "$\phi(x) + c$", where "c" is an arbitrary constant. *Thus, on the differential level, we have a useful interpretation of antiderivative of "f".*

Since we have $\int f(x)dx = \phi(x)$, we can say that *an antiderivative* of "f" is a function "ϕ", whose differential $\phi'(x)dx$ equals $f(x)dx$. Thus, we can say that in the symbol $\int f(x)dx$, *the expression "$f(x)dx$" is the differential of some function $\phi(x)$.*

Remark: Equation (2) suggests that differentiation and antidifferentiation (or integration) are inverse processes of each other. (We shall come back to this discussion again in Chapter 6a).

Leibniz introduced the convention of writing *the differential of a function after the integral symbol* "$\int$". The advantage of using the differential in this manner will be apparent to the reader later when we compute antiderivatives by the method of substitution—to be studied later in Chapters 3a and 3b. Whenever we are asked to evaluate the integral $\int f(x)dx$, we are required to find a function $\phi(x)$, satisfying the condition $\phi'(x) = f(x)$. *But how can we find the function $\phi(x)$?*

Because of certain practical difficulties, it is not possible to formulate a set of rules by which any function may be integrated. However, certain methods have been devised for integrating certain types of functions.

- The knowledge of these methods,
- good grasp of differentiation formulas, and
- necessary practice, should help the students to integrate most of the commonly occurring functions.

The methods of integration, in general, consist of certain mathematical operations applied to the integrand so that it assumes some known form(s) of which the integrals are known. Whenever it is possible to express the integrand in any of the known forms (which we call *standard forms*), *the final solution becomes a matter of recognition and inspection.*

Remark: It is important to remember that in the integral $\int f(x)dx$, the variable in the integrand "$f(x)$" and in the differential "dx" must be same (Here it is "x" in both). Thus, $\int \cos y \, dx$ *cannot be evaluated as it stands.* It would be necessary, if possible, to express $\cos y$ as a function of x. Any other letter may be used to represent the independent variable besides x. Thus, $\int t^2 dt$ indicates that t^2 is to be integrated (wherein t is the independent variable), and we need to integrate it with respect to t (which appears in dt).

Note: Integration has one advantage that the result can always be checked by differentiation. If the function obtained by integration is differentiated, we should get back the original function.

1.2 USEFUL SYMBOLS, TERMS, AND PHRASES FREQUENTLY NEEDED

TABLE 1.1 Useful Symbols, Terms, and Phrases Frequently Needed

Symbols/Terms/Phrases	Meaning
$f(x)$ in $\int f(x)dx$	Integrand
The expression $f(x)dx$ in $\int f(x)dx$	The element of integration
$\int f(x)dx$	Integral of $f(x)$ with respect to x. Here, x in "dx" is the variable of integration
Integrate	Find the indefinite integral (i.e., find an antiderivative and add an arbitrary constant to it)[a]
An integral of $f(x)$	A function $\phi(x)$, such that $\phi'(x) = f(x)$
Integration	The process of finding the integral
Constant of integration	An arbitrary real number denoted by "c" (or any other symbol) and considered as a constant.

[a] The term integration also stands for the process of computing the definite integral of $f(x)$, to be studied in Chapter 5.

1.3   TABLE(S) OF DERIVATIVES AND THEIR CORRESPONDING INTEGRALS

TABLE 1.2a   Table of Derivatives and Corresponding Integrals

S. No.	Differentiation Formulas Already Known to us $\dfrac{d}{dx}[f(x)] = f'(x)$	Corresponding Formulas for Integrals $\int f'(x)dx = f(x) + c$ (Antiderivative with Arbitrary Constants)
1.	$\dfrac{d}{dx}(x^n) = n\,x^{n-1},\, n \in \mathbf{R}$	$\int n x^{n-1} dx = x^n + c,\, n \in \mathbf{R}$
2.	$\dfrac{d}{dx}\left[\dfrac{x^{n+1}}{n+1}\right] = x^n,\, n \neq -1$	$\int x^n dx = \dfrac{x^{n+1}}{n+1} + c,\, n \neq -1,\, n \in \mathbf{R}.$ This form is more useful
3.	$\dfrac{d}{dx}(e^x) = e^x$	$\int e^x dx = e^x + c$
4.	$\dfrac{d}{dx}(a^x) = a^x \cdot \log_e a\,(a > 0)$ or $\dfrac{d}{dx}\left(\dfrac{a^x}{\log_e a}\right) = a^x\,(a > 0)$	$\int a^x \cdot \log_e a\, dx = a^x + c\,(a > 0)$ $\therefore \int a^x\, dx = \dfrac{a^x}{\log_e a} + c$
5.	$\dfrac{d}{dx}(\log_e x) = \dfrac{1}{x}\,(x > 0)$	$\int \dfrac{1}{x} dx = \log_e |x| + c,\, x \neq 0^*$

*This formula is discussed at length in Remark (2), which follows.

From the formulas of derivatives of functions, we can write down directly the corresponding formulas for integrals. The formulas for integrals of the important functions given on the right-hand side of the Table 1.2a are referred to as standard formulas which will be used to find integrals of other (similar) functions.

Remark (1):   We make *two comments about formula (2) mentioned in Table 1.2a.*

(i) It is meant to include the case when $n = 0$, that is,

$$\int x^0 dx = \int 1 dx = x + c$$

(ii) Since no interval is specified, *the conclusion is understood to be valid for any interval* on which x^n *is defined. In particular, if $n < 0$, we must exclude any interval containing the origin.* (Thus, $\int x^{-3} dx = (x^{-2}/-2) = -(1/2x^2)$, which is valid in any interval not containing zero.)

Remark (2):   Refer to formula (5) mentioned in Table 1.2a. *We have to be careful when considering functions whose domain is not the whole real line.* For instance, when we say $d/dx(\log_e x) = 1/x$, it is obvious that in this equality $x \neq 0$. However, it is important to remember that, $\log_e x$ *is defined only for positive x.*[5]

[5] Recall that $y = e^x \Leftrightarrow \log_e y = x$. Note that $e^x (=y)$ is always a positive number. It follows that $\log_e y$ is defined only for positive numbers. In fact, in any equality involving the function $\log_e x$ (to any base), it is assumed that $\log x$ is defined only for positive values of x.

In view of the above, the derivative of $\log_e x$ must also be *considered only for positive values of x*. Further, when we write $\int 1/x(dx) = \log_e x$, one must remember that in this equality *the function 1/x is to be considered only for positive values of x*.

Note: Observe that, though *the integrand 1/x is defined for negative values of x, it will be wrong to say that, since 1/x is defined for all nonzero values of x, the integral of 1/x (which is* $\log_e x$*) may be defined for negative values of x*. To overcome this situation, we write

$$\int \frac{1}{x} dx = \log_e |x|, \quad x \neq 0. \; Let\;us\;prove\;this.$$

For $x > 0$, we have,

$$\frac{d}{dx}(\log_e x) = \frac{1}{x},$$

and, for $x < 0$,

$$\frac{d}{dx}[\log_e(-x)] = \frac{1}{-x}(-1) = \frac{1}{x}$$

[Note that for $x < 0$, $(-x) > 0$]. *Combining these two results*, we get,

$$\frac{d}{dx}(\log_e |x|) = \frac{1}{x}, \quad x \neq 0 \qquad \therefore \quad \int \frac{1}{x} dx = \log_e |x| + c, \quad x \neq 0.$$

From this point of view, it is *not appropriate* to write

$$\int \frac{1}{x} dx = \log_e x, \quad x \neq 0. \quad \text{(Why?)}$$

The correct statement is:

$$\int \frac{1}{x} dx = \log_e x, \quad x > 0. \tag{A}$$

$$\text{or } \int \frac{1}{x} dx = \log_e |x|, \quad x \neq 0 \tag{B}$$

Note that both the equalities at (A) and (B) above clearly indicate that $\log_e x$ is defined only for positive values of x.

In solving problems involving log functions, generally the base "e" is assumed. It is convenient and saves time and effort, both (To avoid confusion, one may like to indicate the base of logarithm, if necessary). Some important formulas for integrals that are directly obtained from the derivatives of certain functions, are listed in Tables 1.2b and 1.2c.

Besides, there are *certain results (formulas) for integration*, which are *not obtained directly* from the formulas for derivatives but obtained indirectly by applying other methods of integration. (These methods will be discussed and developed in subsequent chapters).

Many important formulas for *integration* (whether obtained directly or indirectly) are treated as *standard formulas for integration*, which means that we can use these results to write the integrals of (other) similar looking functions.

TABLE 1.2b   Table of Derivatives and Corresponding Integrals

S. No.	Differentiation Formulas Already Known to us $\frac{d}{dx}[f(x)] = f'(x)$	Corresponding Formulas for Indefinite Integrals $\int f'(x)dx = f(x) + c$
6.	$\frac{d}{dx}(\sin x) = \cos x$	$\int \cos x\, dx = \sin x + c$
7.[a]	$\frac{d}{dx}(\cos x) = -\sin x$	$\int(-\sin x)dx = \cos x + c$ $\therefore\ \int \sin x\, dx = -\cos x + c$
8.	$\frac{d}{dx}(\tan x) = \sec^2 x$	$\int \sec^2 x\, dx = \tan x + c$
9.[a]	$\frac{d}{dx}(\cot x) = -\mathrm{cosec}^2 x$	$\int(-\mathrm{cosec}^2 x)dx = \cot x + c$ $\therefore\ \int \mathrm{cosec}^2 x\, dx = -\cot x + c$
10.	$\frac{d}{dx}(\sec x) = \sec x \cdot \tan x$	$\int \sec x \cdot \tan x\, dx = \sec x + c$
11.[a]	$\frac{d}{dx}(\mathrm{cosec}\, x) = -\mathrm{cosec}\, x \cdot \cot x$	$\int(-\mathrm{cosec}\, x \cdot \cot x)dx = \mathrm{cosec}\ x + c$ $\therefore\ \int \mathrm{cosec}\, x \cdot \cot x\, dx = -\mathrm{cosec}\, x + c$

[a] Observe that derivatives of trigonometric functions starting with "co," (i.e., cos x, cot x, and cosec x) are with *negative* sign. Accordingly, the corresponding integrals are also with *negative* sign.

Important Note: *The main problem in evaluating an integral lies in expressing the integrand in the standard form.* For this purpose, we may have to use algebraic operations and/or trigonometric identities. For certain integrals, we may have to change the variable of integration by using the *method of substitution*, to be studied later, in Chapters 3a and 3b. In such cases, *the element of integration is changed to a new element of integration, in which the integrand* (in a new variable) may be in the standard form. *Once the integrand is expressed in the standard*

TABLE 1.2c   Derivatives of Inverse Trigonometric Functions and Corresponding Formulas for Indefinite Integrals

S. No.	Differentiation Formulas Already Known to us $\frac{d}{dx}[f(x)] = f'(x)$	Corresponding Formulas for Indefinite Integrals $\int f'(x)dx = f(x) + c$
12.	$\frac{d}{dx}(\sin^{-1} x) = \frac{1}{\sqrt{1-x^2}}$ $\frac{d}{dx}(\cos^{-1} x) = \frac{-1}{\sqrt{1-x^2}}$	$\int \frac{dx}{\sqrt{1-x^2}} = \begin{cases} \sin^{-1} x + c \\ \text{or} \\ -\cos^{-1} x + c \end{cases}$
13.	$\frac{d}{dx}(\tan^{-1} x) = \frac{1}{1+x^2}$ $\frac{d}{dx}(\cot^{-1} x) = \frac{-1}{1+x^2}$	$\int \frac{dx}{1+x^2} = \begin{cases} \tan^{-1} x + c \\ \text{or} \\ -\cot^{-1} x + c \end{cases}$
14.	$\frac{d}{dx}(\sec^{-1} x) = \frac{1}{x \cdot \sqrt{x^2-1}}$ $\frac{d}{dx}(\mathrm{cosec}^{-1} x) = \frac{-1}{x \cdot \sqrt{x^2-1}}$	$\int \frac{dx}{x \cdot \sqrt{x^2-1}} = \begin{cases} \sec^{-1} x + c \\ \text{or} \\ -\mathrm{cosec}^{-1} x + c \end{cases}$

form, evaluating the integral depends only on recognizing the form and remembering the table of integrals.

Remark: *Thus, integration as such is not at all difficult. The real difficulty lies in applying the necessary algebraic operations and using trigonometric identities needed for converting the integrand to standard form(s).*

1.3.1   Table of Integrals of tan x, cot x, sec x, and cosec x

Now consider Table 1.2b.

Note: Table 1.2b does not include the integrals of tan x, cot x, sec x, and cosec x. The *integrals of these functions will be established by using the method of substitution* (to be studied later in Chapter 3a). However, we list below these results for convenience.

1.3.2   Results for the Integrals of tan x, cot x, sec x, and cosec x

(i) $\int \tan x \, dx = \log_e |\sec x| + c = \log(\sec x) + c$

(ii) $\int \cot x \, dx = \log_e |\sin x| + c = \log(\sin x) + c$

(iii) $\int \sec x \, dx = \log(\sec x + \tan x) + c = \log\left(\tan\left(\frac{x}{2} + \frac{\pi}{4}\right)\right) + c$

(iv) $\int \cosec x \, dx = \log(\cosec x - \cot x) + c = \log\left(\tan \frac{x}{2}\right) + c$

These four integrals are also treated as *standard integrals*.
Now, we consider Table 1.2c.

Remark: Derivatives of inverse circular functions are certain algebraic functions. In fact, *there are only three types of algebraic functions whose integrals are inverse circular functions.*

1.4   INTEGRATION OF CERTAIN COMBINATIONS OF FUNCTIONS

There are some *theorems of differentiation that have their counterparts in integration*. These theorems state the properties of "indefinite integrals" and can be easily proved using the definition of antiderivative. Almost every theorem is proved with the help of differentiation, *thus stressing the concept of antidifferentiation*. To integrate a given function, we shall need these theorems of integration, in addition to the above standard formulas. We give below these results without proof.

(a) $\int [f(x) + g(x)] dx = \int f(x) dx + \int g(x) dx$
In words, *"an integral of the sum of two functions, is equal to the sum of integrals of these two functions"*. The above rule can be extended to the sum of a *finite* number of functions. The result also holds good, if the *sum* is replaced by the *difference*. Hence, integration can be extended to the sum or difference of a finite number of functions.

(b) $\int c \cdot f(x) dx = c \cdot \int f(x) dx$, where *c is a real number.*
Note that result (b) follows from result (a).

Thus, *a constant can be taken out of the integral sign*. The theorem can also be extended as follows:

Corollary:

$$\int [k_1 f(x) + k_2 g(x)] dx = k_1 \int f(x) dx + k_2 \int g(x) dx, \text{ where } k_1 \text{ and } k_2 \text{ are real numbers}$$

(c) If $\int f(x) dx = F(x) + c$

$$\text{then} \quad \int f(x + b) dx = F(x + b) + c.$$

Example:

$$\int \cos(x + 3) dx = \sin(x + 3) + c$$

(d) If $\int f(x) dx = F(x) + c$,

$$\text{then} \quad \int f(ax + b) dx = \frac{1}{a} F(ax + b) + c.$$

This result is easily proved, by differentiating both the sides.

Proof: It is given that $\int f(x) dx = F(x) + c$

$$\therefore \quad F'(x) = f(x) \quad \text{(By definition)}$$

To prove the desired result, we will show that the derivatives of both the sides give the same function.

Now consider,

$$\text{LHS:} \quad \frac{d}{dx} \left[\int f(ax + b) dx \right] = f(ax + b)^{(6)}$$

$$\text{RHS:} \quad \frac{d}{dx} \left[\frac{1}{a} F(ax + b) + c \right] = \frac{1}{a} F'(ax + b) \cdot a$$

$$= F'(ax + b)$$

$$= f(ax + b)$$

$$\therefore \quad \text{L.H.S.} = \text{R.H.S.}$$

Note: *This result is very useful since it offers a new set of "standard forms of integrals", wherein "x" is replaced by a linear function $(ax + b)$. Later on, we will show that this result is more conveniently proved by the method of substitution,* to be studied in Chapter 3a.

Let us now evaluate the integrals of some functions using the above theorems, and *the standard formulas given in Tables 1.2a–1.2c.*

---

[6] We know that the process of differentiation is the inverse of integration (and vice versa). Hence, differentiation nullifies the integration, and we get the integrand as the result. (Detailed explanation on this is given in Chapter 6a).

Examples: We can write,

(i) $\displaystyle\int \sin(5x+7)dx = -\frac{1}{5}\cos(5x+7) + c$

$$\left[\because \int \sin x\, dx = -\cos x + c\right]$$

Similarly,

(ii) $\displaystyle\int e^{3x-2}dx = \frac{1}{3}e^{3x-2} + c$

(iii) $\displaystyle\int \left[\sin(2x+1) + \frac{x-2}{x} - 4^x\right]dx$

$$= \int \sin(2x+1)dx + \int 1 dx - 2\int \frac{1}{x}dx - \int 4^x\, dx$$

$$= -\frac{1}{2}\cos(2x+1) + x - 2\log_e x - \frac{4^x}{\log_e 4} + c \quad \text{Ans.}$$

(iv) $\displaystyle\int \left[x(x+3) - 5\sec^2 x - 3e^{6x-1}\right]$

$$= \int x^2 dx + 3\int x\, dx - 5\int \sec^2 x\, dx - 3\int e^{6x-1}dx$$

$$= \frac{x^3}{3} + \frac{3x^2}{2} - 5\tan x - \frac{3e^{6x-1}}{6} + c$$

$$= \frac{x^3}{3} + \frac{3}{2}x^2 - 5\tan x - \frac{1}{2}e^{6x-1} + c \quad \text{Ans.}$$

(v) $\displaystyle\int \frac{x + 2\sqrt{x} + 7}{\sqrt{x}}dx = I \quad \text{(say)}$

$$\therefore I = \int \left(x^{1/2} + 2 + 7\cdot x^{-1/2}\right)dx$$

$$= \frac{x^{3/2}}{3/2} + 2x + 7\frac{x^{1/2}}{1/2} + c$$

$$= \frac{2}{3}x^{3/2} + 2x + 14x^{1/2} + c \quad \text{Ans.}$$

Here, *the integrand is in the form of a ratio,* which can be easily reduced to a sum of functions in the standard form and hence their antiderivatives can be written, using the tables.

(vi) $\displaystyle\int \frac{3x+1}{x-3}dx = \int \frac{3x - 9 + 9 + 1}{x - 3}dx$

$$= \int \frac{3(x-3) + 10}{x-3}dx$$

$$= 3\int dx + \int \frac{10}{x-3}dx$$

$$= 3x + 10\log_e(x-3) + c \quad \text{Ans.}$$

Here again, *the integrand is in the form of a ratio, which can be easily reduced to the standard form.* If the degree of numerator and denominator is same, *then creating the same factor as the denominator* (as shown above) is a quicker method than actual division.

(vii) $\int (2x + 3)\sqrt{x - 4}\ dx = I$ (say)

$$\therefore\quad I = \int (2x - 8 + 11)\sqrt{x - 4}\ dx$$

$$= \int [2(x - 4) + 11] \cdot \sqrt{x - 4}\ dx$$

$$= 2\int (x - 4)^{3/2}dx + 11\int (x - 4)^{1/2}dx$$

$$= 2 \cdot \frac{(x - 4)^{5/2}}{5/2} + 11 \cdot \frac{(x - 4)^{3/2}}{3/2}$$

$$= \frac{4}{5}(x - 4)^{5/2} + \frac{22}{3}(x - 4)^{3/2} + c \qquad \text{Ans.}$$

Here, the *integrand is in the form of a product*, which can be easily reduced to the standard forms, as indicated above.

In solving the above problems, it has been possible to evaluate the integrals in the form of quotients and products of functions, simply because the integrands *can be converted to standard forms, by applying certain algebraic operations.* In fact, there are different methods for handling integrals involving quotients and products and so on. For example, consider the following integrals.

(a) $\int (3x^2 - 5)^{100} x\ dx$

(b) $\int \sin^3 x \cos x\ dx$

(c) $\int \frac{\sin x}{1 + \sin x}dx$

(d) $\int \frac{1 - \cos 2x}{1 + \cos 2x}\ dx$

(e) $\int x^2 \sin x\ dx$

(f) $\int \frac{3x - 4}{x^2 - 3x + 2}dx$

The above integrals are not in the standard form(s), but they *can be reduced to the standard forms, by using algebraic operations, trigonometric identities, and some special methods to be studied later.*

Note: We emphasize that the main problem in evaluating integrals lies in converting the given integrals into standard forms. *Some integrands can be reduced to standard forms by using* *algebraic operations and trigonometric identities.* For instance, consider $\int \sin^2 x\, dx$. Here, the integrand $\sin^2 x$ is not in the standard form. But, we know the trigonometric identity $\cos 2x = 1 - 2 \sin^2 x$. $\therefore \sin^2 x = (1 - \cos 2x)/2$.

Thus, $\int \sin^2 x\, dx = \int ((1 - \cos 2x)/2) dx = 1/2 \int dx - (1/2) \int \cos 2x\, dx$, where the integrands are in the standard form and so their (indefinite) integrals can be written easily. Note that, here we could express the integrand in a standard form by using a trigonometric identity. Similarly, we can show that

$$\int \frac{\sin x}{1 + \sin x} dx = \int (\sec x \cdot \tan x - \sec^2 x + 1) dx$$

and

$$\int \frac{1 - \cos 2x}{1 + \cos 2x} dx = \int (\sec^2 x - 1) dx$$

wherein, the integrands on the right-hand side are in the standard form(s). A good number of such integrals, *involving trigonometric functions, are evaluated in Chapter 2, using trigonometric identities and algebraic operations.* Naturally, the variable of integration remains unchanged in these operations.

Now, consider the integral $\int f(x) dx = \int \sin^3 x \cos x\, dx$. Here, again the integrand is not in the standard form. Moreover, *it is not possible to convert it to a standard form by using algebraic operations and/or trigonometric identities. However, it is possible to convert it into a standard form as follows*:

We put $\sin x = t$ and differentiate both sides of this equation with respect to t to obtain $\cos x\, dx = dt$. Now, by using these relations in the expression for the element of integration, we get $\int \sin^3 x \cos x\, dx = \int t^3 dt$, which can be easily evaluated. We have

$$\int t^3 dt = \frac{t^4}{4} + c = \frac{\sin^4 x}{4} + c$$

Note that, in the process of *converting the above integrand into a standard form, we had to change the variable of integration* from x to t. This method is known as the *method of substitution* which is to be studied later.

The method of substitution is a very useful method for integration, *associated with the change of variable of integration.* Besides these there are other methods of integration. In this book, our interest is restricted to study the following methods of integration.

(a) Integration of certain trigonometric functions by using algebraic operations and/or trigonometric identities.

(b) *Method of substitution.* This method involves the change of variable.

(c) *Integration by parts.* This method is applicable for integrating product(s) of two different functions. It is also used for evaluating integrals of powers of trigonometric functions (reduction formula). Finer details of this method will be appreciated only while solving problems in Chapters 4a and 4b

(d) *Method of integration by partial fractions.* For integrating rational functions like $\int (3x - 4)/(x^2 - 3x + 2) dx$.

The purpose of each method is to reduce the integrad into the standard form.

Before going for discussions about the above methods of integration, it is useful to realize and appreciate the following points related to the processes of *differentiation* and *integration*, in connection with the similarities and differences in these operations.

1.5   COMPARISON BETWEEN THE OPERATIONS OF DIFFERENTIATION AND INTEGRATION

(1) Both operate on functions.

(2) Both satisfy the property of linearity, that is,

(i) $\dfrac{\mathrm{d}}{\mathrm{d}x}[k_1 f_1(x) + k_2 f_2(x)] = k_1 \dfrac{\mathrm{d}}{\mathrm{d}x} f_1(x) + k_2 \dfrac{\mathrm{d}}{\mathrm{d}x} f_2(x)$, where k_1 and k_2 are constants.

(ii) $\displaystyle\int [k_1 f_1(x) + k_2 f_2(x)]\mathrm{d}x = k_1 \int f_1(x)\mathrm{d}x + k_2 \int f_2(x)\mathrm{d}x$, where k_1 and k_2 are constants.

(3) We have seen that *all functions are not differentiable*. Similarly, *all functions are not integrable*. We will learn about this later in Chapter 5.

(4) The derivative of a function (when it exists) is a unique function. The integral of a function is not so. However, integrals are *unique up to an additive constant, that is, any two integrals of a function differ by a constant.*

(5) When a polynomial function P is differentiated, the result is a polynomial whose degree is one less than the degree of P. When a polynomial function P is *integrated*, the result is a polynomial *whose degree is one more than that of P.*

(6) We can speak of the derivative at a point. We do not speak of an integral at a point. We speak of an integral over an interval on which the integral is defined. (This will be seen in the Chapter 5).

(7) The derivative of a function has a *geometrical meaning*, namely *the slope of the tangent to the corresponding curve at a point.* Similarly, *the indefinite integral of a function represents geometrically, a family of curves placed parallel to each other having parallel tangents at the points of intersection of the curve* by the family of lines perpendicular to the axis representing the variable of integration. (Definite integral has a geometrical meaning as an area under a curve).

(8) The derivative is used to find some physical quantities such as the *velocity of a moving particle*, when the distance traversed at any time t is known. *Similarly, the integral is used in calculating the distance traversed, when the velocity at time t is known.*

(9) Differentiation is the process involving limits. *So is the process of integration*, as will be seen in Chapter 5. Both processes deal with situations where the quantities vary.

(10) The process of differentiation and integration are inverses of each other as will be clear in Chapter 6a.

2 Integration Using Trigonometric Identities

2.1 INTRODUCTION

The main problem in evaluating integrals lies in converting the integrand to some standard form. When the integrand involves *trigonometric functions, it is sometimes possible to convert the integrand into a standard form, by applying algebraic operations and/or trigonometric identities. Obviously, in such cases, the integrand can be changed to a standard form, without changing the variable of integration.* Once this is done, we can easily write the final result, using the standard formulas.

2.1.1 Illustrative Examples

Example (1): To evaluate $\int \sqrt{1 + \sin 2x}\, dx$

Solution: Let $I = \int \sqrt{1 + \sin 2x}\, dx$

Here, the integrand is *not* in the standard form.
We consider,

$$1 + \sin 2x = \sin^2 x + \cos^2 x + 2 \sin x \cdot \cos x$$
$$= (\sin x + \cos x)^2$$
$$\therefore \quad I = \int \sqrt{(\sin x + \cos x)^2}\, dx = \int (\sin x + \cos x)\, dx$$

Now, the constituent functions in the integrand are in the standard form (since, we have formulas for the integrals of $\sin x$ and $\cos x$). Therefore, by applying the theorem on the integral of a sum and the standard formulas for the integrals of $\sin x$ and $\cos x$, we have,

$$I = \int (\sin x + \cos x)\, dx = -\cos x + \sin x + c$$
$$= \sin x - \cos x + c \qquad \text{Ans.}$$

2-Integration of certain trigonometric functions using trigonometric identities and applying algebraic manipulations, to express them in some standard form(s).

Introduction to Integral Calculus: Systematic Studies with Engineering Applications for Beginners, First Edition.
Ulrich L. Rohde, G. C. Jain, Ajay K. Poddar, and A. K. Ghosh.
© 2012 John Wiley & Sons, Inc. Published 2012 by John Wiley & Sons, Inc.

Example (2): To evaluate $\int \frac{1}{\sqrt{1+\sin 2x}}\,dx$

Solution: We have seen in Example (1), that $1 + \sin 2x = \sin x + \cos x$.

$$\text{Now,}\quad \sin x + \cos x = \sqrt{2}\left(\frac{\sin x + \cos x}{\sqrt{2}}\right) = \sqrt{2}\left[\sin x\frac{1}{\sqrt{2}} + \cos x\frac{1}{\sqrt{2}}\right]$$

$$= \sqrt{2}\left[\sin x\cdot\cos\frac{\pi}{4} + \cos x\cdot\sin\frac{\pi}{4}\right], \quad \left(\because \cos\frac{\pi}{4} = \sin\frac{\pi}{4} = \frac{1}{\sqrt{2}}\right)$$

$$= \sqrt{2}\sin\left[x + \frac{\pi}{4}\right]$$

$$\therefore \quad \int\frac{1}{\sqrt{1+\sin 2x}}\,dx = \int\frac{1}{\sqrt{2}\sin(x + (\pi/4))}\,dx$$

$$= \frac{1}{\sqrt{2}}\int\operatorname{cosec}\left(x + \frac{\pi}{4}\right)dx = \frac{1}{\sqrt{2}}\log\left[\tan\left(\frac{x}{2} + \frac{\pi}{8}\right)\right] + c$$

$$\left[\because \int\operatorname{cosec} x\,dx = \log\left(\tan\left(\frac{x}{2}\right) + c\right)\right]^{(1)} \quad \text{Ans.}$$

Example (3): To evaluate $\int\frac{\sin x}{1+\sin x}\,dx$

Solution: Let $I = \int\frac{\sin x}{1+\sin x}\,dx$

Observe that the integrand $(\sin x)/(1 + \sin x)$ is *not in the standard form*.
Now consider,

$$\frac{\sin x}{1 + \sin x} = \frac{\sin x}{1 + \sin x} \times \frac{1 - \sin x}{1 - \sin x} = \frac{\sin x - \sin^2 x}{1 - \sin^2 x}$$

$$= \frac{\sin x - \sin^2 x}{\cos^2 x} \quad (\because 1 - \sin^2 x = \cos^2 x)$$

$$= \frac{\sin x}{\cos^2 x} - \frac{\sin^2 x}{\cos^2 x} = \sec x\cdot\tan x - \tan^2 x$$

$$= \sec x\cdot\tan x - (\sec^2 x - 1), \quad [\because \tan^2 x = \sec^2 x - 1]$$

$$= \sec x\cdot\tan x - \sec^2 x + 1$$

$$\therefore \quad I = \int\sec x\cdot\tan x\,dx - \int\sec^2 x\,dx + \int dx$$

$$\therefore \quad I = \sec x - \tan x + x + c \quad \text{Ans.}$$

Note that here we had to use *the identity* $\tan^2 x + 1 = \sec^2 x$, to express $\tan^2 x$ in the *standard form*.

---

(1) We have already listed this formula in Chapter 1, to meet such requirements. However, its proof is given only in Chapter 3a.

Example (4): To evaluate $\int \frac{\sin x}{\cos^2 x} dx$

Solution: Let $I = \int \frac{\sin x}{\cos^2 x} dx$

Consider

$$\frac{\sin x}{\cos^2 x} = \frac{1}{\cos x} \cdot \frac{\sin x}{\cos x} = \sec x \cdot \tan x$$

$$\therefore \quad I = \int \sec x \cdot \tan x \, dx$$

$$= \sec x + c \qquad \text{Ans.}$$

Example (5): To evaluate $\int \sin^2 nx \, dx$

Solution: We know that

$$\cos 2x = \begin{cases} \cos^2 x - \sin^2 x & (1) \\ 2\cos^2 x - 1 & (2) \\ 1 - 2\sin^2 x & (3) \end{cases}$$

Therefore, using the identity (3), we get

$$\sin^2 x = \frac{1 - \cos 2x}{2}$$

$$\therefore \quad \int \sin^2 nx \, dx = \int \frac{1 - \cos 2nx}{2} dx$$

$$= \frac{1}{2} \int dx - \frac{1}{2} \int \cos 2nx \, dx$$

$$= \frac{1}{2} x - \frac{1}{2} \left[\frac{1}{2n} \sin 2nx \right] + c$$

$$= \frac{1}{2} x - \frac{1}{4n} \sin 2nx + c \qquad \text{Ans.}$$

Similarly, using the identity (2), we can write

$$\int \cos^2 x \, dx = \frac{1}{2} \int (\cos 2x + 1) dx$$

$$= \frac{1}{2} \int \cos 2x \, dx + \frac{1}{2} \int dx$$

$$= \frac{1}{4} \sin 2x + \frac{1}{2} x + c \qquad \text{Ans.}$$

Next consider, $\int (\cos x + \sin x)(\cos x - \sin x) dx = \int (\cos^2 x - \sin^2 x) \, dx$

$$= \int \cos 2x \, dx$$

$$= \frac{1}{2} \sin 2x + c \qquad \text{Ans.}$$

Now, we give below some examples of trigonometric functions and show how easily they can be converted into standard form(s) by using simple algebraic operations and trigonometric identities.[2]

The basic idea behind these operations is to simplify the given integrand (to the extent it is possible) and then express it in some standard form. Once this is done, the only requirement is to make use of *the standard formulas for integration.*

While simplifying the expressions, it will be observed that depending on some (type of) similarity in expressions, certain steps are naturally repeated. Besides, *the simplified expressions so obtained are not only useful for integration but also equally important for computing their derivatives.* This will be pointed out wherever necessary.

S. No.	Given Trigonometric Function(s)	Operations Involved in Converting the Function(s) to the Standard Form
1.	$\dfrac{\sin x}{\cos^2 x}$	$= \dfrac{1}{\cos x} \cdot \dfrac{\sin x}{\cos x} = \sec x \cdot \tan x$
2.	$\dfrac{\cos x}{\sin^2 x}$	$= \dfrac{1}{\sin x} \cdot \dfrac{\cos x}{\sin x} = \operatorname{cosec} x \cdot \cot x$
3.	$\dfrac{1}{1 + \sin x}$	$= \dfrac{1}{1 + \sin x} \cdot \dfrac{1 - \sin x}{1 - \sin x} = \dfrac{1 - \sin x}{\cos^2 x}$ $= \sec^2 x - \sec x \cdot \tan x$
4.	$\dfrac{1}{1 - \sin x}$	$= \dfrac{1 + \sin x}{\cos^2 x} = \sec^2 x + \sec x \cdot \tan x$
5.	$\dfrac{1}{1 + \cos x}$	$= \dfrac{1}{1 + \cos x} \cdot \dfrac{1 - \cos x}{1 - \cos x} = \dfrac{1 - \cos x}{\sin^2 x}$ $= \operatorname{cosec}^2 x - \operatorname{cosec} x \cdot \cot x$
6.	$\dfrac{1}{1 - \cos x}$	$= \dfrac{1 + \cos x}{\sin^2 x} = \operatorname{cosec}^2 x + \operatorname{cosec} x \cdot \cot x$
7.	$\dfrac{\sin x}{1 + \sin x}$	$= \dfrac{\sin x}{1 + \sin x} \cdot \dfrac{1 - \sin x}{1 - \sin x}$ $= \dfrac{\sin x - \sin^2 x}{\cos^2 x} = \dfrac{\sin x}{\cos^2 x} - \tan^2 x$ $= \sec x \cdot \tan x - (\sec^2 x - 1)$ $= \sec x \cdot \tan x - \sec^2 x + 1$
8.	$\dfrac{\cos x}{1 + \cos x}$	$= \dfrac{\cos x(1 - \cos x)}{1 - \cos^2 x} = \dfrac{\cos x - \cos^2 x}{\sin^2 x}$ $= \operatorname{cosec} x \cdot \cot x - \cot^2 x$ $= \operatorname{cosec} x \cdot \cot x - (\operatorname{cosec}^2 x - 1)$ $= \operatorname{cosec} x \cdot \cot x - \operatorname{cosec}^2 x + 1$

Note: We have already shown at S. Nos. (5) and (6) respectively, that $(1/(1 + \cos x)) = \operatorname{cosec}^2 x - \operatorname{cosec} c \cdot \cot x$ and $(1/(1 - \cos x)) = \operatorname{cosec}^2 x + \operatorname{cosec} c \cdot \cot x.$

[2] In general, the trigonometric identities listed in Chapter 5 of Part I are sufficient to meet our requirements.

These expressions can also be put in a *simpler "standard form"* as follows:

$$\frac{1}{1 + \cos x} = \frac{1}{1 + (2\cos^2(x/2) - 1)} = \frac{1}{2}\sec^2\frac{x}{2}$$

and

$$\frac{1}{1 - \cos x} = \frac{1}{1 - (1 - 2\sin^2(x/2))} = \frac{1}{2}\operatorname{cosec}^2\frac{x}{2}$$

$$\therefore \quad \int \frac{1}{1 + \cos x}dx = \frac{1}{2}\int \sec^2\frac{x}{2}dx = \frac{1}{2}\cdot\frac{[\tan(x/2)]}{1/2} + c$$

$$= \tan x/2 + c$$

Similarly,

$$\int \frac{1}{1 - \cos x}dx = \frac{1}{2}\int \operatorname{cosec}^2\frac{x}{2}dx = -\frac{1}{2}\frac{\cot(x/2)}{1/2}$$

$$= -\cot(x/2)$$

Note: Whenever expressions like $1 \pm \cos x$, $1 \pm \cos 2x$, $1 \pm \cos 3x$, and so on, occur in any function, we generally replace $\cos x$, $\cos 2x$, and so on, using one of the identities, $\cos 2x = 2\cos^2 x - 1$ or $\cos 2x = 1 - 2\sin^2 x$, keeping in mind that the number 1 (in $1 \pm \cos x$, $1 \pm \cos 2x, \ldots$, etc.) must be removed by using the correct identity. Besides, for any constant "a", we write $a = a(\sin^2 x + \cos^2 x) = a(\sin^2(x/2) + \cos^2(x/2))$, and so on. Also, we write $\sin x = 2\sin(x/2)\cdot\cos(x/2)$, $\sin 2x = 2\sin x\cdot\cos x$, and so on, as per the requirement. Other useful relations to be remembered are $\sin x = \cos((\pi/2) - x)$ and $\cos x = \sin((\pi/2) - x)$. Now, we consider some more functions and express them in the standard forms.

S. No.	Given Trigonometric Function(s)	Operations Involved in Converting the Function(s) to the Standard Form
9.	$\dfrac{\sin x}{1 + \cos x}$	$= \dfrac{2\sin(x/2)\cdot\cos(x/2)}{1 + 2\cos^2(x/2) - 1} = \tan(x/2)$
10.	$\dfrac{\cos x}{1 + \sin x}$	$= \dfrac{\sin((\pi/2) - x)}{1 + \cos((\pi/2) - x)}$ [Imp. step] $= \tan(1/2)((\pi/2) - x) = \tan((\pi/4) - (x/2))$
11.	$\dfrac{1 + \sin x}{1 + \cos x}$	$= \dfrac{1}{1 + \cos x} + \dfrac{\sin x}{1 + \cos x}$ $= \dfrac{1}{2\cos^2(x/2)} + \dfrac{2\sin(x/2)\cdot\cos(x/2)}{2\cos^2(x/2)}$ $= \dfrac{1}{2}\sec^2\dfrac{x}{2} + \tan\dfrac{x}{2}$
12.	$\dfrac{1 + \cos x}{1 + \sin x}$	$= \dfrac{1 + \cos x}{1 + \sin x} = \dfrac{1}{1 + \sin x} + \dfrac{\cos x}{1 + \sin x}$ $= \dfrac{(1 - \sin x)}{1 - \sin^2 x} + \dfrac{\sin((\pi/2) - x)}{1 + \cos((\pi/2) - x)}$ $= \sec^2 x + \sec x\cdot\tan x + \tan\left(\dfrac{\pi}{4} - \dfrac{x}{2}\right)$

(continued)

(*Continued*)

S. No.	Given Trigonometric Function(s)	Operations Involved in Converting the Function(s) to the Standard Form
13.	$\dfrac{1 + \sin x}{1 - \sin x}$	$= \dfrac{1 + \sin x}{1 - \sin x} \cdot \dfrac{1 + \sin x}{1 + \sin x}$ [Note this step]
		$= \dfrac{(1 + \sin x)^2}{1 - \sin^2 x} = \dfrac{\sin^2 x + 2 \sin x + 1}{\cos^2 x}$ (A)
		$= \tan^2 x + 2 \sec x \cdot \tan x + \sec^2 x$
		$= (\sec^2 x - 1) + 2 \sec x \cdot \tan x + \sec^2 x$
		$= 2 \sec^2 x + 2 \sec x \cdot \tan x - 1$
14.	Similarly, $\dfrac{1 - \sin x}{1 + \sin x}$	$= \dfrac{1 - \sin x}{1 + \sin x} \cdot \dfrac{1 - \sin x}{1 - \sin x}$ [Note this step]
		$= \dfrac{\sin^2 x - 2 \sin x + 1}{\cos^2 x}$ (B)
		$= \tan^2 x - 2 \sec x \cdot \tan x + \sec^2 x$
		$= 2 \sec^2 x - 2 \sec x \cdot \tan x - 1$

Observe that each term in the expressions (A) and (B) is in the standard form. Now, we shall express (13) and (14) in other *new forms, which are frequently useful for both integration and differentiation*. They might appear complicated, but in reality they pose no difficulty, once we learn the operations involved in expressing them in desired forms.

S. No.	Given Trigonometric Function(s)	Operations Involved in Converting the Function(s) to the Standard Form
13a.	$\dfrac{1 + \sin x}{1 - \sin x}$	$= \dfrac{\sin^2(x/2) + \cos^2(x/2) + 2 \sin(x/2) \cdot \cos(x/2)}{\sin^2(x/2) + \cos^2(x/2) - 2 \sin(x/2) \cdot \cos(x/2)}$
		$= \left[\dfrac{\cos(x/2) + \sin(x/2)}{\cos(x/2) - \sin(x/2)} \right]^2$ (C)
		Dividing N^r and D^r by $\cos(x/2)$, we get
		$= \left[\dfrac{1 + \tan(x/2)}{1 - \tan(x/2)} \right]^2$ (D)
		$= \left[\dfrac{\tan(\pi/4) + \tan(x/2)}{1 - \tan(\pi/4) \cdot \tan(x/2)} \right]^2, \quad \left(\because \tan \dfrac{\pi}{4} = 1 \right)$ (E)
		$= \left[\tan \left(\dfrac{\pi}{4} + \dfrac{x}{2} \right) \right]^2 = \tan^2 \left(\dfrac{\pi}{4} + \dfrac{x}{2} \right)$
		$= \left[\tan \left(\dfrac{\pi}{4} + x \right) \right]^2$, and so on
	Similarly, $\dfrac{1 + \sin 2x}{1 - \sin 2x}$	

On similar lines as above, it is easy to express the following function in the standard form.

S. No.	Given Trigonometric Function(s)	Operations Involved in Converting the Function(s) to the Standard Form
14a.	$\dfrac{1 - \sin x}{1 + \sin x}$	$= \left[\dfrac{\cos(x/2) - \sin(x/2)}{\cos(x/2) + \sin(x/2)}\right]^2 = \left[\dfrac{1 - \tan(x/2)}{1 + \tan(x/2)}\right]^2$
		$= \left[\tan\left(\dfrac{\pi}{4} - \dfrac{x}{2}\right)\right]^2 = \tan\left(\dfrac{\pi}{4} - \dfrac{x}{2}\right)$
		$= \tan^2\left(\dfrac{\pi}{4} - x\right)$, and so on
	and $\dfrac{1 - \sin 2x}{1 + \sin 2x}$	

Now, let us consider the following examples wherein the integrands can be easily expressed in the standard form(s) using the steps shown above in achieving the results at (C), (D), and (E).

Applications	Integration
$\displaystyle\int \sqrt{\dfrac{1 + \sin 6x}{1 - \sin 6x}}\,dx$	$= \displaystyle\int \sqrt{\tan^2\left(\dfrac{\pi}{4} + 3x\right)}\,dx$
	$= \displaystyle\int \tan\left(\dfrac{\pi}{4} + 3x\right)dx^{\,a}$
	$= \dfrac{1}{3}\log\left[\sec\left(\dfrac{\pi}{4} + 3x\right)\right] + c$
$\displaystyle\int \dfrac{1 - \tan 2x}{1 + \tan 2x}\,dx$	$= \displaystyle\int \tan\left(\dfrac{\pi}{4} - 2x\right)dx$
	$= \dfrac{1}{-2}\log\left[\sec\left(\dfrac{\pi}{4} - 2x\right)\right] + c^{\,a}$
$\displaystyle\int \dfrac{\cos x + \sin x}{\cos x - \sin x}\,dx$	$= \displaystyle\int \dfrac{1 + \tan x}{1 - \tan x}\,dx$
	$= \displaystyle\int \tan\left(\dfrac{\pi}{4} + x\right)dx = \log\left[\sec\left(\dfrac{\pi}{4} + x\right)\right] + c^{\,a}$
$\displaystyle\int \dfrac{1 + \sin 2x}{1 - \sin 2x}\,dx$	$= \displaystyle\int \tan^2\left(\dfrac{\pi}{4} + x\right)dx$
	$= \displaystyle\int \left[\sec^2\left(\dfrac{\pi}{4} + x\right) - 1\right]dx^{\,a}$
	$= \tan\left(\dfrac{\pi}{4} + x\right) - x + c$
$\displaystyle\int \dfrac{1 - \sin 6x}{1 + \sin 6x}\,dx$	$= \displaystyle\int \tan^2\left(\dfrac{\pi}{4} - 3x\right)dx$
	$= \displaystyle\int \left[\sec^2\left(\dfrac{\pi}{4} - 3x\right) - 1\right]dx^{\,a}$
	$= -\dfrac{1}{3}\tan\left(\dfrac{\pi}{4} - 3x\right) - x + c$

^{a}These results should not be used as formulas. They must be properly derived before use.

S. No.	Given Trigonometric Function(s)	Operations Involved in Converting the Function(s) to the Standard Form
15.	$\dfrac{1 + \cos x}{1 - \cos x}$	$= \dfrac{1 + (2\cos^2(x/2) - 1)}{1 - (1 - 2\sin^2(x/2))} = \dfrac{2\cos^2(x/2)}{2\sin^2(x/2)}$ $= \cot^2\left(\dfrac{x}{2}\right) = \operatorname{cosec}^2\left(\dfrac{x}{2}\right) - 1$
16.	$\dfrac{1 - \cos 2x}{1 + \cos 2x}$	$= \dfrac{2\sin^2 x}{2\cos^2 x} = \tan^2 x = \sec^2 x - 1$
17.	$\dfrac{1 + \sin 2x}{1 + \cos 2x}$	$= \dfrac{1 + 2\sin x \cdot \cos x}{1 + (2\cos^2 x - 1)}$ $= \dfrac{1}{2\cos^2 x} + \dfrac{2\sin x \cdot \cos x}{2\cos^2 x} = \dfrac{1}{2}\sec^2 x + \tan x$
18.	$\dfrac{1 + \cos x}{1 + \sin x}$	$= \dfrac{1}{1 + \sin x} + \dfrac{\cos x}{1 + \sin x}$ $= \dfrac{1 - \sin x}{1 - \sin^2 x} + \dfrac{\sin((\pi/2) - x)}{1 + \cos((\pi/2) - x)}$ [Imp. step] $= \dfrac{1 - \sin x}{\cos^2 x} + \dfrac{2\sin((\pi/4) - (x/2)) \cdot \cos((\pi/4) - (x/2))}{2\cos^2((\pi/4) - (x/2))}$ $\sec^2 x - \sec x \tan x + \tan\left(\dfrac{\pi}{4} - \dfrac{x}{2}\right)$

We give below some more examples of different types of trigonometric functions, which can be easily converted to the standard form(s).[3]

S. No.	Given Trigonometric Function(s)	Operations Involved in Converting the Function(s) to the Standard Form
19.	$\sqrt{1 + \sin x}$	$= \sqrt{\cos^2\dfrac{x}{2} + \sin^2\dfrac{x}{2} + 2\sin\dfrac{x}{2} \cdot \cos\dfrac{x}{2}}$ $= \sqrt{\left(\cos\dfrac{x}{2} + \sin\dfrac{x}{2}\right)^2} = \cos\dfrac{x}{2} + \sin\dfrac{x}{2}$ [a]
20.	$\sqrt{1 - \sin x}$	$= \sqrt{\left(\cos\dfrac{x}{2} - \sin\dfrac{x}{2}\right)^2} = \cos\dfrac{x}{2} - \sin\dfrac{x}{2}$ or $\sin\dfrac{x}{2} - \cos\dfrac{x}{2}$ [a]

Note:
$$\cos\frac{x}{2} - \sin\frac{x}{2} = \sqrt{2}\left[\frac{\cos(x/2) - \sin(x/2)}{\sqrt{2}}\right]$$
$$= \sqrt{2}\left[\sin\frac{\pi}{4}\cos\frac{x}{2} - \cos\frac{\pi}{4}\cdot\sin\frac{x}{2}\right] \quad \left(\because \sin\frac{\pi}{4} = \frac{1}{\sqrt{2}} = \cos\frac{\pi}{4}\right)$$
$$= \sqrt{2}\sin\left(\frac{\pi}{4} - \frac{x}{2}\right) = \sqrt{2}\cos\left(\frac{\pi}{4} + \frac{x}{2}\right) \quad \left(\because \sin\theta = \cos\left(\frac{\pi}{2} - \theta\right)\right)$$
[a]These expressions are in the standard form(s) for integration.

[3] At this stage, the reader's attention is drawn to the following functions:
$$\frac{1+\sin x}{1+\cos x} \cdot e^x, \quad \frac{1-\sin x}{1-\cos x} \cdot e^x, \quad \frac{1+\sin 2x}{1+\cos 2x} \cdot e^{2x}, \quad \frac{x}{1+\sin x}, \quad \text{and} \quad \frac{x}{1+\cos x}.$$

Note that these functions are different than from those listed at S. Nos. (11) and (12). These functions can be converted to *certain products of functions*, which can be easily integrated using the method of *Integration by Parts* (to be studied later in Chapters 4a and 4b).

S. No.	Given Trigonometric Function(s)	Operations Involved in Converting the Function(s) to the Standard Form
21.	$\dfrac{1}{\sqrt{1+\sin x}}$	$=\dfrac{1}{\cos(x/2)+\sin(x/2)}$ $\left.\begin{array}{l}=\dfrac{1}{\sqrt{2}\,\sin((x/2)+(\pi/4))}=\dfrac{1}{\sqrt{2}}\,\mathrm{cosec}\left(\dfrac{x}{2}+\dfrac{\pi}{4}\right)\\[2mm]=\dfrac{1}{\sqrt{2}}\,\sec\left(\dfrac{\pi}{4}-\dfrac{x}{2}\right)\end{array}\right\}a$
	Similarly,	
22.	$\dfrac{1}{\sqrt{1+\sin 2x}}$	$\left.=\dfrac{1}{\sqrt{2}}\,\mathrm{cosec}\left(x+\dfrac{\pi}{4}\right)=\dfrac{1}{\sqrt{2}}\,\sec\left(\dfrac{\pi}{4}-x\right)\right\}a$
23.	$\dfrac{1}{\sqrt{1-\sin 2x}}$	$\left.\begin{array}{l}=\dfrac{1}{\sqrt{2}}\,\mathrm{cosec}\left(\dfrac{\pi}{4}-x\right),\quad\text{or}\\[2mm]=\dfrac{1}{\sqrt{2}}\,\sec\left(\dfrac{\pi}{4}+x\right)\end{array}\right\}a$
24.	$\sqrt{1+\cos x}$	$=\sqrt{1+\left(2\cos^2\dfrac{x}{2}-1\right)}=\sqrt{2}\cos\dfrac{x}{2}{}^{a}$
25.	$\sqrt{1-\cos x}$	$=\sqrt{1-\left(1-2\sin^2\dfrac{x}{2}\right)}=\sqrt{2}\sin\dfrac{x}{2}{}^{a}$
26.	$\dfrac{1}{\sqrt{1+\cos 2x}}$	$\dfrac{1}{\sqrt{2}}\sec x{}^{a}$
27.	$\dfrac{1}{\sqrt{1-\cos 2x}}$	$\dfrac{1}{\sqrt{2}}\,\mathrm{cosec}\,x{}^{a}$

[a]These expressions are in the standard form for integration.

The next four functions are in the standard form with regard to integration, but if they appear in their reciprocal form, they have to be converted to a standard form for integration.

S. No.	Given Trigonometric Function(s)	Operations Involved in Converting the Function(s) to the Standard Form
28.	$\sec x+\tan x$	$=\dfrac{1}{\cos x}+\dfrac{\sin x}{\cos x}=\dfrac{1+\sin x}{\cos x}$ $=\dfrac{\sin^2(x/2)+\cos^2(x/2)+2\sin(x/2)\cdot\cos(x/2)}{\cos^2(x/2)-\sin^2(x/2)}{}^{a}$ $=\dfrac{\cos(x/2)+\sin(x/2)}{\cos(x/2)-\sin(x/2)}=\dfrac{1+\tan(x/2)}{1-\tan(x/2)}=\tan\left(\dfrac{\pi}{4}+\dfrac{x}{2}\right)$
	$\therefore\quad\dfrac{1}{\sec x+\tan x}$	$=\dfrac{1}{\tan((\pi/4)+(x/2))}=\cot((\pi/4)+(x/2))=\tan((\pi/4)-(x/2))^{a}$
29.	$\sec x-\tan x$	$=\dfrac{1-\sin x}{\cos x}=\dfrac{\cos(x/2)-\sin(x/2)}{\cos(x/2)-\sin(x/2)}=\dfrac{1-\tan(x/2)}{1+\tan(x/2)}=\tan\left(\dfrac{\pi}{4}-\dfrac{x}{2}\right)$ $=\dfrac{1}{\tan((\pi/4)-(x/2))}=\cot\left(\dfrac{\pi}{4}-\dfrac{x}{2}\right)$
	$\therefore\quad\dfrac{1}{\sec x-\tan x}$	$=\tan\left(\dfrac{\pi}{4}+\dfrac{x}{2}\right){}^{b}$

(*continued*)

(Continued)

S. No.	Given Trigonometric Function(s)	Operations Involved in Converting the Function(s) to the Standard Form
30.	$\csc x + \cot x$	$= \dfrac{1 - \cos x}{\sin x} = \dfrac{2\cos^2(x/2)}{2\sin(x/2)\cdot\cos(x/2)} = \cot\dfrac{x}{2}$
	$\therefore \dfrac{1}{\csc x + \cot x}$	$= \dfrac{1}{\cot(x/2)} = \tan\dfrac{x}{2}^b$
31.	$\csc x - \cot x$	$= \dfrac{1 - \cos x}{\sin x} = \dfrac{2\sin^2(x/2)}{2\sin(x/2)\cdot\cos(x/2)} = \tan\dfrac{x}{2}$
	$\therefore \dfrac{1}{\csc x - \cot x}$	$= \dfrac{1}{\tan(x/2)} = \cot\dfrac{x}{2} = \tan\left(\dfrac{\pi}{2} - \dfrac{x}{2}\right)^b$

[a]These forms frequently appear in problems for *differentiation and integration*. For example,
$\frac{d}{dx}[\tan^{-1}(\sec x + \tan x)] = \frac{d}{dx}\tan^{-1}\left[\tan\left(\frac{\pi}{4} + \frac{x}{2}\right)\right] = \frac{d}{dx}\left(\frac{\pi}{4} + \frac{x}{2}\right) = \frac{1}{2}$. Also, $\int\frac{1+\sin x}{\cos x}dx = \int\tan\left(\frac{\pi}{4} + \frac{x}{2}\right)dx$. Similarly,
$\int\frac{1}{\sec x + \tan x}dx = \int\tan\left(\frac{\pi}{4} - \frac{x}{2}\right)dx$ and so on. These expressions are in the standard form.
[b]These are in the standard forms convenient for integration.

Integration of trigonometric functions involving *certain* higher powers and those involving *certain* products:

S. No.	Given Trigonometric Function(s)	Operations Involved in Converting the Function(s) to the Standard Form
32.	$\begin{cases} \sin^2 x \\ \cos^2 x \end{cases}$	Consider the identities $\cos 2x = 2\cos^2 x - 1 = 1 - 2\sin^2 x$ $\left.\begin{array}{l}\therefore\ \sin^2 x = \dfrac{1 - \cos 2x}{2} \\ \text{and}\ \ \cos^2 x = \dfrac{1 + \cos 2x}{2}\end{array}\right\}^a$
33.	$\begin{cases} \sin^3 x \\ \cos^3 x \end{cases}$	$\sin 3x = 3\sin x - 4\sin^3 x$ $\therefore\ \sin^3 x = \dfrac{3\sin x - \sin 3x}{4}{}_{a\text{-}}$ $\cos 3x = 4\cos^3 x - 3\cos x$ $\therefore\ \cos^3 x = \dfrac{\cos 3x + 3\cos x}{4}{}_a$
34.	$\begin{cases} \tan^2 x \\ \cot^2 x \end{cases}$	$\sin^2 x + \cos^2 x = 1$ $\therefore \tan^2 x + 1 = \sec^2 x$ $\therefore \tan^2 x = \sec^2 x - 1^a$ and $1 + \cot^2 x = \csc^2 x$ $\therefore \cot^2 x = \csc^2 x - 1^a$
35.	$\begin{cases} \left(\cos\dfrac{x}{2} \pm \sin\dfrac{x}{2}\right)^2 \\ \text{Similarly,} \\ (\cos 3x \pm \sin 3x)^2 \end{cases}$	$= \cos^2\dfrac{x}{2} + \sin^2\dfrac{x}{2} \pm 2\sin\dfrac{x}{2}\cos\dfrac{x}{2}{}_a$ $= 1 \pm 2\sin\dfrac{x}{2}\cos\dfrac{x}{2} = 1 \pm \sin x$ $= 1 \pm \sin 6x$
36.	$\begin{cases} (\sec x \pm \tan x)^2 \\ \\ (\csc x \pm \cot x)^2 \end{cases}$	$= \sec^2 x + \tan^2 x \pm 2\sec x \cdot \tan x$ $= 2\sec^2 x \pm 2\sec x \cdot \tan x - 1$ ${}_a$ $(\because \tan^2 x = \sec^2 x - 1)$ $= \csc^2 x + \cot^2 x \pm 2\csc x \cdot \cot x{}_a$ $= 2\csc^2 x \pm 2\csc x \cdot \cot x - 1$

(*Continued*)

S. No.	Given Trigonometric Function(s)	Operations Involved in Converting the Function(s) to the Standard Form
37.	$\sin^2 x \cdot \cos^2 x$	$= (\sin x \cdot \cos x)^2 = \left[\dfrac{1}{2}\sin 2x\right]^2{}^a$ $= \dfrac{1}{4}\sin^2 2x = \dfrac{1 - \cos 4x}{4}$
	$\dfrac{1}{\sin^2 x \cdot \cos^2 x}$	$= \dfrac{4}{\sin^2 2x} = 4\cosec^2 2x^a$
	Also $\dfrac{1}{\sin^2 x \cdot \cos^2 x}$	$= \dfrac{\sin^2 x + \cos^2 x}{\sin^2 x \cdot \cos^2 x} = \dfrac{1}{\cos^2 x} + \dfrac{1}{\sin^2 x}$
	$\therefore\quad \sec^2 x \cdot \cosec^2 x$	$= \sec^2 x + \cosec^2 x^a$ $= \sec^2 x + \cosec^2 x = 4\cosec^2 2x^a$

^{a}These are in standard form for integration.

S. No.	Given Trigonometric Function(s)	Operations Involved in Converting the Function(s) to the Standard Form
38.	$\sin A \cdot \cos B$ $\therefore\quad \sin 5x \cdot \cos x$ and $\sin x \cdot \cos 5x$	$= \dfrac{1}{2}[\sin(A+B) + \sin(A-B)]$ $= \dfrac{1}{2}[\sin 6x + \sin 4x]$ $= \dfrac{1}{2}[\sin 6x + \sin(-4x)] = \dfrac{1}{2}[\sin 6x - \sin 4x]^a$ $[\because\quad \sin(-\theta) = -\sin\theta]$
39.	$\cos A \cdot \sin B$ $(= \sin B \cdot \cos A)$ $\therefore\quad \cos 7x \cdot \sin 3x$	$= \dfrac{1}{2}[\sin(A+B) - \sin(A-B)]_a$ $= \dfrac{1}{2}[\sin 10x - \sin 4x]$
40.	$\cos A \cdot \cos B$ $\therefore\quad \cos 5x \cdot \cos 3x$	$= \dfrac{1}{2}[\cos(A+B) + \cos(A-B)]_a$ $= \dfrac{1}{2}[\cos 8x + \cos 2x]$
	$\left[= \sin\left(\dfrac{\pi}{2} - 5x\right) \cdot \cos 3x\right]$	$= \dfrac{1}{2}\left[\sin\left(\dfrac{\pi}{2} - 2x\right) + \sin\left(\dfrac{\pi}{2} - 8x\right)\right]$ $= \dfrac{1}{2}[\cos 2x + \cos 8x]^a$
41.	$\sin A \cdot \sin B$ $\therefore\quad \sin 3x \cdot \sin 5x$	$= \dfrac{1}{2}[\cos(A-B) - \cos(A+B)]$ [Note this formula] $= \dfrac{1}{2}[\cos(-2x) - \cos(8x)]$ $\qquad a$ $= \dfrac{1}{2}[\cos 2x - \cos 8x]$
	$\left[= \sin 3x \cdot \cos\left(\dfrac{\pi}{2} - 5x\right)\right]$	$= \dfrac{1}{2}\left[\sin\left(\dfrac{\pi}{2} - 2x\right) + \sin\left(8x - \dfrac{\pi}{2}\right)\right]$ $= \dfrac{1}{2}\left[\cos 2x - \sin\left(\dfrac{\pi}{2} - 8x\right)\right]$ $= \dfrac{1}{2}[\cos 2x - \cos 8x]^a$

^{a}These are in the standard form. Note that identities at S. Nos. (38–41) can be easily remembered, if we carefully remember the formula at (38).

Sometimes, trigonometric identities can be directly used for expressing the given trigonometric function in the standard form.

---

42. $\because \quad \sin 2x = 2 \sin x \cdot \cos x = \dfrac{2 \tan x}{1 + \tan^2 x}$,

we have $\displaystyle\int \dfrac{2 \tan x}{1 + \tan^2 x} dx = \int \sin 2x \, dx$

---

43. $\left.\begin{array}{c} \because \quad \cos 2x = 2 \cos^2 x - 1 \\ = 1 - 2 \sin^2 x \\ = \cos^2 x - \sin^2 x \end{array}\right\}$

$= \dfrac{1 - \tan^2 x}{1 + \tan^2 x}$, we have

$\displaystyle\int \dfrac{1 - \tan^2 x}{1 + \tan^2 x} dx = \int \cos 2x \, dx$

---

44. $\because \quad \tan 2x = \dfrac{2 \tan x}{1 - \tan^2 x} \displaystyle\int \dfrac{2 \tan x}{1 - \tan^2 x} dx = \int \tan 2x \, dx$

2.1.2 Illustrative Examples

Example (6): To evaluate $I = \displaystyle\int \dfrac{\tan x}{\sec x + \tan x} dx$

Solution: Consider $\dfrac{\tan x}{\sec x + \tan x} = \dfrac{\sin x / \cos x}{1 / \cos x + \sin x / \cos x}$

$= \dfrac{\sin x}{1 + \sin x} = \dfrac{\sin x (1 - \sin x)}{(1 + \sin x)(1 - \sin x)} = \dfrac{\sin x - \sin^2 x}{1 - \sin^2 x}$

$= \dfrac{\sin x}{\cos^2 x} - \dfrac{\sin^2 x}{\cos^2 x} = \sec x \cdot \tan x - \tan^2 x$

$= \sec x \cdot \tan x (\sec^2 x - 1)$

$\therefore \quad I = \displaystyle\int \sec x \cdot \tan x \, dx - \int \sec^2 x \, dx + \int dx$

$= \sec x - \tan x + x + c \qquad$ Ans.

Example (7): To evaluate $I = \displaystyle\int (\tan x + \cot x)^2 dx$

Solution: Consider $(\tan x + \cot x)^2$

$= \left(\dfrac{\sin x}{\cos x} + \dfrac{\cos x}{\sin x}\right)^2 = \left[\dfrac{\sin^2 x + \cos^2 x}{\sin x \cdot \cot x}\right]^2$

$= \dfrac{1}{\sin^2 x \cdot \cos^2 x} = \dfrac{\sin^2 x + \cos^2 x}{\sin^2 x \cdot \cos^2 x}, \quad (\because \quad \sin^2 x + \cos^2 x = 1)$

$= \dfrac{1}{\cos^2 x} + \dfrac{1}{\sin^2 x} = \sec^2 x + \cosec^2 x$

$\therefore \quad I = \displaystyle\int \sec^2 x \, dx + \int \cosec^2 x \, dx$

$= \tan x - \cot x + c \quad$ Ans.

Also, note that

$$(\tan x + \cot x)^2 = \tan^2 x + \cot^2 x + 2$$
$$= (\sec^2 x - 1) + (\operatorname{cosec}^2 x - 1) + 2$$
$$= \sec^2 x + \operatorname{cosec}^2 x$$

Example (8): To evaluate $I = \int \cos 3x \cdot \cos 2x \cdot \cos x \, dx$

Solution: Consider $\cos 3x \cdot \cos 2x \cdot \cos x^{(4)}$

$$= \frac{1}{2}[\cos 5x + \cos x]\cos x$$

$$= \frac{1}{2}\left[\cos 5x \cdot \cos x + \cos^2 x\right]$$

$$= \frac{1}{4}[\cos 6x + \cos 4x] + \frac{1}{2}\cos^2 x$$

$$\therefore \quad I = \frac{1}{4}\int \cos 6x \, dx + \frac{1}{4}\int \cos 4x \, dx + \frac{1}{2}\int \cos^2 x \, dx$$

$$= \frac{1}{24}\sin 6x + \frac{1}{16}\sin 4x + \frac{1}{2}\left[\int \frac{\cos 2x + 1}{2} dx\right]$$

$$= \frac{1}{24}\sin 6x + \frac{1}{16}\sin 4x + \frac{1}{4}\cdot\frac{\sin 2x}{2} + \frac{1}{4}x + c$$

$$= \frac{1}{24}\sin 6x + \frac{1}{16}\sin 4x + \frac{1}{8}\sin 2x + \frac{1}{4}x + c \qquad \text{Ans.}$$

Example (9): To evaluate $I = \int \sin 3x \cdot \sin x \, dx$

Solution: Consider $\sin 3x \cdot \sin x = \sin 3x \cdot \cos\left(\frac{\pi}{2} - x\right)^{(5)}$

$$= \frac{1}{2}\left[\sin\left(2x + \frac{\pi}{2}\right) + \sin\left(4x - \frac{\pi}{2}\right)\right]$$

$$= \frac{1}{2}\left[\sin\left(\frac{\pi}{2} + 2x\right) - \sin\left(\frac{\pi}{2} - 4x\right)\right]$$

$$= \frac{1}{2}\cos 2x - \frac{1}{2}\cos 4x$$

$$\therefore \quad I = \frac{1}{2}\int \cos 2x \, dx - \frac{1}{2}\int \cos 4x \, dx$$

$$= \frac{1}{4}\sin 2x - \frac{1}{8}\sin 4x + c \qquad \text{Ans.}$$

---

(4) $\cos A \cdot \cos B = \frac{1}{2}[\cos(A+B) + \cos(A-B)]$

(5) $\sin A \cdot \cos B = \frac{1}{2}[\sin(A+B) + \sin(A-B)]$

Example (10): To evaluate $I = \int \frac{5\cos^3 x + 7\sin^3 x}{2\sin^2 x \cdot \cos^2 x} dx$

Solution: Consider $\dfrac{5\cos^3 x + 7\sin^3 x}{2\sin^2 x \cdot \cos^2 x}$

$$= \frac{5}{2} \cdot \frac{\cos x}{\sin^2 x} + \frac{7}{2} \cdot \frac{\sin x}{\cos^2 x}$$

$$= \frac{5}{2} \cot x \cdot \operatorname{cosec} x + \frac{7}{2} \tan x \cdot \sec x$$

$$\therefore \quad I = \frac{5}{2} \int \cot x \cdot \operatorname{cosec} x \, dx + \frac{7}{2} \int \tan x \cdot \sec x \, dx$$

$$= \frac{7}{2} \int \sec x \cdot \tan x \, dx + \frac{5}{2} \int \operatorname{cosec} x \cot x \, dx$$

$$= \frac{7}{2} \sec x - \frac{5}{2} \operatorname{cosec} x + c \qquad \text{Ans.}$$

Example (11): To evaluate $I = \int \sec^2 x \operatorname{cosec}^2 x \, dx$

Solution: Consider $\sec^2 x \cdot \operatorname{cosec}^2 x$

$$= \frac{1}{\cos^2 x} \cdot \frac{1}{\sin^2 x} = \frac{\sin^2 x + \cos^2 x}{\sin^2 x \cdot \cos^2 x}$$

$$= \frac{1}{\cos^2 x} + \frac{1}{\sin^2 x} = \sec^2 x + \operatorname{cosec}^2 x$$

$$\therefore \quad I = \int \sec^2 x \, dx + \int \operatorname{cosec}^2 x \, dx$$

$$= \tan x - \cot x + c \qquad \text{Ans.}$$

Example (12): To evaluate $I = \int \tan^{-1}\left(\frac{\sin x}{1+\cos x}\right) dx$

Solution: Consider $\dfrac{\sin x}{1 + \cos x}$

$$= \frac{2\sin(x/2) \cdot \cos(x/2)}{1 + (2\cos^2(x/2) - 1)} = \frac{\sin(x/2) \cdot \cos(x/2)}{\cos^2(x/2)} = \tan\frac{x}{2}$$

$$\therefore \quad I = \int \tan^{-1}\left(\tan\frac{x}{2}\right) dx = \int \frac{x}{2} dx, \qquad \left[\because \quad \tan^{-1}(\tan t) = t\right]$$

$$= \frac{1}{2} \int x \, dx = \frac{1}{2} \cdot \frac{x^2}{2} + c$$

$$= \frac{1}{4}x^2 + c \qquad \text{Ans.}$$

Example (13): To evaluate $I = \int \tan^{-1}\left(\frac{\cos x}{1+\sin x}\right) dx$

Solution: Consider $\dfrac{\cos x}{1 + \sin x}$

$$= \frac{\sin((\pi/2) - x)}{1 - \cos((\pi/2) - x)} = \tan\frac{1}{2}\left(\frac{\pi}{2} - x\right) \quad \text{[As in } \textbf{Example (7)]}$$

$$= \tan\left(\frac{\pi}{4} - \frac{x}{2}\right)$$

$$\therefore \quad I = \int \tan^{-1}\left[\tan\left(\frac{\pi}{4} - \frac{x}{2}\right)\right] dx = \int \left(\frac{\pi}{4} - \frac{x}{2}\right) dx$$

$$= \frac{\pi}{4}x - \frac{x^2}{4} + c \qquad \text{Ans.}$$

Example (14): To evaluate $I = \int \sqrt{\frac{1+\sin 2x}{1-\sin 2x}}\, dx$

Solution: $\dfrac{1 + \sin 2x}{1 - \sin 2x} = \left[\tan\left(\frac{\pi}{4} + x\right)\right]^2 \qquad \text{[see: S. No. (13)]}$

$$\therefore \quad I = \int \tan\left(\frac{\pi}{4} + x\right) dx = \log\left[\sec\left(\frac{\pi}{4} + x\right)\right] + c \qquad \text{Ans.}$$

Note: $\displaystyle \int \sqrt{\frac{1 + \sin x}{1 - \sin x}}\, dx = \int \tan\left(\frac{\pi}{4} + \frac{x}{2}\right) dx = \frac{1}{2}\log\left[\sec\left(\frac{\pi}{4} + \frac{x}{2}\right)\right] + c$

and $\displaystyle \int \tan^{-1}\left[\sqrt{\frac{1 - \sin 6x}{1 + \sin 6x}}\right] dx = \int \tan^{-1}\left[\tan\left(\frac{\pi}{4} - 3x\right)\right] dx$

$$= \int \left(\frac{\pi}{4} - 3x\right) dx = \frac{\pi}{4}x - \frac{3}{2}x^2 + c \quad \text{Ans.}$$

Example (15): To evaluate $I = \int \tan^{-1}\left[\frac{1+\tan 4x}{1-\tan 4x}\right] dx$

Solution: $I = \displaystyle \int \tan^{-1}\left[\tan\left(\frac{\pi}{4} + 4x\right)\right] dx$

$$= \int \left(\frac{\pi}{4} + 4x\right) dx = \frac{\pi}{4}x + 2x^2 + c \quad \text{Ans.}$$

Example (16): To evaluate $I = \int \sin^{-1}(\cos x) dx$

Solution: $I = \displaystyle \int \sin^{-1}\left[\sin\left(\frac{\pi}{2} - x\right)\right] dx$

$$= \int \left(\frac{\pi}{2} - x\right) dx = \frac{\pi}{2}\int dx - \int x\, dx$$

$$= \frac{\pi}{2}x - \frac{x^2}{2} + c \qquad \text{Ans.}$$

Similarly,

(a) $\displaystyle\int \cos^{-1}(\cos^2 x - \sin^2 x)dx$

$\displaystyle\qquad = \int \cos^{-1}(\cos 2x)dx = \int 2x\,dx$

$\displaystyle\qquad = 2 \cdot \frac{x^2}{2} + c = x^2 + c$

(b) $\displaystyle\int \tan^{-1}\left(\frac{2\tan x}{1 - \tan^2 x}\right)dx = \int \tan^{-1}(\tan 2x)dx$

$\displaystyle\qquad = \int 2x\,dx = x^2 + c. \quad \left[\text{Recall:}\quad \tan 2x = \frac{2\tan x}{1 - \tan^2 x}\right]$

(c) $\displaystyle\int \sin^{-1}\left(\frac{2\tan 2x}{1 + \tan^2 2x}\right)dx = \int \sin^{-1}(\sin 4x)dx$

$\displaystyle\qquad = \int 4x\,dx = 2x^2 + c. \quad \left[\text{Recall:}\quad \sin 2x = \frac{2\tan x}{1 + \tan^2 x}\right]$

(d) $\displaystyle\int \cos^{-1}\left(\frac{1 - \tan^2 x}{1 + \tan^2 x}\right)dx = \int \cos^{-1}(\cos 2x)dx$

$\displaystyle\qquad = \int 2x\,dx = x^2 + c. \quad \left[\text{Recall: } \cos 2x = \frac{1 - \tan^2 x}{1 + \tan^2 x}\right]$

(e) $\displaystyle\int \sin^{-1}\left(\frac{1 - \tan^2 x}{1 + \tan^2 x}\right)dx = \int \sin^{-1}(\cos 2x)dx$

$\displaystyle\qquad = \int \sin^{-1}\left[\sin\left(\frac{\pi}{2} - 2x\right)\right]dx, \quad \left[\because\quad \cos = \sin\left(\frac{\pi}{2} - \theta\right)\right]$

$\displaystyle\qquad = \int \left(\frac{\pi}{2} - 2x\right)dx = \frac{\pi}{2}x - x^2 + c$

Example (17): To evaluate $I = \int \cos mx \cos nx\,dx$, where m and n are positive integers and $m \neq n$. What will happen if $m = n$?

Solution: We have the identity

$$\cos mx \cdot \cos nx = \frac{1}{2}[\cos(m + n)x + \cos(m - n)x]$$

$$\therefore\quad I = \frac{1}{2}\left[\int \cos(m + n)x\,dx + \int \cos(m - n)x\,dx\right]$$

$$= \frac{1}{2}\left[\frac{\sin(m + n)x}{m + n} + \frac{\sin(m - n)x}{m - n}\right] + c. \quad (\text{since } m \neq n, \text{ as given})$$

When $m = n$, $\cos(m - n)x = \cos 0 = 1$, and therefore, its integral is x.
    Also, $\cos(m + n)x = \cos 2mx$, and its integral is $(\sin 2mx)/(2m)$. Thus, we get

$$I = \frac{1}{2}\left[\frac{\sin 2mx}{2m} + x\right] + c$$

Also, note that when $m = n$,

$$\int \cos mx \cos nx \, dx = \int \cos^2 mx \, dx$$

$$= \int \frac{1 + \cos 2mx}{2} \, dx = \frac{1}{2} \left[\int (1 + \cos 2mx) dx \right]$$

$$= \frac{1}{2} \left[x + \frac{\sin 2mx}{2m} \right] + c \qquad \text{Ans.}$$

Exercise

Integrate the following with respect to x:

(1) $\dfrac{3 - 2 \sin x}{\cos^2 x}$

Ans. $3 \tan x - 2 \sec x + c$

(2) $\tan^{-1} \sqrt{\dfrac{1 - \cos 2x}{1 + \cos 2x}}$

Ans. $\dfrac{x^2}{2} + c$

(3) $\sqrt{1 + \sin 2x}$

Ans. $\sin x - \cos x + c$

(4) $\sqrt{1 + \cos 2x}$

Ans. $\sqrt{2} \sin x + c$

(5) $\dfrac{1}{\sqrt{1 + \cos 2x}}$

Ans. $\dfrac{1}{\sqrt{2}} \log \left[\tan \left(\dfrac{x}{2} + \dfrac{\pi}{4} \right) \right] + c$

(6) $\tan^{-1} \left[\dfrac{\cos 2x - \sin 2x}{\cos 2x + \sin 2x} \right]$

Ans. $\dfrac{\pi}{4} x - x^2 + c$

(7) $\dfrac{1 - \tan^2 3x}{1 + \tan^2 3x}$

Ans. $\dfrac{1}{6} \sin 6x$

[Hint: $(1 - \tan^2 3x)/(1 + \tan^2 3x) = \cos 6x$]

(8) $\left(\dfrac{1 - \tan 3x}{1 + \tan 3x} \right)^2$

Ans. $-\dfrac{1}{3} \tan \left(\dfrac{\pi}{4} - 3x \right) - x + c$

[Hint: $((1 - \tan 3x)/(1 + \tan 3x))^2 = \tan^2((\pi/4) - 3x) = \sec^2((\pi/4) - 3x) - 1$]

(9) $\sin x \cdot \sin 2x \cdot \sin 3x$

Ans. $\dfrac{\cos 6x}{24} - \dfrac{\cos 2x}{8} - \dfrac{\cos 4x}{6} + c$

[Hint:
$\sin A \cdot \sin B = (1/2)[\cos(A - B) - \cos(A + B)]$
or $\sin 3x \cdot \sin x = \sin 3x \cdot \cos((\pi/2) - x) = (1/2)[\sin(2x + (\pi/2)) + \sin(4x - (\pi/2))]]$

true for identities like, $\sin mx \cdot \sin nx$, where m and n are *distinct positive integers.*

and $\sin mx \cdot \cos nx$, where m and n are *distinct positive integers.*

Remark: In this chapter, we have been able to integrate functions such as $\sin^{-1}(\sin x)$, $\cos^{-1}(\cos x)$, $\tan^{-1}(\tan x)$, and so on, *because they can be reduced to simple algebraic functions in the standard form.*

On the other hand, by using the methods learnt so far, it is not possible to integrate inverse trigonometric functions (i.e., $\sin^{-1} x$, $\cos^{-1} x$, $\tan^{-1} x$, etc.). Integration of these functions is discussed later under the *method of integration by parts* in Chapters 4a and 4b.

2.2 SOME IMPORTANT INTEGRALS INVOLVING $\sin x$ AND $\cos x$

Certain trigonometric and algebraic manipulations are required to convert the following types of integrals into standard forms.

2.2.1 Integrals of the Form

(i) $\displaystyle \int \frac{\sin x}{\sin(x + a)}\,dx,$

(ii) $\displaystyle \int \frac{\sin(x - a)}{\cos(x + a)}\,dx,$

(iii) $\displaystyle \int \frac{\sin(x - a)}{\sin(x - b)}\,dx,$

(iv) $\displaystyle \int \frac{1}{\sin(x - a)\cos(x - b)}\,dx,$

(v) $\displaystyle \int \frac{1}{\cos(x + a)\cos(x + b)}\,dx,$ and so on

Example (18): Evaluate $\int \frac{\sin x}{\sin(x+a)}\,dx = I$ (say)

Method: We express the variable x *(in the numerator) in terms of the variable* $(x + a)$*, which is in the denominator.*

Thus, $x = (x+a) - a^{(6)}$

$$\therefore \quad I = \int \frac{\sin(\overline{x+a} - a)}{\sin(x+a)} dx$$

$$= \int \frac{\sin(x+a) \cdot \cos a - \cos(x+a) \cdot \sin a}{\sin(x+a)} dx$$

$$= \cos a \int dx - \sin a \int \cot(x+a) dx$$

$$= x \cos a - (\sin a)\log[\sin(x+a)] + c \qquad \text{Ans.}$$

(Note that $\cos a$ and $\sin a$ are constants.)

Example (19): Evaluate $\int \frac{\sin(x-a)}{\cos(x+a)} dx = I$ (say)

Note: $x - a = (x+a) - 2a$

$$\therefore \quad I = \int \frac{\sin(\overline{x+a} - 2a)}{\cos(x+a)} dx$$

$$= \int \frac{\sin(x+a) \cdot \cos 2a - \cos(x+a) \cdot \sin 2a}{\cos(x+a)} dx$$

$$\therefore \quad I = \cos 2a \int \tan(x+a) dx - \sin 2a \int dx$$

$$= \cos 2a \cdot \log[\sec(x+a)] - \sin 2a \cdot x + c \qquad \text{Ans.}$$

Now, show that $\int \frac{\sin(x-a)}{\sin(x+a)} dx$

$$= (\cos 2a)x - (\sin 2a)\log[\sin(x+a)] + c$$

Example (20): Evaluate $\int \frac{\sin(x-a)}{\sin(x-b)} dx = I$ (say)

Note: $x - a = (x-b) + b - a$

$$= (x-b) + (b-a)$$

$$\therefore \quad I = \int \frac{\sin[(x-b) + (b-a)]}{\sin(x-b)} dx$$

$$= \int \frac{\sin(x-b) \cdot \cos(b-a) + \cos(x-b) \cdot \sin(b-a)}{\sin(x-b)} dx$$

$$= \cos(b-a) \int dx + \sin(b-a) \int \cot(x-b) dx$$

$$= \cos(b-a) \cdot x + \sin(b-a) \cdot \log[\sin(x-b)] + c \qquad \text{Ans.}$$

---

[6] Note that for evaluating $\int \frac{\sin(x\pm a)}{\sin x} dx$, $\int \frac{\cos(x\pm a)}{\cos x} dx$, or $\int \frac{\cos(x\pm a)}{\sin x} dx$, no such adjustments are needed.

Now, show that $\int \dfrac{\cos(x-a)}{\cos(x-b)} \, dx$

$$= x\cos(b-a) - \sin(b-a)\log[\sec(x-b)] + c$$

$$= x\cos(b-a) + \sin(b-a)\log[\cos(x-b)] + c$$

Example (21): Evaluate $\int \frac{1}{\sin(x-a)\cdot\cos(x-b)}\,dx = I$ (say)

Note: In such cases, we observe that $(x-a) - (x-b) = (b-a)$, *which is a constant.*

Now consider $\dfrac{1}{\sin(x-a)\cdot\cos(x-b)}$ [7]

$$= \frac{1}{\sin(x-a)\cdot\cos(x-b)} \cdot \frac{\cos(b-a)}{\cos(b-a)}$$

$$= \frac{1}{\cos(b-a)} \cdot \frac{\cos(b-a)}{\sin(x-a)\cdot\cos(x-b)}$$

$$\therefore \quad I = \frac{1}{\cos(b-a)} \int \frac{\cos[(x-a)-(x-b)]}{\sin(x-a)\cdot\cos(x-b)}\,dx$$

$$= \frac{1}{\cos(b-a)} \int \frac{\cos(x-a)\cos(x-b) + \sin(x-a)\sin(x-b)}{\sin(x-a)\cdot\cos(x-b)}\,dx$$

$$= \frac{1}{\cos(b-a)} \int [\cot(x-a) + \tan(x-b)]\,dx$$

$$= \frac{1}{\cos(b-a)} [\log[\sin(x-a)] + \log[\sec(x-b)]] + c$$

$$= \frac{1}{\cos(b-a)} [\log\sin(x-a) - \log\cos(x-b)] + c$$

$$= \sec(b-a)\left[\log\frac{\sin(x-a)}{\cos(x-b)}\right] + c \qquad \text{Ans.}$$

Example (22): Evaluate $\int \frac{1}{\cos(x+a)\cos(x+b)}\,dx$

Note: $(x+a) - (x+b) = (a-b)$

Consider $\dfrac{1}{\cos(x+a)\cos(x+b)} \cdot \dfrac{\sin(a-b)}{\sin(a-b)}$

---

[7] In this case, we multiply the integrand by the number $(\cos(b-a)/\cos(b-a))(=1)$ and then expand the N^r by expressing it suitably. Here, we should not multiply the integrand by $(\sin(b-a)/\sin(b-a))(=1)$. However, if the integrand contains the product $\sin(x-a)\cdot\sin(x-b)$ or $\cos(x-a)\cdot\cos(x-b)$, we must choose the quantity $(\sin(b-a)/\sin(b-a))(=1)$ for multiplying with the integrand. These choices are important for converting the integrand into standard form. Check this.

$$\therefore \quad I = \frac{1}{\sin(a-b)} \int \frac{\sin[(x+a)-(x+b)]}{\cos(x+a) \cdot \cos(x+b)} dx$$

$$= \frac{1}{\sin(a-b)} \int \frac{\sin(x+a) \cdot \cos(x+b) - \cos(x+a)\sin(x+b)}{\cos(x+a) \cdot \cos(x+b)} dx$$

$$= \operatorname{cosec}(a-b) \int [\tan(x+a) - \tan(x+b)] dx$$

$$= \operatorname{cosec}(a-b)[\log \sec(x+a) - \log \sec(x+b)] + c$$

$$= \operatorname{cosec}(a-b)\left[\log \frac{\sec(x+a)}{\sec(x+b)}\right] + c \quad \text{Ans.}$$

Now, evaluate the following integrals:

(1) $\displaystyle \int \frac{1}{\cos(x-a)\sin(x-b)} dx$

Ans. $[\sec(b-a)]\log\left[\dfrac{\sin(x-b)}{\cos(x-a)}\right] + c$

(2) $\displaystyle \int \frac{1}{\cos(x-a)\cos(x-b)} dx$

Ans. $[\operatorname{cosec}(b-a)]\log\left[\dfrac{\cos(x-b)}{\cos(x-a)}\right] + c$

(3) $\displaystyle \int \frac{1}{\sin(x-a)\sin(x-b)} dx$

Ans. $[\operatorname{cosec}(b-a)]\log\left[\dfrac{\sin(x-b)}{\sin(x-a)}\right] + c$

2.3 INTEGRALS OF THE FORM ∫(dx/(a sin x + b cos x)), WHERE a, b ∈ r

Method: Consider the expression: $a \sin x + b \cos x$.
This can be converted into a single trigonometric quantity.

$$\text{Put } a = r \cos \alpha \text{ and } b = r \sin \alpha.$$

Then, $r^2 = a^2 + b^2 \therefore r = \sqrt{a^2 + b^2}$ and $\alpha = \tan^{-1}(b/a)$

$$\therefore \quad a \sin x + b \cos x = r \sin x \cos \alpha + r \cos x \sin \alpha$$
$$= r(\sin x \cos \alpha + \cos x \sin \alpha)$$
$$= r \sin(x + \alpha), \text{ where } r \text{ and } \alpha \text{ are defined above.}$$

$$\therefore \quad \int \frac{dx}{a \sin x + b \cos x} = \frac{1}{r} \int \frac{dx}{\sin(x+\alpha)}$$

$$= \frac{1}{r} \int \operatorname{cosec}(x+\alpha) dx^{(8)}$$

$$= \frac{1}{r} \log \left[\tan \left(\frac{x+\alpha}{2} \right) \right] + c$$

$$= \frac{1}{r} \log \left[\tan \left(\frac{x}{2} + \frac{1}{2}(\alpha) \right) \right] + c$$

$$= \frac{1}{\sqrt{a^2 + b^2}} \log \left[\tan \left(\frac{x}{2} + \frac{1}{2} \tan^{-1} \frac{b}{a} \right) \right] + c \qquad \text{Ans.}$$

Now, let us consider some expressions of the form $a \sin x + b \cos x$, and convert them into a single trigonometric quantity.

(i) $\dfrac{a \sin x + b \cos x}{\sqrt{a^2 + b^2}} = E$ (say)

Put $a = r \cos \alpha$ and $b = r \sin \alpha$

$$\therefore \quad r^2 = a^2 + b^2 \qquad \therefore \quad r = \sqrt{a^2 + b^2}$$

and

$$\alpha = \tan^{-1} \frac{b}{a}$$

$$\therefore \quad E = \frac{r \sin(x+\alpha)}{r} = \sin(x+\alpha)$$

where

$$\alpha = \tan^{-1} \frac{b}{a}$$

(ii) $\dfrac{a \cos x + b \sin x}{\sqrt{a^2 + b^2}} = E$ (say)

Put $a = r \sin \alpha$ and $b = r \cos \alpha$

$$\therefore \quad r^2 = a^2 + b^2 \qquad \therefore \quad r = \sqrt{a^2 + b^2}$$

[8] We know that

$\int \operatorname{cosec} x \, dx = \log(\operatorname{cosec} x - \cot x) + c$ (i)

$\qquad = \log\left(\tan \dfrac{x}{2}\right) + c$ (ii)

It is always better to use form (ii) of the integral, since it is convenient to write. Also, it is easier to compute. Recall that in evaluating $\int \operatorname{cosec} x \, dx$, we have to use the method of substitution.

and

$$\alpha = \tan^{-1}\frac{a}{b}$$

$$\therefore \quad E = \frac{r\sin(\alpha + x)}{r} = \sin(x + \alpha)$$

where

$$\alpha = \tan^{-1}\frac{a}{b} \quad \text{(Note this)}$$

(iii) $x \cos \alpha + \sqrt{1 - x^2} \sin \alpha = E \quad$ (say)

Put $x = \sin t \quad \therefore \quad \sqrt{1 - x^2} = \cos t$

and $\tan t = \dfrac{x}{\sqrt{1 - x^2}} \quad \therefore \quad t = \tan^{-1}\dfrac{x}{\sqrt{1 - x^2}}$

$\therefore \quad E = \sin t \cos \alpha + \cos t \sin \alpha$

$\qquad = \sin(t + \alpha)$

where

$$t = \tan^{-1}\frac{x}{\sqrt{1 - x^2}}$$

(iv) $\dfrac{\sin x + \cos x}{\sqrt{2}} = E \quad$ (say)

We know that $\sin\dfrac{\pi}{4} = \dfrac{1}{\sqrt{2}} = \cos\dfrac{\pi}{4}$

$$\therefore \quad E = \sin x \cdot \cos\frac{\pi}{4} + \cos x \cdot \sin\frac{\pi}{4}$$

$$= \sin\left(x + \tfrac{\pi}{4}\right)$$

To evaluate integrals of the type, $\int \dfrac{dx}{a \sin x + b \cos x} = I \quad$ (say)

Example (23): $\int \dfrac{dx}{2 \cos x + 3 \sin x}$

Consider the expression $2 \cos x + 3 \sin x$.

Let $2 = r \sin \alpha$ and $3 = r \cos \alpha$

$\therefore \quad r^2 = 2^2 + 3 \quad \therefore \quad r = \sqrt{4 + 9} = \sqrt{13}$

and $\qquad \tan \alpha = \dfrac{2}{3} \quad \therefore \quad \alpha = \tan^{-1}\dfrac{2}{3}$

$$\therefore \quad I = \int \frac{dx}{r \sin \alpha \cos x + r \cos \alpha \sin x}$$

$$= \frac{1}{r}\int \frac{dx}{\sin(\alpha + x)} = \frac{1}{r}\int \cosec(x + \alpha)dx$$

$$= \frac{1}{r}\log\left[\tan\frac{x + \alpha}{2}\right] + c$$

$$= \frac{1}{\sqrt{13}}\log\left[\tan\left(\frac{x}{2} + \frac{1}{2}\tan^{-1}\frac{2}{3}\right)\right] + c \qquad \text{Ans.}$$

Example (24):

$$\int \frac{dx}{\cos x + \sin x} = \int \frac{1}{\sqrt{2}((1/\sqrt{2})\cos x + (1/\sqrt{2})\sin x)} dx$$

$$= \frac{1}{\sqrt{2}} \int \frac{dx}{\sin(\pi/4 + x)} = \frac{1}{\sqrt{2}} \int \frac{dx}{\sin((x + \pi/4))}$$

$$= \frac{1}{\sqrt{2}} \int \operatorname{cosec}(x + \pi/4)dx \quad \left[\because \int \operatorname{cosec} x \, dx = \log\left(\tan \frac{x}{2}\right)\right]$$

$$= \frac{1}{\sqrt{2}} \log\left[\tan\left(\frac{x}{2} + \frac{\pi}{8}\right)\right] + c \quad \text{Ans.}$$

Example (25): $\int \dfrac{1}{5\cos x - 12 \sin x} dx = I \quad \text{(say)}$

Method (I): Let $5 = r \sin \alpha$ and $12 = r \cos \alpha$

$$\therefore \quad r^2 = 5^2 + (12)^2 \quad \therefore \quad r = \sqrt{25 + 144} = \sqrt{169} = 13$$

and $$\tan \alpha = \frac{5}{12} \quad \therefore \quad \alpha = \tan^{-1}\frac{5}{12}$$

$$\therefore \quad I = \frac{1}{r} \int \frac{1}{\sin(\alpha - x)} dx = \frac{1}{r} \int \operatorname{cosec}(\alpha - x)dx$$

$$= \frac{1}{r} \log\left[\tan\frac{(\alpha - x)}{2}\right] + c = \frac{1}{r} \log\left[\tan\left(\frac{\alpha}{2} - \frac{x}{2}\right)\right] + c$$

$$= \frac{1}{13} \log\left[\tan\left(\frac{1}{2}\tan^{-1}\left(\frac{5}{12}\right) - \frac{x}{2}\right)\right] + c \quad \text{Ans.}$$

Method (II): Let $5 = r \cos \alpha$ and $12 = r \sin \alpha$

$$\therefore \quad r^2 = 5^2 + 12^2 \quad \therefore \quad r = \sqrt{169} = 13$$

and $\tan \alpha = 12/5 \quad \therefore \quad \alpha = \tan^{-1} 12/5$[18]

$$\therefore \quad I = \frac{1}{r} \int \frac{1}{\cos x \cos \alpha - \sin x \sin \alpha} dx \qquad {}^{(9)}$$

$$= \frac{1}{r} \int \frac{1}{\cos(x + \alpha)} dx$$

$$= \frac{1}{r} \int \sec(x + \alpha)dx$$

$$= \frac{1}{r} \log\left[\tan\left(\frac{x + \alpha}{2} + \frac{\pi}{4}\right)\right] + c$$

$$= \frac{1}{13} \log\left[\tan\left\{\left(\frac{x}{2} + \frac{1}{2}\tan^{-1}\frac{12}{5}\right) + \frac{\pi}{4}\right\}\right] + c \text{ Ans.}$$

$${}^{(9)} \cos A \cdot \cos B - \sin A \cdot \sin B = \cos(A + B)$$

$$\int \sec x \, dx = \log\left[\tan\left(\frac{x}{2} + \frac{\pi}{4}\right)\right] + c$$

Example (26): $\int \dfrac{1}{2 \cos x + 3 \sin x} dx = I$ (say)

Consider the expression $2 \cos x + 3 \sin x$

Method (I): Let $2 = r \sin \alpha$ and $3 = r \cos \alpha$

$$\therefore \quad r^2 = 2^2 + 3^2 = 13 \quad \therefore \quad r = \sqrt{13}$$

and $\tan \alpha = 2/3 \quad \therefore \quad \alpha \tan^{-1} 2/3$

$$\therefore \quad I = \frac{1}{r} \int \frac{dx}{\sin(\alpha + x)} = \frac{1}{r} \int \frac{dx}{\sin(x + \alpha)} = \frac{1}{r} \int \cosec(x + \alpha) dx$$

$$= \frac{1}{r} \log \left[\tan \frac{(x + \alpha)}{2} \right] + c$$

$$= \frac{1}{\sqrt{13}} \log \left[\tan \left(\frac{x}{2} + \frac{1}{2} \tan^{-1} \frac{2}{3} \right) \right] + c \quad \text{Ans.}$$

Method (II): Let $2 = r \cos \alpha$ and $3 = r \sin \alpha$

$$\therefore \quad r^2 = 2^2 + 3^2 = 13 \quad \therefore \quad r = \sqrt{13}$$

and $\tan \alpha = 3/2 \quad \therefore \quad \alpha = \tan^{-1} 3/2$

$$\therefore \quad I = \frac{1}{r} \int \frac{dx}{\cos \alpha \cos x + \sin \alpha \sin x} dx$$

$$= \frac{1}{r} \int \frac{dx}{\cos x \cos \alpha + \sin x \sin \alpha} dx$$

$$= \frac{1}{r} \int \frac{dx}{\cos(x - \alpha)}$$

$$= \frac{1}{r} \int \sec(x - \alpha) dx$$

$$= \frac{1}{r} \log \left[\tan \left(\frac{x - \alpha}{2} + \frac{\pi}{4} \right) \right] + c$$

$$= \frac{1}{\sqrt{13}} \log \left[\tan \left\{ \frac{x}{2} - \frac{1}{2} \tan^{-1} \frac{3}{2} + \frac{\pi}{4} \right\} \right] + c \quad \text{Ans.}$$

Remark: Observe that the integral is in simpler form, *if the expression $(a \sin x \pm b \cos x)$ in question is expressed in the form $\sin(\alpha \pm x)$ instead of $\cos(\alpha \pm x)$.*

2.3.1 Converting the Non-Standard Formats to the Standard form of the Integral $(dx/(a \sin x + b \cos x))$

Certain integrals can be expressed in the form $\int (dx/a \sin x + b \cos x)$. By identifying such integrals, we can easily integrate them as done in the above solved examples. Such examples are important. One such example is given below, which may be evaluated.

$$\int \frac{\sec x}{\sqrt{3} + \tan x} \, dx = \int \frac{dx}{\sqrt{3} \cos x + \sin x}$$

$$\frac{1}{2} \log \left[\tan \left(\frac{x}{2} + \frac{\pi}{6} \right) \right] + c \quad \text{Ans.}$$

The following integrals involving trigonometric functions, $\sin x$ and $\cos x$, appear to be simple, but they cannot be converted to the standard form(s) by trigonometric and algebraic manipulations

$$\int \frac{dx}{a + b \sin x}, \quad \int \frac{dx}{a + b \cos x}, \quad \int \frac{dx}{a \sin x + b \cos x + c}$$

where a, b, and c are integers. (These integrals should not be confused with those discussed in Section 2.3).

We shall introduce a very simple substitution, which can be uniformly used in evaluating all such integrals. But, *as a prerequisite, it is necessary to first establish the following standard integrals*, since the above integrals are reduced to quadratic algebraic functions, due to substitution:

(1) $\int \frac{1}{x^2 + a^2} \, dx = \frac{1}{a} \tan^{-1} \left(\frac{x}{a} \right) + c$

(2) $\int \frac{1}{x^2 - a^2} \, dx = \frac{1}{2a} \log \left(\frac{x - a}{x + a} \right) + c, \quad x > a$

(3) $\int \frac{1}{a^2 - x^2} \, dx = \frac{1}{2a} \log \left(\frac{a + x}{a - x} \right) + c, \quad x < a$

Using these *standard integrals*, we can also evaluate integrals of the form $\int (dx/(ax^2 + bx + c))$.

Details are discussed in Chapter 3b.

3a Integration by Substitution: Change of Variable of Integration

3a.1 INTRODUCTION

So far we have evaluated integrals of functions, which are of *standard forms* and those, *which can be reduced to standard forms* by simple algebraic operations or trigonometric simplification methods including the use of trigonometric identities. Many integrals cannot be reduced to standard forms by these methods. We must, therefore, learn other techniques of integration.

In this chapter, we shall discuss the method of substitution, which is applicable in reducing to standard forms, the integrals involving composite functions. It will be observed that *this method involves change of variable of integration* as against the earlier methods, wherein the variable of integration remains unchanged. Before introducing the theorem which governs the rule of *integration by substitution*, let us recall the *chain rule for differentiation, as applied to a power of a function*. If $u = f(x)$ is a differentiable function and r is a rational number, then

$$\frac{d}{dx}\left[\frac{u^{r+1}}{r+1}\right] = \frac{(r+1)u^r}{(r+1)} \cdot \frac{du}{dx} = u^r \cdot \frac{d}{dx}(u), (r \text{ is rational}, r \neq -1)$$

$$\text{or } \frac{d}{dx}\left(\frac{[f(x)]^{r+1}}{r+1}\right) = [f(x)]^r \cdot f'(x)^{(1)}$$

From the above result, *we obtain the following important rule for indefinite integrals.*

3a.2 GENERALIZED POWER RULE

Let f be a *differentiable function* and "r" be a *rational number* other than "-1". Then

$$\int [f(x)]^r f'(x)dx = \frac{[f(x)]^{r+1}}{r+1} + c, (r \text{ rational}, r \neq -1)$$

Note that, in the above statement, we have simply used the definition of an antiderivative (or an integral).

Let us apply the above rule for evaluating the following integrals.

3a-Integration by substitution (Change of variable of integration)

[1] It follows that $\int [f(x)]^r f'(x)dx = \frac{[f(x)]^{r+1}}{r+1} + c$, ($r$ rational, $r \neq -1$). In particular, $\int x^n dx = \frac{x^{n+1}}{n+1} + c$, $n \neq -1$.

Introduction to Integral Calculus: Systematic Studies with Engineering Applications for Beginners, First Edition.
Ulrich L. Rohde, G. C. Jain, Ajay K. Poddar, and A. K. Ghosh.
© 2012 John Wiley & Sons, Inc. Published 2012 by John Wiley & Sons, Inc.

Example (1): Find

(a) $\int (x^3 + 2x)^{25}(3x^2 + 2)dx$

(b) $\int \sin^{12} x \cos x \, dx$

Solution:

(a) To evaluate $\int (x^3 + 2x)^{25}(3x^2 + 2)dx$,

we observe that $\dfrac{d}{dx}(x^3 + 2x) = 3x^2 + 2.$

Let $f(x) = x^3 + 2x$,
$\therefore f'(x) = 3x^2 + 2.$
Thus, by the above theorem,

$$\int (x^3 + 2x)^{25}(3x^2 + 2)dx = \int [f(x)]^{25} \cdot f'(x)dx$$
$$= \frac{[f(x)]^{26}}{26} + c$$
$$= \frac{(x^3 + 2x)^{26}}{26} + c$$

(b) To evaluate $\int \sin^{12} x \cos x \, dx$, we observe that $(d/dx)(\sin x) = \cos x.$
Let $f(x) = \sin x$, then $f'(x) = \cos x.$ Thus,

$$\int \sin^{12} x \cos x \, dx = \int [f(x)]^{12} f'(x)dx$$
$$= \frac{[f(x)]^{13}}{13} + c$$
$$= \frac{\sin^{13} x}{13} + c$$

Now, *we can see why Leibniz used the differential dx in his notation* $\int \ldots dx$. If we put $u = f(x)$,
then $du = f'(x)dx$. Therefore, *the conclusion of result (1) is that*

$$\int u^r du = \frac{u^{r+1}}{r+1} + c, \quad r \neq -1, \quad \text{which is the ordinary power rule, } with \text{ "}u\text{" } as the variable.$$

Thus, the generalized power rule is just the ordinary power rule applied to functions. But in
applying the power rule to functions, we must make sure that we have du to go with u^r.

In the integral of Example 1(a), the function $f(x) = (x^3 + 2x)$ and its differential $f'(x)$
$dx = (3x^2 + 2)dx$, *both appear in the element of integration*. Similarly, in Example 1(b) the
function sin x and its *differential* cos xdx both appear in the element of integration.
Such integrals are easily expressed in the standard form(s) by substituting $f(x) = u$ *and
replacing* $f'(x)dx$ *by* du.

The following examples will make this point clearer.

Example (2): Evaluate the following integrals:

(a) $\int (x^3 + 6x)^5 (6x^2 + 12)dx$

(b) $\int (x^2 + 4)^{10} x \, dx$

(c) $\int \left(\frac{x^2}{2} + 3\right)^2 x^2 \, dx$

(d) $\int (2x + 3)\cos(x^2 + 3x)dx$

(a) To evaluate $\int (x^3 + 6x)^5 (6x^2 + 12) dx$

Let $x^3 + 6x = u$

$\therefore \quad (3x^2 + 6)dx = du$

$\therefore \quad (6x^2 + 12)dx = 2(3x^2 + 6)dx = 2du.$

$\therefore \quad \int (x^3 + 6x)^5 (6x^2 + 12)dx = \int u^5 \cdot 2 du$

$$= 2\int u^5 du = 2\left[\frac{u^6}{6} + c\right] = \frac{u^6}{3} + 2c$$

$$= \frac{u^6}{3} + k.$$

Here, *two things must be noted* about our solution.

(i) Note that $(6x^2 + 12)dx = 2du$ (instead of "du"). The factor 2 could be moved in front of the integral sign as shown above.

(ii) The constant "$2c$" obtained above is still *an arbitrary constant* and we may call it k.

(b) To evaluate $\int (x^2 + 4)^{10} x\, dx$

Let $x^2 + 4 = u$

$\therefore \quad 2x\, dx = du$

$\therefore \quad x\, dx = \frac{1}{2}du.$ Thus,

$\therefore \quad \int (x^2 + 4)^{10} x \quad dx = \int u^{10} \cdot \frac{1}{2} \cdot du$

$$= \frac{1}{2}\int u^{10} du$$

$$= \frac{1}{2}\left[\frac{u^{11}}{11} + c\right] = \frac{(x^2 + 4)^{11}}{22} + k.$$

(c) To evaluate $\int \left(\frac{x^2}{2} + 3\right)^2 x^2\, dx$

Let $\frac{x^2}{2} + 3 = u,\quad \therefore x \cdot dx = du$

Here, *the method illustrated in (a) and (b) fails because* $x^2\, dx = x(x\, dx) = x\, du$, *and* x *cannot be passed in front of the integral sign* (that can be done only with a constant factor). However, by ordinary algebra, we can express the given integral, as

$$\int \left(\frac{x^2}{2} + 3\right)^2 x^2\, dx = \int \left(\frac{x^4}{4} + 3x^2 + 9\right)x^2\, dx$$

$$= \int \left(\frac{x^6}{4} + 3x^4 + 9x^2\right)dx$$

$$= \frac{x^7}{28} + \frac{3x^5}{5} + 3x^3 + c.$$

(d) To evaluate $\int (2x + 3)\cos(x^2 + 3x)dx$

If we put $x^2 + 3x = u$, then we get $(2x + 3)dx = du$.

By using these relations, the given integral transforms to $\int \cos u\, du$. *Note that this integral is not a power function.* However, formally we get

$$\int (2x + 3)\cos(x^2 + 3x)dx = \int \cos u\, du$$

$$= \sin u + c$$

$$= \sin(x^3 + 3x) + c.$$

Remark: We observe that *the method of substitution, introduced above for power functions, extends for beyond that use.* Now, we introduce the theorem, which governs the *rule of integration by substitution.*

3a.3   THEOREM

If $x = \phi(t)$ is a differentiable function of t and $f[\phi(t)]$ exists, then

$$\int f(x)dx = \int f(x) \cdot \frac{dx}{dt} \cdot dt = \int f[\phi(t)]\phi'(t)dt.$$

Proof: It is given that
$x = \phi(t)$ is a differentiable function of t.

$$\therefore \quad \frac{dx}{dt} = \phi'(t).$$

Let $\int f(x)dx = F(x)$ \hfill (1)

$$\therefore \text{ By definition of an integral. } \frac{d}{dx}[F(x)] = f(x), \hfill (2)$$

Now, by chain rule, we have

$$\therefore \quad \frac{d}{dt}[F(x)] = \frac{d}{dx}F(x) \cdot \frac{dx}{dt}$$

$$= f(x) \cdot \frac{dx}{dt}, [\text{using}(2)]$$

$\therefore$ By the definition of integration,

$$F(x) = \int f(x)\frac{dx}{dt} \cdot dt,$$

$$\text{or } \int f(x)dx = \int f(x)\frac{dx}{dt} \cdot dt \hfill (3)$$

$$= \int f[\phi(t)]\phi'(t)dt \left[\begin{array}{l} \because x = \phi(t) \\ \therefore \dfrac{dx}{dt} = \phi'(t) \end{array}\right.$$

Note (1): We have seen that if $x = \phi(t)$ is a differentiable function of t and $f[\phi(t)]$ exists then,

$$\int f(x)dx = \int f(x)\frac{dx}{dt} \cdot dt$$

$$= \int f[\phi(t)]\phi'(t)dt.$$

From this statement, we see that if in $\int f(x)dx$, we substitute $x = \phi(t)$, then "dx" gets replaced by $\phi'(t)dt$. Thus, if $x = \phi(t)$, then $dx = \phi'(t)dt$.[2]

---

[2] In the case of any composite function $y = f(x) = f[\phi(t)]$ the role of independent variable is played by the function $\phi(t)$ and we have seen that its differential [i.e., $\phi'(t)dt$] replaces the differential dx.

Note (2): Again consider the result (3). We have

$$\int f[\phi(t)]\phi'(t)\mathrm{d}t = \int f(x)\mathrm{d}x, \quad \text{where } \phi(t) = x$$

Now, interchanging the roles of x and t, we get,

$$\left. \begin{array}{c} \int f[\phi(x)]\phi'(x)\mathrm{d}x = \int f(t)\mathrm{d}t, \\[2mm] \text{or} \quad t = \phi(x), \quad \therefore \quad \phi'(x)\mathrm{d}x = \mathrm{d}t \end{array} \right\} \tag{4}$$

3a.3.1 Corollaries from the Rule of Integration by Substitution

Observe that the integrand on left-hand side of Equation (4) is complicated. In this integrand, we have $f[\phi(x)]$ (*as a part of the integrand*), which is a function of a function. If we substitute $\phi(x) = t$, then $\phi'(x)\mathrm{d}x$ gets replaced by $\mathrm{d}t$. Thus, the substitution $\phi(x) = t$ simplifies the integrand.

The new integral with changed variable may be in the standard form. In using this method, it is important to recognize the form $f[\phi(x)] \cdot \phi'(x)$ in the integrand. (In other words, it is helpful to decide the most convenient substitution, if we can identify a function and its derivative in the integrand). Only then can we find a suitable substitution, viz. $\phi(x) = t$. *Essentially, this method of integration reduces to finding out what kind of substitution has to be performed for the given integrand.*

Also, remember that the differential $\phi'(x)\mathrm{d}x$ may at times be expressed in the form $k \cdot \mathrm{d}t$, where k is a constant. Using the rule of integration by substitution, given by (4), we can easily prove the following results.

Corollary (1): $\displaystyle \int [f(x)]^n \cdot f'(x)\mathrm{d}x = \frac{[f(x)]^{n+1}}{n+1} + c, \quad n \neq -1$

Corollary (2): $\displaystyle \int \frac{f'(x)}{[f(x)]^n}\mathrm{d}x = \frac{-1}{(n-1)[f(x)]^{n-1}} + c, \quad n \neq 1$

Corollary (3): $\displaystyle \int \frac{f'(x)}{\sqrt{f(x)}}\mathrm{d}x = 2\sqrt{f(x)} + c$

Corollary (4): $\displaystyle \int \frac{f'(x)}{f(x)}\mathrm{d}x = \log_e[f(x)] + c, \quad f(x) > 0$

$\qquad\qquad$ [More correctly $= \log_e|f(x)| + c$]

Corollary (5): $\displaystyle \int e^{f(x)} \cdot f'(x)\mathrm{d}x = e^{f(x)} + c$

Corollary (6): $\displaystyle \int a^{f(x)} \cdot f'(x)\mathrm{d}x = \frac{a^{f(x)}}{\log_e a} + c, \quad a > 0$

Corollary (7): If $\int f(x)dx = \phi(x)$, then $\int f(ax+b)dx = \dfrac{\phi(ax+b)}{a} + c$

Note: *These corollaries should be treated as individual problems and not as formulas.* When solving problems we must use *only the standard formulas,* which are necessary for writing the integral(s), in the new variable "*t*" (say). Later on *t* must be replaced by *f(x)* while writing the final result(s).

To get a feel, how simple it is to establish these corollaries, we prove Corollaries (2), (4), (6), and (7).

Corollary (2): $\displaystyle \int \frac{f'(x)}{[f(x)]^n}\,dx = \frac{-1}{(n-1)[f(x)]^{n-1}} + c, \quad n \neq -1$

Proof: Let $I = \displaystyle \int \frac{f'(x)}{[f(x)]^n}\,dx$

Put $f(x) = t$ $\therefore$ $f'(x)\,dx = dt$

$$\therefore \quad I = \int \frac{dt}{t^n} = \int t^{-n}\,dt$$

$$= \frac{t^{-n+1}}{-n+1} = \frac{t^{-(n-1)}}{-(n-1)}$$

$$= \frac{1}{-(n-1)t^{n-1}} + c, \quad n \neq 1$$

$$= \frac{-1}{(n-1)[f(x)]^{n-1}} + c, \quad n \neq 1 \quad \text{Ans.}$$

Corollary (4): $\displaystyle \int \frac{f'(x)}{f(x)}\,dx = \log_e|f(x)| + c$

Here, it is convenient to put $f(x) = t$, so that, $f'(x)dx = dt$,

$$\therefore \quad \int \frac{f'(x)dx}{f(x)} = \int \frac{dt}{t} = \log_e|t| = \log_e|f(x)| + c \quad \text{Ans.}$$

Corollary (6): $\displaystyle \int a^{f(x)} \cdot f'(x)dx = \frac{a^{f(x)}}{\log_e a} + c, \quad a > 0$

Proof: Let $I = \int a^{f(x)} \cdot f'(x)dx$

Put $f(x) = t$ $\therefore$ $f'(x)\mathrm{d}x = \mathrm{d}t$

$$\therefore \quad I = \int a^t\,\mathrm{d}t = \frac{a^t}{\log_e a} + c = \frac{a^{f(x)}}{\log_e a} + c \quad \text{Ans.}$$

Corollary (7):   If $\int f(x)\mathrm{d}x = \phi(x)$, then $\int f(ax+b)\mathrm{d}x = \dfrac{\phi(ax+b)}{a} + c$

Proof: It is given that

$$\int f(x)\mathrm{d}x = \phi(x) \tag{5}$$

Let $\int f(ax+b)\mathrm{d}x = I$

Put $ax + b = t$

$\therefore$ $a\,\mathrm{d}x = \mathrm{d}t$.

$\therefore$ $\mathrm{d}x = (1/a)\mathrm{d}t$

$$\therefore \quad I = \int f(t)\frac{1}{a}\mathrm{d}t = \frac{1}{a}\int f(t)\mathrm{d}t = \frac{1}{a}\phi(t), \text{by (5)}$$

$$= \tfrac{1}{a}\phi(ax+b) + c \qquad \text{Ans.}$$

3a.3.2   Importance of Corollary (7)

The result of Corollary (7) tells us that *corresponding to every standard integral of the type* $\int f(x)\mathrm{d}x$, *we can at once write one more standard result for* $\int f(ax+b)\mathrm{d}x$. *In fact,* $f(ax+b)$ *is an extended form of* $f(x)$. Here, we must remember two important things:

(i) If x in $f(x)$ is replaced by *a linear expression* $(ax+b)$, then the corresponding integral is expressed by *writing the linear expression in place of* x *in the standard formula* divided by the coefficient of x. Note that if x in $f(x)$ is replaced by any expression other than the linear one, then we do not get any standard result. Check this.

(ii) *The integrals of such extended forms are also treated as standard results and hence they can be used as formulas.*

Now, we give a list *of some standard results,* for extended forms.

3a.3.2.1   Standard Results for Extended Forms

(1) $\int \sin(ax+b)\mathrm{d}x = -\tfrac{1}{a}\cos(ax+b) + c$

(2) $\int \cos(ax+b)\mathrm{d}x = \tfrac{1}{a}\sin(ax+b) + c$

(3) $\int \sec^2(ax+b)\mathrm{d}x = \tfrac{1}{a}\tan(ax+b) + c$

(4) $\int \mathrm{cosec}^2(ax+b)\mathrm{d}x = -\tfrac{1}{a}\cot(ax+b) + c$

(5) $\int \sec(ax+b)\tan(ax+b)\mathrm{d}x = \tfrac{1}{a}\sec(ax+b) + c$

(6) $\int \mathrm{cosec}(ax+b)\cot(ax+b)\mathrm{d}x = -\tfrac{1}{a}\mathrm{cosec}(ax+b) + c$

(7) $\int e^{ax+b}\mathrm{d}x = \tfrac{1}{a}e^{ax+b} + c$

(8) $\int m^{ax+b}\mathrm{d}x = \tfrac{1}{a} \cdot \dfrac{m^{ax+b}}{\log_e m} + c, \quad m > 0$ [see Corollary (6)]

(9) $\int (ax+b)^n dx = \dfrac{1}{a} \cdot \dfrac{(ax+b)^{n+1}}{n+1} + c, \quad n \neq -1$

(10) $\int \dfrac{1}{ax+b} dx = \dfrac{1}{a} \log_e (ax+b), \quad (ax+b) > 0$

3a.3.3   Importance of Corollary (4)

Corollary (4) helps in *finding the integrals of tan x, cot x, sec x, and cosec x.*

(1) $\int \tan x \, dx = \int \dfrac{\sin x}{\cos x} dx = I$ (say)

> **Method (1):** Put $\cos x = t$
>
> $\therefore \quad -\sin x \, dx = dt$
>
> or $\sin x \, dx = -dt$
>
> $\therefore \quad I = -\int \dfrac{dt}{t} = -\log_e t = \log_e t^{-1}$
>
> $\qquad = \log_e (\cos x)^{-1} = \log_e (\sec x) + c, \quad \sec x > 0$
>
> $\boxed{\therefore \quad \int \tan x \, dx = \log_e |\sec x| + c^{(3)}}$ Ans.

(2) $\int \cot x \, dx = \int \dfrac{\cos x}{\sin x} dx = I$ (say)

> Put $\sin x = t$
>
> $\therefore \quad \cos x \, dx = dt$
>
> $\therefore \quad I = \int \dfrac{dt}{t} = \log_e t = \log_e (\sin x) + c, \quad \sin x > 0$
>
> $\boxed{\therefore \quad \int \cot x = \log_e |\sin x| + c}$ Ans.

(3) $\int \cosec x \, dx = \int \dfrac{1}{\sin x} dx = I$ (say)

> $\therefore \quad I = \int \dfrac{1}{2 \sin(x/2) \cdot \cos(x/2)} dx$
>
> Dividing N^r and D^r by $\cos^2 \dfrac{x}{2}$, we get $I = \int \dfrac{\sec^2(x/2)}{2 \tan(x/2)}$
>
> $= \int \dfrac{(1/2) \cdot \sec^2(x/2)}{\tan(x/2)} dx^{(4)}$

[3] **Method (2)** $\int \tan x \, dx = \int \dfrac{\sec x \tan x}{\sec x} dx$

$\qquad\qquad\qquad\qquad = \int \dfrac{(d/dx)(\sec x)}{\sec x} dx$

$\qquad\qquad\qquad\qquad = \log_e (\sec x) + c, \quad \sec x > 0$

$\qquad\qquad\qquad\qquad = \log_e |\sec x| + c$ Ans.

[4] Here, we have expressed "cosec x" in the form $(\sec^2(x/2))/(2\tan(x/2))$, which is convenient for integration by the method of substitution. (It can also be expressed in other useful forms.)

Method (1): Put $\tan\dfrac{x}{2} = t$

$$\therefore \frac{1}{2}\sec^2\frac{x}{2}\,dx = dt$$

$$\therefore \quad I = \int \frac{dt}{t} = \log_e(t)$$

$$= \log_e\left(\tan\frac{x}{2}\right) + c, \quad \tan\frac{x}{2} > 0$$

$$\boxed{\therefore \ \int \operatorname{cosec} x\,dx = \log_e\left|\tan\frac{x}{2}\right| + c^{(5)}} \quad \text{Ans.}$$

(4) $\displaystyle\int \sec x\,dx = \int \frac{1}{\cos x}\,dx = I \quad$ (say)

$$\therefore \quad I = \int \frac{1}{\cos^2(x/2) - \sin^2(x/2)}\,dx$$

Dividing N^r and D^r by $\cos^2(x/2)$, we get

$$I = \int \frac{\sec^2(x/2)}{1 - \tan^2(x/2)}\,dx$$

Method (1): Put $\tan(x/2) = t \quad \therefore \quad \frac{1}{2}\sec^2(x/2) \cdot dx = dt$

$$\therefore \quad I = \int \frac{2\,dt}{1 - t^2}$$

$$= \int \left[\frac{1}{1+t} + \frac{1}{1-t}\right]dt, \quad \left\{ \begin{array}{l} \text{Note that} \dfrac{1}{1+t} + \dfrac{1}{1-t} = \dfrac{(1-t)+(1+t)}{1-t^2} = \dfrac{2}{1-t^2} \\ \text{We shall learn about this technique, later.} \end{array} \right.$$

$$= \log_e(1+t) - \log_e(1-t) + c$$

$$= \log_e\frac{1+t}{1-t} + c$$

$$= \log_e\left[\frac{1+\tan(x/2)}{1-\tan(x/2)}\right] = \log_e\left[\frac{\tan(x/2) + \tan(\pi/4)}{1 - \tan(x/2) \cdot \tan(\pi/4)}\right]$$

$$= \log_e[\tan((x/2) + (\pi/4))] + c \quad [\because \tan(\pi/4) = 1]$$

(5) **Method (2):** $\displaystyle\int \operatorname{cosec} x\,dx = \int \frac{\operatorname{cosec} x(\operatorname{cosec} x - \cot x)}{(\operatorname{cosec} x - \cot x)}\,dx$

$$= \int \frac{\operatorname{cosec}^2 x - \operatorname{cosec} x \cot x}{\operatorname{cosec} x - \cot x}\,dx$$

$$= \int \frac{(d/dx)(\operatorname{cosec} x - \cot x)}{\operatorname{cosec} x - \cot x}\,dx = \log_e(\operatorname{cosec} x - \cot x) + c$$

$$\boxed{\therefore \int \operatorname{cosec} x\,dx = \log_e|\operatorname{cosec} x - \cot x| + c}$$

Further note that

$$\operatorname{cosec} x - \cot x = \frac{1}{\sin x} - \frac{\cos x}{\sin x} = \frac{1 - \cos x}{\sin x}$$

$$= \frac{1 - [1 - 2\sin^2(x/2)]}{2\sin(x/2) \cdot \cos(x/2)} = \frac{2\sin^2(x/2)}{2\sin(x/2) \cdot \cos(x/2)}$$

$$= \tan\frac{x}{2}.$$

$$\int \sec x \, dx = \log_e \left| \tan \left(\frac{x}{2} + \frac{\pi}{4} \right) \right| + c^{(6)} \quad \text{Ans.}$$

Note: The four integrals obtained above are treated as *standard integrals*. Hence, we add the following results to our list of standard formulae.

(1) $\int \tan x \, dx = \log_e |\sec x| + c = \log(\sec x) + c$

(2) $\int \cot x \, dx = \log_e |\sin x| + c = \log(\sin x) + c^{(7)}$

(3) $\int \mathrm{cosec}\, x \, dx = \log(\mathrm{cosec}\, x - \cot x) + c$

$\qquad = \log\left(\tan\frac{x}{2}\right) + c$

(4) $\int \sec x \, dx = \log(\sec x + \tan x) + c$

$\qquad = \log\left(\tan\left(\frac{x}{2} + \frac{\pi}{4}\right)\right) + c$

Further, in view of the Corollary (7)

$$\left[\text{i.e.,} \int f(x)dx = \phi(x) \Rightarrow \int f(ax+b)dx = \frac{1}{a}\phi(ax+b) \right]$$

We also have the following *standard results*:

(1A)$\int \tan(ax+b)dx = \frac{1}{a}\log(\sec(ax+b)) + c$

(1B)$\int \cot(ax+b)dx = \frac{1}{a}\log(\sin(ax+b)) + c$

(1C) $\int \mathrm{cosec}(ax+b)dx = \frac{1}{a}\log\left(\tan\left(\frac{ax+b}{2}\right)\right) + c$

$\qquad = \frac{1}{a}\log(\mathrm{cosec}(ax+b) - \cot(ax+b)) + c$

(1D) $\int \sec(ax+b)dx = \frac{1}{a}\log\left[\tan\left(\frac{ax+b}{2} + \frac{\pi}{4}\right)\right] + c$

$\qquad = \frac{1}{a}\log[\sec(ax+b) + \tan(ax+b)] + c$

[6] **Method (2):**

$$\int \sec x \, dx = \int \frac{\sec x(\sec x + \tan x)}{(\sec x + \tan x)} dx$$

$$= \int \frac{\sec^2 x + \sec x \tan x}{\sec x + \tan x} = \int \frac{(d/dx)(\sec x + \tan x)}{(\sec x + \tan x)} dx$$

$$= \log_e |\sec x + \tan x| + c$$

Further, note that $\sec x + \tan x = \dfrac{1}{\cos x} + \dfrac{\sin x}{\cos x} = \dfrac{1 + \sin x}{\cos x}$

$$= \frac{\sin^2(x/2) + \cos^2(x/2) + 2\sin(x/2)\cdot\cos(x/2)}{\cos^2(x/2) - \sin^2(x/2)} = \frac{(\cos(x/2) + \sin(x/2))^2}{\cos^2(x/2) - \sin^2(x/2)}$$

$$= \frac{\cos(x/2) + \sin(x/2)}{\cos(x/2) - \sin(x/2)} = \frac{1 + \tan(x/2)}{1 - \tan(x/2)} = \tan\left(\frac{x}{2} + \frac{\pi}{4}\right)$$

[7] **Important Note:** Here onwards, we agree to use the notation $\log[\phi(x)]$ to mean $\log_e|\phi(x)|$. This is done for saving time and effort. However, the importance of the base "e" and that of the symbol for absolute value must always be remembered.

3a.3.4   Some Solved Examples

In using the method of substitution, it is important to see carefully the form of the element of integration. *Usually, we make a substitution for a function whose derivative also occurs in the integrand.* This will be clear from the following examples.

Example (3):   Find $I = \int x^3 \sin x^4 \, dx$

Put $x^4 = t$

$\therefore$ $4x^3 \, dx = dt$ $\therefore x^3 \, dx = \frac{1}{4} \, dt$

$\therefore$ $I = \frac{1}{4} \int \sin t \, dt$

$= \frac{1}{4}(-\cos t) + c = -\frac{1}{4}\cos x^4 + c$ Ans.

Example (4):   Find $I = \int \dfrac{\sin x}{1 + \cos x} \, dx$

Put $1 + \cos x = t$

$\therefore$ $-\sin x \, dx = dt$ $\therefore \sin x \, dx = -dt$

$\therefore$ $I = -\int \dfrac{dt}{t} = -\log t + c$

$= -\log(1 + \cos x) + c$ Ans.

Example (5):   Find $I = \int \dfrac{\cos(\log x)}{x} \, dx$

Put $\log x = t$ $\therefore (1/x) \, dx = dt$

$\therefore$ $I = \int \cos t \, dt = \sin t + c$

$= \sin(\log x) + c$ Ans.

Example (6):   Find $I = \int \dfrac{1}{x \log x [\log(\log x)]} \, dx$

Put $\log (\log x) = t$ $\therefore$ $\dfrac{1}{\log x} \cdot \dfrac{1}{x} dx = dt$

i.e., $\dfrac{1}{x \log x} dx = dt$

$\therefore$ $I = \int \dfrac{dt}{t} = \log t + c = \log[\log(\log x)] + c$ Ans.

Example (7):   Find $I = \int x \, a^{x^2} \, dx$

Put $x^2 = t$ $\therefore$ $2x \, dx = dt$

$\therefore$ $x \, dx = \frac{1}{2} \, dt$

$\therefore$ $I = \frac{1}{2} \int a^t \, dt = \frac{1}{2} \cdot \dfrac{a^t}{\log a} + c$

$= \frac{1}{2} \cdot \dfrac{a^{x^2}}{\log a} + c$ Ans.

Example (8): Evaluate $I = \displaystyle\int \dfrac{2 \sin x \cdot \cos x}{\sin^4 x + \cos^4 x} dx$

Method (1) Dividing N^r and D^r by $\cos^4 x$, we get

$$I = \int \frac{2 \tan x \cdot \sec^2 x}{\tan^4 x + 1} dx. \qquad \begin{cases} \dfrac{2 \sin x \cdot \cos x}{\cos^4 x} \\ = 2\dfrac{\sin x}{\cos x} \cdot \dfrac{1}{\cos^2 x} \\ = 2 \tan x \cdot \sec^2 x \end{cases}$$

Now, put $\tan^2 x = t$ $\quad \therefore \quad 2 \tan x \sec^2 x \, dx = dt$

$$I = \int \frac{dt}{1 + t^2} = \tan^{-1} t + c = \tan^{-1}(\tan^2 x) + c \quad \text{Ans.}$$

Method (2)

$$I = \int \frac{2 \sin x \cos x}{\sin^4 x + \cos^4 x} dx = \int \frac{2 \sin x \cos x}{(\sin^2 x)^2 + (1 - \sin^2 x)^2} dx$$

Put $\sin^2 x = t$ $\quad \therefore 2 \sin x \cos x \, dx = dt$

$$\therefore \quad I = \int \frac{dt}{t^2 + (1 - t)^2} = \int \frac{dt}{t^2 + 1 - 2t + t^2}$$

$$= \int \frac{dt}{2t^2 - 2t + 1} = \frac{1}{2} \int \frac{dt}{t^2 - t + (1/2)} \qquad \begin{cases} \text{Consider } 2t^2 - 2t + 1 \\ = 2(t^2 - t + (1/2)) \\ \text{Now, } t^2 - t + \dfrac{1}{2} \\ = t^2 - 2\left(\dfrac{1}{2}\right)t + \dfrac{1}{4} - \dfrac{1}{4} + \dfrac{1}{2} \\ = \left(t - \dfrac{1}{2}\right)^2 + \left(\dfrac{1}{2}\right)^2 \end{cases}$$

$$= \frac{1}{2} \int \frac{dt}{(t - (1/2))^2 + (1/2)^2}$$

$$= \frac{1}{2} \tan^{-1} \frac{t - (1/2)}{1/2} + c = \frac{1}{2} \tan^{-1}(2t - 1) + c$$

$$= \frac{1}{2} \tan^{-1}(2 \sin^2 x - 1) + c \quad \text{Ans.}$$

Example (9): Evaluate $I = \displaystyle\int \dfrac{x^{e-1} + e^{x-1}}{x^e + e^x} dx$

Put $x^e + e^x = t$

$\therefore \quad (e \, x^{e-1} + e^x) \, dx = dt$

$\therefore \quad e \, (x^{e-1} + e^{x-1}) \, dx = dt \qquad$ [Note this step]

$\therefore \quad I = \dfrac{1}{e} \displaystyle\int \dfrac{dt}{t} = \dfrac{1}{e} \log t + c = \dfrac{1}{e} \log(x^e + e^x) + c$

Example (10): Evaluate $I = \displaystyle\int \frac{4\,e^x}{7 - 3\,e^x}\,dx$

$$I = \int \frac{4\,e^x}{7 - 3\,e^x}\,dx \qquad \begin{cases} \text{Put}(7 - 3\,e^x) = t \\[2mm] \therefore \quad -3\,e^x\,dx = dt \\[2mm] \therefore \quad e^x\,dx = -\dfrac{1}{3}\,dt \end{cases}$$

$$-\frac{4}{3}\log(7 - 3\,e^x) + c \quad \text{Ans.}$$

Next, observe that the integral $I = \int \frac{1}{e^x + e^{-x}}\,dx$ can also be expressed in a form that is convenient for substitution.

We have, $e^x + e^{-x} = e^x + \dfrac{1}{e^x} = \dfrac{e^{2x} + 1}{e^x}$.

$$\therefore \quad I = \int \frac{1}{e^x + e^{-x}}\,dx = \int \frac{e^x}{e^{2x} + 1}\,dx = \int \frac{e^x}{1 + (e^x)^2}\,dx$$

Now, put $e^x = t$ $\therefore$ $e^x\,dx = dt$

$$\therefore \quad I = \int \frac{dt}{1 + t^2} = \tan^{-1} t + c = \tan^{-1}(e^x) + c \quad \text{Ans.}$$

Similarly, $\int \frac{1}{3 + 4e^{-x}}\,dx = \int \frac{e^x}{3e^x + 4}\,dx$, which is of the same form as in the Example (8).

Now, look at the following *integral which appears to be of similar type, but it cannot be integrated so easily.*

Example (11): $\displaystyle\int \frac{1}{3 + 4e^x}\,dx$

$$\int \frac{1}{3 + 4\,e^x}\,dx \qquad [\text{Here, } N^r \text{ is a constant and } D^r \text{ is of the form}(ae^x + b)]$$

(Note that here no substitution is possible without changing the integrand, as shown below.) In such cases, we divide N^r and D^r by e^x, so that we get

$I = \int \frac{e^{-x}}{3\,e^{-x} + 4}\,dx$, which is now of the type at Example (8).

Now put $3e^{-x} + 4 = t$

$\therefore$ $-3e^{-x}\,dx = dt$

$\therefore$ $e^{-x}\,dx = -1/3\ dt$

$$\therefore I = -\frac{1}{3}\int \frac{dt}{t} = -\frac{1}{3}\log t + c$$

$$= -\frac{1}{3}\log(3e^{-x} + 4) + c \quad \text{Ans.}$$

Remark: Note that, it is comparatively simpler to evaluate the integral $\int \frac{1}{3 + 4e^{-x}}\,dx$, than to evaluate the integral $\int \frac{1}{3 + 4e^x}\,dx$.

Example (12): Evaluate $I = \int \dfrac{e^{2x} - 1}{e^{2x} + 1} dx$

Method (1): In such cases, we may break up the integral into two parts: one of the type at Example (10) and the other of the type at Example (11), or else divide N^r and D^r by half the power given to "e".

Dividing N^r and D^r by e^x, we get

$$I = \int \frac{e^x - e^{-x}}{e^x + e^{-x}} dx$$

Put $(e^x + e^{-x}) = t$ $\therefore (e^x - e^{-x}) dx = dt$

$$\therefore I = \int \frac{dt}{t} = \log t + c = \log(e^x - e^{-x}) + c \qquad \text{Ans.}$$

Method (2): $I = \int \dfrac{e^{2x} - 1}{e^{2x} + 1} dx$

$$= \int \frac{e^{2x}}{e^{2x} + 1} dx - \int \frac{1}{e^{2x} + 1} dx$$

$$= I_1 - I_2 \text{ (say).}$$

Consider I_1

Put $e^{2x} + 1 = t \therefore 2 e^{2x} dx = dt \therefore e^{2x} dx = \frac{1}{2} dt$

$$\therefore I_1 = \frac{1}{2} \int \frac{dt}{t} = \frac{1}{2} \log t + c_1 = \frac{1}{2} \log(e^{2x} + 1) + c_1$$

Consider $I_2 = \int \dfrac{1}{e^{2x} + 1} dx = \int \dfrac{e^{-2x}}{1 + e^{-2x}} dx$

Put $1 + e^{-2x} = t \therefore -2e^{-2x} dx = dt$

$$\therefore e^{-2x} dx = -\frac{1}{2} dt$$

$$I_2 = -\frac{1}{2} \int \frac{dt}{t} = -\frac{1}{2} \log t = -\frac{1}{2} \log(1 + e^{-2x}) + c_2$$

$$\therefore I = I_1 - I_2 = \frac{1}{2} \left[\log(e^{2x} + 1) + \log(1 + e^{-2x}) \right] + c$$

$$= \frac{1}{2} \log \left[e^{2x} + 2 + e^{-2x} \right] + c$$

$$= \frac{1}{2} \log[e^x + e^{-x}]^2 + c$$

$$= \log(e^x + e^{-x}) + c \quad \text{Ans.}$$

Let us solve one more example of the above type.

Example (13): Evaluate $I = \int \dfrac{e^x - 1}{e^x + 1} dx$

Dividing N^r and D^r by $e^{x/2}$, we get

$$I = \int \frac{e^{x/2} - e^{-x/2}}{e^{x/2} + e^{-x/2}} dx$$

Now, put $e^{x/2} + e^{-x/2} = t$ $\therefore$ $(e^{x/2} - e^{-x/2})\,dx = 2dt$

$$\therefore \quad I = 2\int \frac{dt}{t} = 2\log t + c$$

$$= 2\log(e^{x/2} + e^{-x/2}) + c \quad \text{Ans.}$$

To evaluate $\displaystyle\int \frac{ae^x + b}{ce^x + d}\,dx$.

Note: Examples (12) and (13) are special cases of $\int \frac{ae^x+b}{ce^x+d}\,dx$. Of course, this type of integral can be easily evaluated by breaking it into two parts as explained in Example (12), Method (2). However, there is a *simpler method* (which is more general and applicable to many other integrals) as explained in Example 14.

Example (14): Evaluate $I = \displaystyle\int \frac{3e^x + 5}{2e^x + 7}\,dx$

We express,

$$N^r = A(D^r) + B\left(\frac{d}{dx}D^r\right)^{(8)}$$

$$\text{i.e., } 3e^x + 5 = A(2e^x + 7) + B(2e^x) \quad \left[\because \frac{d}{dx}(2e^x + 7) = 2e^x\right]$$

$$= (A + B)2e^x + 7A.$$

Now, comparing terms and their coefficients, on both sides we get $A = 5/7$ and hence $B = 11/14$.

$$\therefore \quad I = \frac{5}{7}\int \frac{2e^x + 7}{2e^x + 7}\,dx + \frac{11}{14}\int \frac{2e^x}{2e^x + 7}\,dx$$

$$= \frac{5}{7}\int dx + \frac{11}{14}\int \frac{(d/dx)(2e^x + 7)}{2e^x + 7}\,dx$$

$$= \frac{5}{7}x + \frac{11}{14}\log(2e^x + 7) + c \quad \text{Ans.}$$

Example (15): Find $I = \displaystyle\int \frac{\sin\sqrt{x}}{\sqrt{x}}\,dx^{(9)}$

$$\text{Put } \sqrt{x} = t \quad \therefore \quad \frac{1}{2\sqrt{x}}\,dx = dt \quad \therefore \quad \frac{dx}{\sqrt{x}} = 2dt$$

$$\therefore \quad I = 2\int \sin t\,dt = -2\cos t + c$$

$$= -2\cos\sqrt{x} + c \quad \text{Ans.}$$

---

(8) N^r = numerator and D^r = denominator.
(9) **Remark**: It is because of the function $\sqrt{x}$, that these integrals are easily evaluated. If we replace $\sqrt{x}$ by x, then the new functions cannot be integrated by substitution. Even the integrals $\int \sin\sqrt{x}\,dx$ and $\int \cos\sqrt{x}\,dx$, and so on, cannot be evaluated by this method. Later on, it will be shown that the integrals $\int \sin\sqrt{x}\,dx$ and $\int \cos\sqrt{x}\,dx$ can be evaluated by the method of "integration by parts", to be studied later.

Now, evaluate the following integrals:

(i) $\int \dfrac{e^{\sqrt{x}}}{\sqrt{x}} dx$

Ans. $2e^{\sqrt{x}} + c$

(ii) $\int \dfrac{\cos\sqrt{x}}{\sqrt{x}} dx$

Ans. $2 \sin \sqrt{x} + c$

(iii) $\int \dfrac{\tan^2 \sqrt{x}}{\sqrt{x}} dx$

Ans. $2[\tan\sqrt{x} - \sqrt{x}] + c$

(iv) $\int \dfrac{\sec^2 \sqrt{x}}{\sqrt{x}} dx$

Ans. $2 \tan \sqrt{x} + c$

(v) $\int \dfrac{\sec \sqrt{x}}{\sqrt{x}} dx$

Ans. $2 \log[\sec\sqrt{x} + \tan\sqrt{x}] + c$

Exercise

Integrate the following with respect to x:

(1) $I = \int \dfrac{(\sin^{-1} x)^2}{\sqrt{1 - x^2}} dx$

Ans. $\dfrac{(\sin^{-1} x)^3}{3} + c$

(2) $I = \int \dfrac{2x}{1 + x^4} dx$

Ans. $\tan^{-1} x^2 + c$

(3) $I = \int \dfrac{1}{x \sin^2 (\log x)} d$

Ans. $-\cot (\log x) + c$

(4) $I = \int e^x \cos(e^x) d$

Ans. $\sin(e^x) + c$

(5) $I = \int \dfrac{(\log x)^2}{x} dx$

Ans. $\dfrac{(\log x)^3}{3} + c$

(6) $I = \int e^{\tan x} \sec^2 x \, dx$

Ans. $e^{\tan x} + c$

(7) $I = \int \dfrac{2x - 5}{\sqrt{x^2 - 5x + 13}} \, dx$

Ans. $2\sqrt{x^2 - 5x + 13} + c$

(8) $I = \int \dfrac{2x \, e^{\tan^{-1} x^2}}{1 + x^4} \, dx$

Ans. $e^{\tan^{-1} x^2} + c$

(9) $I = \int \dfrac{\sin 2x}{3 \sin^2 x + 5 \cos^2 x} \, dx$

Ans. $-\dfrac{1}{2} \log(5 - 2 \sin^2 x) + c$

(10) $I = \int \dfrac{\cos 2x}{\sin^2 x + \cos^2 x + 2 \sin x \cos x} \, dx$

Ans. $\dfrac{1}{2} \log(1 + \sin 2x) + c$

(11) $I = \int \dfrac{\sin 2x}{a^2 \cos^2 x + b^2 \sin^2 x} \, dx$

Ans. $\dfrac{1}{b^2 - a^2} \log[a^2 + (b^2 - a^2)\sin^2 x] + c$

(12) $I = \int \dfrac{1}{e^x + 1} \, dx$

Ans. $-\log(1 + e^{-x}) + c$ or $\log\left(\dfrac{e^x}{e^x + 1}\right) + c$

(13) $I = \int \dfrac{1}{3 + 4e^{2x}} \, dx$

Ans. $-\dfrac{1}{6} \log(3e^{-2x} + 4) + c$

(14) $I = \int \dfrac{2e^x + 3}{4e^x + 5} \, dx$

Ans. $\dfrac{3}{5} x - \dfrac{1}{10} \log(4e^x + 5) + c$

Note: Solutions to the above problems are available at the end of this chapter.

3a.4   TO EVALUATE INTEGRALS OF THE FORM $\int \frac{a\sin x+b\cos x}{c\sin x+d\cos x}dx$, WHERE A, B, C, AND D ARE CONSTANT

These integrals are handled as the integrals of the form $\int((ae^x+b)/(ce^x+d))dx$.

Example (16):   To find $I = \displaystyle\int \frac{\sin x + 2\cos x}{3\sin x + 4\cos x}dx$

we express $N^r = A(D^r) + B\frac{d}{dx}(D^r)$

$$\frac{d}{dx}(D^r) = \frac{d}{dx}[3\sin x + 4\cos x] = 3\cos x - 4\sin x$$

Now, $\sin x + 2\cos x = A(3\sin x + 4\cos x) + B(3\cos x - 4\sin x)$
$$= (3A-4B)\sin x + (4A+3B)\cos x.$$

Equating coefficients on both sides, we get
$$3A - 4B = 1 \quad \text{(i)}$$
$$4A + 3B = 2 \quad \text{(ii)}$$

(i) $\times$ 3: $9A - 12B = 3$
(ii) $\times$ 4: $16A + 12B = 8$
- - - - - - - - - - - - -
$\qquad 25A \qquad = 11 \quad \therefore A = 11/25$

Now (ii) $\times$ 3: $12A + 9B = 6$
and   (i) $\times$ 4: $12A - 16B = 4$
$\qquad\quad - \qquad + \qquad -$
- - - - - - - - - - - - -
$\qquad\qquad 25B = 2 \qquad \therefore B = 2/25$[10]

$$\therefore I = \frac{11}{25}\int dx + \frac{2}{25}\int \frac{(d/dx)(3\sin x + 4\cos x)}{3\sin x + 4\cos x)} + c$$

$$= \frac{11}{25}x + \frac{2}{25}\log(3\sin x + 4\cos x) + c \quad \text{Ans.}$$

Example (17):   Find $I = \displaystyle\int \frac{1}{1 + \tan x}dx$

Consider $\dfrac{1}{1 + \tan x} = \dfrac{1}{1 + (\sin x/\cos x)} = \dfrac{\cos x}{\cos x + \sin x}$

We express $N^r = A(D^r) + B\, d/dx\, (D^r)$

$$\frac{d}{dx}(D^r) = \frac{d}{dx}(\cos x + \sin x) = -\sin x + \cos x$$
$$= \cos x - \sin x$$

Let $\cos x = A(\cos x + \sin x) + B(\cos x - \sin x) = (A+B)\cos x + (A-B)\sin x$

---

[10] **Note:** Value of B could also be found by putting the value of A in any of the Equations (i) or (ii).

Equating coefficients, on both sides, we get

$$A + B = 1 \quad (6) \atop A - B = 0 \quad (7)\Bigg\} \quad \therefore \quad 2A = 1 \quad \therefore \quad A = 1/2 \quad \text{and} \quad 2B = 1 \quad \therefore \quad B = 1/2$$

$$\therefore \quad I = \int \frac{(1/2)(\cos x + \sin x) + (1/2)(\cos x - \sin x)}{\cos x + \sin x} dx$$

Put $\cos x + \sin x = t$ $\quad \therefore (\cos x - \sin x)\, dx = dt$

$$\therefore \quad I = \frac{1}{2}\int dx + \frac{1}{2}\int \frac{dt}{t} = \frac{x}{2} + \frac{1}{2}\log t + c$$

$$= \frac{x}{2} + \frac{1}{2}\log(\cos x + \sin x) + c \quad \text{Ans.}$$

Now evaluate the following integrals:

(i) $\displaystyle\int \frac{1}{1 + \cot x}\,dx$

Ans. $\quad \dfrac{1}{2}x - \dfrac{1}{2}\log(\sin x + \cos x) + c$

(ii) $\displaystyle\int \frac{1}{3 + 2\tan x}\,dx$

Ans. $\quad \dfrac{3}{13}x + \dfrac{2}{13}\log(3\cos x + 2\sin x) + c$

(iii) $\displaystyle\int \frac{3\sin x + 4\cos x}{2\sin x + \cos x}\,dx$

Ans. $\quad 2x + \log(2\sin x + \cos x) + c$

Information in Advance: We can also handle the integrals of the form $\int((a + b\cos x + c\sin x)/(\alpha + \beta\cos x + \gamma\sin x))dx$. Here, we shall express the N^r (i.e., $a + b\cos x + c\sin x$) as $l(D^r) + m((d/dx)\text{of } D^r) + n$, where l, m, n will be found by comparing the coefficients of corresponding terms (i.e., $\cos x$, $\sin x$, and constant terms). By this method, the given integral reduces to the sum of three integrals. The first two integrals can be easily evaluated, whereas the third integral is of the form $\int(n/(\alpha + \beta\cos x + \gamma\sin x))dx$. *We shall learn about this integral in the next chapter. Obviously, evaluation of such integrals is time consuming.*

Solutions to the Problems of the Exercise

Q. (1): Find $I = \displaystyle\int \frac{(\sin^{-1} x)^2}{\sqrt{1 - x^2}}\,dx$

Put $\sin^{-1}x = t$ $\quad \therefore \quad \dfrac{1}{\sqrt{1 - x^2}}dx = dt$

$$\therefore \quad I = \int t^2\, dt = \frac{t^3}{3} + c = \frac{(\sin^{-1} x)^3}{3} + c \quad \text{Ans.}$$

Q. (2): Find $I = \int \dfrac{2x}{1+x^4} dx$

$$I = \int \dfrac{2x}{1+(x^2)^2} dx$$

Put $x^2 = t$ $\therefore$ $2x\, dx = dt$

$$\therefore \quad I = \int \dfrac{dt}{1+t^2} = \tan^{-1} t + c$$

$$= \tan^{-1} x^2 + c \quad \text{Ans.}$$

Q. (3): Find $I = \int \dfrac{1}{x \sin^2(\log x)} dx$

Put $\log x = t$ $\therefore$ $(1/x)dx = dt$

$$\therefore \quad I = \int \dfrac{dt}{\sin^2 t} = \int \operatorname{cosec}^2 t\, dt = -\cot t + c$$

$$= -\cot(\log x) + c \quad \text{Ans.}$$

Q. (4): Find $I = \int e^x \cos(e^x)dx$

Put $Q = t$ $\therefore$ $Q\, dx = dt$

$$\therefore \quad I = \int \cos t\, dt = \sin t + c$$

$$= \sin(e^x) + c \quad \text{Ans.}$$

Now evaluate the following integrals:

(i) $\int \dfrac{1}{\sqrt{1-x^2} \cdot \sin^{-1} x} dx$

Ans. $\log(\sin^{-1} x) + c$

(ii) $\int \dfrac{2x}{\sqrt{1-x^4}} dx$

Ans. $\sin^{-1} x^2 + c$

(iii) $\int \dfrac{1+\sin 2x}{\sqrt{x+\sin^2 x}} dx$

Ans. $2\sqrt{x+\sin^2 x} + c$

Q. (5): Find $I = \int \dfrac{(\log x)^2}{x} dx$

Put $\log x = t$ $\therefore$ $(1/x)dx = dt$

$$\therefore \quad I = \int t^2\, dt = \dfrac{t^3}{3} + c$$

$$= \dfrac{(\log x)^3}{3} + c \quad \text{Ans.}$$

Now evaluate the following integrals:

(i) $\int \dfrac{\log(x+1) - \log x}{x(x+1)} dx$

Ans. $\dfrac{-[\log(x+1) - \log x]^2}{2} + c$

[Hint: Put $\log(x+1) - \log x = t$]

(ii) $\int \dfrac{(1+x)(\log x + x)}{2x} dx$

Ans. $\dfrac{(\log x + x)^2}{4} + c$

[Hint: Put $\log x + x = t$]

(iii) $\int \dfrac{\sin(a + b \log x)}{x} dx$

Ans. $\dfrac{-1}{b} \cos(a + b \log x) + c$

[Hint: Put $(a + b \log x) = t$]

Q. (6): Find $I = \int e^{\tan x} \sec^2 x \, dx$

Put $\tan x = t$ $\therefore$ $\sec^2 x \, dx = dt$

$\therefore I = \int e^t \, dt = e^t + c = e^{\tan x} + c$ Ans.

Q. (7): Find $I = \int \dfrac{2x - 5}{\sqrt{x^2 - 5x + 13}} dx$

Put $x^2 - 5x + 13 = t$ $\therefore$ $(2x - 5)dx = dt$

$\therefore$ $I = \int \dfrac{dt}{\sqrt{t}} = \dfrac{t^{1/2}}{1/2} + c = 2\sqrt{x^2 - 5x + 13} + c$ Ans.

Q. (8): Find $I = \int \dfrac{2x \, e^{\tan^{-1} x^2}}{1 + x^4} dx$

Put $\tan^{-1} x^2 = t$ $\therefore$ $\dfrac{1}{1 + (x^2)^2} 2x \, dx = dt$

i.e., $\dfrac{2x \, dx}{1 + x^4} = dt$

$\therefore$ $I = \int e^t \, dt = e^t + c$

$= e^{\tan^{-1} x^2 + c}$ Ans.

Q. (9): Find $I = \displaystyle\int \frac{\sin 2x}{3 \sin^2 x + 5 \cos^2 x} \, dx$

$$I = \int \frac{2 \sin x \cdot \cos x}{3 \sin^2 x + 5(1 - \sin^2 x)} \, dx$$

$$= \int \frac{2 \sin x \cdot \cos x}{5 - 2 \sin^2 x} \, dx$$

Method (1): Put $5 - 2 \sin^2 x = t$

$\therefore \quad -4 \sin x \cdot \cos x \, dx = dt$

$\therefore \quad 2 \sin x \cdot \cos x \, dx = -\dfrac{1}{2} dt$

$$\therefore \quad I = -\frac{1}{2} \int \frac{dt}{t} = -\frac{1}{2} \log t + c$$

$$= -\frac{1}{2} \log(5 - 2 \sin^2 x) + c \quad \text{Ans.}$$

Method (2): Put $3 \sin^2 x + 5 \cos^2 x = t$

$\therefore \quad [6 \sin x \cos x - 10 \cos x \sin x] dx = dt$

or $-4 \sin x \cos x \, dx = dt$

or $-2 \sin 2x \, dx = dt$ or $\sin 2x \, dx = -\frac{1}{2} dt$

$$\therefore \quad I = -\frac{1}{2} \int \frac{dt}{t} = -\frac{1}{2} \log t + c$$

$$= -\frac{1}{2} \log(3 \sin^2 x + 5 \cos^2 x) + c \quad \text{Ans.}$$

Note: We have seen above that $3 \sin^2 x + 5 \cos^2 x = 5 - 2 \sin^2 x$.

Q. (10): Find $I = \displaystyle\int \frac{\cos 2x}{\sin^2 x + \cos^2 x + 2 \sin x \cos x} \, dx$

$$I = \int \frac{\cos 2x}{\sin^2 x + \cos^2 x + 2 \sin x \cos x} \, dx$$

$$= \int \frac{\cos 2x}{1 + \sin 2x} \, dx \qquad \left[\begin{array}{l} \because \quad \sin^2 x + \cos^2 x = 1 \text{ and} \\[4pt] \quad 2 \sin x \cos x = \sin 2x \end{array} \right.$$

$$I = \int \frac{\cos 2x}{(\sin x + \cos x)^2} \, dx = \int \frac{\cos^2 x - \sin^2 x}{(\cos x + \sin x)^2} \, dx$$

$$= \int \frac{\cos x - \sin x}{\cos x + \sin x} \, dx = \int \frac{(d/dx)(\cos x + \sin x)}{\cos x + \sin x} \, dx$$

$$= \log[\cos x + \sin x] + c \quad \text{Ans.}$$

Also $I = \displaystyle\int \frac{1 - \tan x}{1 + \tan x} = \int \tan\left(\frac{\pi}{4} - x\right) dx = -\log\left[\sec\left(\frac{\pi}{4} - x\right)\right] + c \quad \text{Ans.}$

Put $1 + \sin 2x = t$ $\therefore$ $2 \cos 2x \, dx = dt$ $\therefore$ $\cos 2x \, dx = \frac{1}{2} dt$

$$\therefore \quad I = \frac{1}{2} \int \frac{dt}{t} = \frac{1}{2} \log t + c = \frac{1}{2} \log(1 + \sin 2x) + c \quad \text{Ans.}$$

Q. (11): Find $I = \displaystyle\int \frac{\sin 2x}{a^2 \cos^2 x + b^2 \sin^2 x} dx$

Consider $a^2 \cos^2 x + b^2 \sin^2 x$

$$= a^2 (1 - \sin^2 x) + b^2 \sin^2 x$$

$$= a^2 + (b^2 - a^2) \sin^2 x$$

$$\therefore \quad I = \int \frac{2 \sin x \cdot \cos x \, dx}{a^2 + (b^2 - a^2) \sin^2 x}$$

Put $a^2 + (b^2 - a^2) \sin^2 x = t$

$\therefore$ $(b^2 - a^2) \, 2 \sin x \cos x \cdot dx = dt$

$\therefore$ $2 \sin x \cos x \, dx = \dfrac{1}{(b^2 - a^2)} dt$

$$\therefore I \quad = \frac{1}{(b^2 - a^2)} \int \frac{dt}{t} = \frac{1}{(b^2 - a^2)} \log t + c$$

$$= \frac{1}{b^2 - a^2} \log\left[[a^2 + (b^2 - a^2)\sin^2 x] + c \right] \quad \text{Ans.}$$

Note: We may also put, $a^2 \cos^2 x + b^2 \sin^2 x = t$

Then $[-2a^2 \cos x \cdot \sin x + 2b^2 \sin x \cdot \cos x] \, dx = dt$

$(2 \sin x \cos x) \cdot (b^2 - a^2) \, dx = dt$

or $\sin 2x \, dx = \dfrac{1}{(b^2 - a^2)} dt$

Q. (12): Evaluate $I = \int \frac{1}{e^x + 1} dx$

Dividing N^r and D^r by e^x, we get

$$I = \int \frac{e^{-x}}{1 + e^{-x}} dx$$

Put $1 + e^{-x} = t$

$\therefore$ $-e^{-x} dx = dt$

$$\therefore \quad I = -\int \frac{dt}{t} = -\log t + c$$

$$= -\log(1 + e^{-x}) + c \quad \text{Ans.}$$

Further simplification

$$I = -\log\left(1 + \frac{1}{e^x}\right) + c$$

$$= -\log\left(\frac{e^x + 1}{e^x}\right) + c = \log\left(\frac{e^x}{e^x + 1}\right) + c$$

Q. (13): Evaluate $I = \int \frac{1}{3 + 4e^{2x}} dx$

Dividing N^r and D^r by e^{2x}

$$I = \int \frac{e^{-2x}}{3\,e^{-2x} + 4} dx$$

Put $3e^{-2x} + 4 = t$

$$\therefore \quad -6e^{-2x}\,dx = dt$$

$$\therefore \quad e^{-2x}\,dx = -1/6\,dt$$

$$\therefore \quad I = -\frac{1}{6}\int \frac{dt}{t} = -\frac{1}{6}\log t + c$$

$$= -\frac{1}{6}\log(3e^{-2x} + 4) + c \quad \text{Ans.}$$

Q. (14): Evaluate $I = \int \frac{2e^x + 3}{4e^x + 5} dx$

We express $N^r = A(D^r) + B\dfrac{d}{dx}(D^r)$

$$\frac{d}{dx}(D^r) = \frac{d}{dx}(4e^x + 5) = 4e^x$$

Let $2Q + 3 = A(4Q + 5) + B(4Q)$

$$= (4A + 4B)Q + 5A$$

Equating coefficients on both sides, we get

$$5A = 3$$

$$\therefore \quad A = 3/5$$

$$4\left(\frac{3}{5}\right) + 4B = 2$$

$$\frac{3}{5} + B = \frac{1}{2}$$

$$\therefore \quad B = \frac{1}{2} - \frac{3}{5} = \frac{5 - 6}{10} = -\frac{1}{10}$$

$$\therefore \quad I = \int \frac{(3/5)(4e^x + 5) - (1/10)(4e^x)}{4e^x + 5} dx$$

$$= \frac{3}{5}\int dx - \frac{1}{10}\int \frac{(d/dx)(4e^x + 5)}{4e^x + 5} dx$$

$$= \frac{3}{5}x - \frac{1}{10}\log(4e^x + 5) + c \quad \text{Ans.}$$

3b Further Integration by Substitution: Additional Standard Integrals

3b.1 INTRODUCTION

We know that the method of *integration by substitution* aims at reducing an integral to a standard form (as in the case of other methods). In fact, *there is no definite rule for choosing a substitution that should convert the given integral to a standard form*. However, as we have seen in the previous Chapter 3a, *if the integrand is a function of a function* [i.e., $f[\phi(x)]$] *and the derivative of $\phi(x)$ also appears in the integrand*, so that the element of integration is in the form $f[\phi(x)] \, \phi'(x) \, dx$, then it is convenient to choose the substitution $\phi(x) = t$.

With this substitution, we get $\phi'(x) \, dx = dt$, and the variable of integration is changed from x to t. The new integral in t is expected to be easier for integration. Also, it is sometimes convenient to choose a change of variable (from x to t) by the substitution $x = f(t)$, where $f(t)$ must be some suitable, trigonometric, algebraic, or hyperbolic function. This statement will be clear from the following trigonometric (and algebraic) substitutions.

[Here, we shall not be considering the hyperbolic substitutions though they are equally convenient for this purpose. We have discussed the hyperbolic functions at length in Chapter 23 of differential Calculus (i.e. Part I)].

3b.1.1 Trigonometric Substitutions

The following integrals can be easily evaluated by trigonometric substitutions as indicated below:

(i) If the integrand involves $\sqrt{a^2 - x^2}$, then put $x = a \sin t$ or $x = a \cos t$.

(ii) If the integrand involves $\sqrt{a^2 + x^2}$ or $a^2 + x^2$, then put $x = a \tan t$.

(iii) If the integrand involves $\sqrt{x^2 - a^2}$, then put $x = a \sec t$.

The idea behind these substitutions is to get rid of the square root sign by using trigonometric identities: $1 - \cos^2 t = \sin^2 t$, $\tan t + 1 = \sec^2 t$ and $\sec^2 t - 1 = \tan^2 t$, as required.

3b-Further integration by substitution

Introduction to Integral Calculus: Systematic Studies with Engineering Applications for Beginners, First Edition.
Ulrich L. Rohde, G. C. Jain, Ajay K. Poddar, and A. K. Ghosh.
© 2012 John Wiley & Sons, Inc. Published 2012 by John Wiley & Sons, Inc.

3b.2   SPECIAL CASES OF INTEGRALS AND PROOF FOR STANDARD INTEGRALS

There are many other function(s) that can be reduced to the standard forms by substitutions. For example, consider the following functions listed in three batches:

Batch (I) $\dfrac{1}{x^2+a^2}, \dfrac{1}{x^2-a^2}, \dfrac{1}{a^2-x^2}$

Batch (II) $\dfrac{1}{\sqrt{x^2+a^2}}, \dfrac{1}{\sqrt{x^2-a^2}}, \dfrac{1}{\sqrt{a^2-x^2}}, \dfrac{1}{x\sqrt{x^2-a^2}}$

Batch (III) $\sqrt{x^2+a^2}, \sqrt{x^2-a^2},$ and $\sqrt{a^2-x^2}$

It is useful to consider the integrals involving the above functions in the batches as listed above. (It will be found helpful in remembering the standard formulae conveniently, in the order they are developed.) Now, we proceed to obtain the formulas for the integrals of Batch (I).

Integral (1) $\displaystyle\int \dfrac{1}{x^2+a^2}\,dx = \dfrac{1}{a}\tan^{-1}\left(\dfrac{x}{a}\right) + c^{(1)}$

Method (1):   Let $I = \displaystyle\int \dfrac{1}{x^2+a^2}\,dx$

Consider $x^2 + a^2 = a^2\left[\dfrac{x^2}{a^2}+1\right]$

$$= a^2\left[1+\dfrac{x^2}{a^2}\right] = a^2\left[1+\left(\dfrac{x}{a}\right)^2\right]$$

$$\therefore\quad I = \dfrac{1}{a^2}\int \dfrac{1}{1+(x/a)^2}\,dx$$

Put $\dfrac{x}{a} = t$ $\therefore\quad \dfrac{1}{a}dx = dt$ $\therefore\quad dx = a\,dt$

$$\therefore\quad I = \dfrac{1}{a^2}\int \dfrac{a\,dt}{1+t^2} = \dfrac{1}{a}\int \dfrac{dt}{1+t^2}$$

$$= \dfrac{1}{a}\tan^{-1}t + c = \dfrac{1}{a}\tan^{-1}\left(\dfrac{x}{a}\right) + c$$

$$\therefore\quad \int \dfrac{1}{x^2+a^2}\,dx = \dfrac{1}{a}\tan^{-1}\left(\dfrac{x}{a}\right) + c \quad \text{Ans.}$$

---

[1] Note that, this result is more general than the result: $\int(1/(1+x^2))dx = \tan^{-1}x + c$, which follows the form $(d/dx)(\tan^{-1}x) = 1/(1+x^2)$. Both these integrals are treated as standard formulas.

Method (2):   Let $I = \displaystyle\int \dfrac{1}{x^2 + a^2} dx$

Put $\qquad\qquad\qquad\qquad\qquad x = at \quad \therefore \quad dx = a\, dt. \quad \text{Also} \quad t = \dfrac{x}{a}$

$$\therefore\ I = \int \frac{1}{a^2t^2 + a^2} \cdot a\, dt = \int \frac{a\, dt}{a^2(t^2 + 1)}$$

$$= \frac{1}{a} \int \frac{dt}{t^2 + 1} = \frac{1}{a} \int \frac{dt}{1 + t^2}$$

$$= \frac{1}{a} \tan^{-1} t + c$$

$$= \frac{1}{a} \tan^{-1}\left(\frac{x}{a}\right) + c \quad \text{Ans.}$$

Method (3):   Let $I = \displaystyle\int \dfrac{1}{x^2 + a^2} dx$

Let $x = a \tan t \quad \therefore \quad dx = a \sec^2 t\, dt$

Also, $\tan t = \dfrac{x}{a} \quad \therefore \quad t = \tan^{-1} \dfrac{x}{a}$

$$\therefore\ I = \int \frac{a \sec^2 t\, dt}{a^2 \tan^2 t + a^2} = \int \frac{a \sec^2 t}{a^2(\tan^2 t + 1)} dt$$

$$= \frac{1}{a} \int \frac{\sec^2 t}{\sec^2 t} dt = \frac{1}{a} \int dt = \frac{1}{a} t + c \qquad\qquad (1)$$

$$= \frac{1}{a} \tan^{-1}\left(\frac{x}{a}\right) + c$$

$$\therefore\ = \int \frac{1}{x^2 + a^2} dx = \frac{1}{a} \tan^{-1}\left(\frac{x}{a}\right) + c$$

Note: To obtain the above result, we expressed the given integral in the standard form $\int (1/(1 + t^2))\, dt$ by using the method of substitution. In the Methods (1) and (2), we used *the same substitution in two different ways, but in Method (3) we used a different substitution which converted the given integral into another standard form.*

Integral (2) $\displaystyle\int \dfrac{1}{x^2 - a^2} dx = \dfrac{1}{2a} \log\left(\dfrac{x-a}{x+a}\right) + c,\ x > a$

Let $I = \displaystyle\int \dfrac{1}{x^2 - a^2} dx$

Consider $\dfrac{1}{x^2 - a^2} = \dfrac{1}{(x-a)(x+a)} = \dfrac{1}{2a}\left[\dfrac{1}{x-a} - \dfrac{1}{x+a}\right]$

$$\therefore\ I = \frac{1}{2a}\int\left[\frac{1}{x-a}-\frac{1}{x+a}\right]dx$$

$$= \frac{1}{2a}[\log(x-a)-\log(x+a)]+c$$

$$= \frac{1}{2a}\log\left(\frac{x-a}{x+a}\right)+c$$

$$\therefore\ \int\frac{1}{x^2-a^2}dx = \frac{1}{2a}\log\left(\frac{x-a}{x+a}\right)+c,\ x>a$$

(2)

or

$$\int\frac{1}{x^2-a^2}dx = \frac{1}{2a}\log\left|\frac{x-a}{x+a}\right|+c$$

Integral (3) $\displaystyle\int\frac{1}{a^2-x^2}dx = \frac{1}{2a}\log\left(\frac{a+x}{a-x}\right)+c,\quad x<a$

Suppose, $\displaystyle I = \int\frac{1}{a^2-x^2}dx = \int\frac{1}{(a-x)(a+x)}dx$

Let $\displaystyle\frac{1}{(a-x)(a+x)} = \frac{A}{a-x}+\frac{B}{a+x}$

$\therefore\ 1 = A(a+x)+B(a-x)$

This is an identity, and hence true for any value of x. To find the numbers A and B:

Put $x=a$, we get $\ 1 = 2a\cdot A \quad \therefore A = (1/2a)$

Now, put $x=-a$, we get $\ 1 = 2a\cdot B \quad \therefore B = (1/2a)$

$$\therefore\ \frac{1}{a^2-x^2} = \frac{1}{(a-x)(a+x)} = \frac{1}{2a}\left[\frac{1}{a-x}+\frac{1}{a+x}\right]$$

$$\therefore\ I = \frac{1}{2a}\int\left[\frac{1}{a-x}+\frac{1}{a+x}\right]dx$$

$$= \frac{1}{2a}[\log(a-x)\cdot(-1)+\log(a+x)]+c$$

$$= \frac{1}{2a}[\log(a+x)-\log(a-x)]+c$$

$$\therefore\ \int\frac{1}{a^2-x^2}dx = \frac{1}{2a}\log\left|\frac{a+x}{a-x}\right|+c$$

(3)

Note: Integral (3) can be deduced from Integral (2) by observing that $\int\frac{dx}{a^2-x^2}=-\int\frac{dx}{x^2-a^2}$ and $\log\left|\frac{a+x}{a-x}\right|=-\log\left|\frac{x-a}{x+a}\right|$.

$$\left[\text{Note that } -\log\left|\frac{x-a}{x+a}\right| = -\log\left|\frac{-(a-x)}{a+x}\right| = -\log\left|\frac{a-x}{a+x}\right| = \log\left|\frac{a+x}{a-x}\right|.\right]$$

Remark: We are now in a position to evaluate integrals of the form,

$$I = \int \frac{dx}{ax^2 + bx + c}$$

To express this integral in the standard form, we express the quadratic expression $ax^2 + bx + c$ in the form of sum or difference of two squares and then apply one of the above three formulas, as applicable. For this purpose, *we consider the following three cases.*

Case (1): When the coefficient of x^2 is *plus one.*

Example (i): $x^2 + 6x + 10 = x^2 + 2(3) \cdot x + 9 + 1$
$$= (x+3)^2 + 1^2$$

$$\therefore \quad \int \frac{dx}{x^2 + 6x + 10} = \int \frac{dx}{(x+3)^2 + 1^2} = \tan^{-1}\left(\frac{x+3}{1}\right) + c$$

Example (ii): $x^2 + 4x - 5 = x^2 + 2(2) \cdot x + 4 - 9$
$$= (x+2)^2 - (3)^2$$

$$\therefore \quad \int \frac{dx}{x^2 + 4x - 5} = \int \frac{dx}{(x+2)^2 - (3)^2} = \frac{1}{2 \cdot 3} \log \frac{(x+2) - 3}{(x+2) + 3} + c$$

$$= \frac{1}{6} \log\left(\frac{x-1}{x+5}\right) + c, \quad x > 1 \quad \text{Ans.}$$

Similarly, $x^2 - 4x + 8 = x^2 - 2(2)x + 4 + 4 = (x-2)^2 + (2)^2$

And, $x^2 + x - 3 = x^2 + 2\left(\frac{1}{2}\right)x + \frac{1}{4} - \frac{1}{4} - 3$

$$= \left(x + \frac{1}{2}\right)^2 - \left(\frac{\sqrt{13}}{2}\right)^2$$

Now consider $x^2 - 12 = x^2 - (2\sqrt{3})^2$

$$\therefore \quad \int \frac{dx}{x^2 - 12} = \int \frac{dx}{x^2 - (2\sqrt{3})^2} = \frac{1}{4\sqrt{3}} \log\left(\frac{x - 2\sqrt{3}}{x + 2\sqrt{3}}\right) + c, x > 2\sqrt{3}$$

$$\left[\because \int \frac{dx}{x^2 - a^2} = \frac{1}{2a} \log\left(\frac{x-a}{x+a}\right) + c, x > a \right]$$

Case (2): When the coefficient of x^2 is *minus one.*

Example (i): $1-x-x^2 = -[x^2+x-1]$

$$= -\left[x^2+2\left(\frac{1}{2}\right)x+\frac{1}{4}-\left(\frac{1}{4}+1\right)\right]$$

$$= -\left[\left(x+\frac{1}{2}\right)^2-\left(\frac{\sqrt{5}}{2}\right)^2\right] = \left[\left(\frac{\sqrt{5}}{2}\right)^2-\left(x+\frac{1}{2}\right)^2\right]$$

$$\therefore \int\frac{dx}{1-x-x^2} = \int\frac{dx}{(\sqrt{5}/2)^2-(x+(1/2))^2}$$

$$= \frac{1}{2(\sqrt{5}/2)}\cdot\log\left(\frac{\sqrt{5}/2+(x+(1/2))}{\sqrt{5}/2-(x+(1/2))}\right)+c$$

$$= \frac{1}{\sqrt{5}}\log\left[\frac{\sqrt{5}+(2x+1)}{\sqrt{5}-(2x+1)}\right]+c$$

$$= \frac{1}{\sqrt{5}}\log\left|\frac{\sqrt{5}+1+2x}{\sqrt{5}-1-2x}\right|+c \quad \text{Ans.}$$

Example (ii): $15+4x-x^2 = -[x^2-4x-15]$

$$= -[x^2-2(2)\,x+4-4-15]$$

$$= -\left[(x-2)^2-(\sqrt{19})^2\right] = (\sqrt{19})^2-(x-2)^2$$

(Note that negative sign of x^2 is absorbed in the final expression.)

$$\therefore \int\frac{dx}{15+4x-x^2} = \int\frac{dx}{(\sqrt{19})^2-(x-2)^2}$$

$$= \frac{1}{2\sqrt{19}}\cdot\log\left[\frac{\sqrt{19}+(x-2)}{\sqrt{19}-(x-2)}\right]+c$$

$$= \frac{1}{2\sqrt{19}}\cdot\log\left[\frac{\sqrt{19}-2+x}{\sqrt{19}+2-x}\right]+c \quad \text{Ans.}$$

Case (3): When the coefficient of x^2 is some constant "a" (other than ±1).

Method: *After making the coefficient of x^2 unity, we express the quadratic expression as sum or difference of two squares.*

Example (i): $3x^2+2x+7 = 3\left(x^2+\frac{2}{3}x+\frac{7}{3}\right)$

$$= 3[x^2+2(\tfrac{1}{3})x+\tfrac{1}{9}-\tfrac{1}{9}+\tfrac{7}{3}]$$

$$= 3\left[(x+\tfrac{1}{3})^2+\tfrac{20}{9}\right] = 3\left[(x+\tfrac{1}{3})^2+\left(\tfrac{2\sqrt{5}}{3}\right)^2\right]$$

$$\therefore \ I = \int \frac{dx}{3x^2 + 2x + 7} = \frac{1}{3} \int \frac{dx}{(x + (1/3))^2 + (2\sqrt{5}/3)^2}$$

$$= \frac{1}{3} \cdot \frac{3}{2\sqrt{5}} \tan^{-1} \frac{(x + (1/3))}{2\sqrt{5}/3} + c$$

$$= \frac{1}{2\sqrt{5}} \tan^{-1} \left(\frac{3x + 1}{2\sqrt{5}} \right) + c \quad \text{Ans.}$$

Similarly,

$$4 - 3x - 2x^2 = -2 \left[x^2 + \frac{3}{2} x - 2 \right]$$

$$= -2 \left[x^2 - 2 \left(\frac{3}{4} \right) x + \frac{9}{16} - \frac{9}{16} - 2 \right]$$

$$= -2 \left[\left(x + \frac{3}{4} \right)^2 - \frac{41}{16} \right] = -2 \left[\left(x + \frac{3}{4} \right)^2 - \left(\frac{\sqrt{41}}{4} \right)^2 \right]$$

$$= 2 \left[\frac{\sqrt{41}}{4} - \left(x + \frac{3}{4} \right)^2 \right]$$

[Note that the negative sign (of $-2x^2$) is absorbed in the final expression. Finally, 2 is kept outside the bracket and not -2.]

Remark: In the quadratic expression, if the coefficient of x^2 is a *positive number* then, it reduces to the forms $v^2 \pm k^2$ and if the coefficient of x^2 is a *negative number*, then it reduces to the form $k^2 - v^2$, where v is in the form $(ax + b)$ and k is a constant.

Note: We have seen that

$$\int f(x) dx = \phi(x) \Rightarrow \int f(ax + b) dx = \frac{1}{a} \phi(ax + b).$$

Thus, it is possible to write the corresponding integrals directly, *without reducing the integrand to the standard form*. Accordingly, one may write the following:

(i) $\displaystyle \int \frac{1}{4x^2 + 9} dx = \int \frac{1}{(2x)^2 + (3)^2} dx = \frac{1}{2} \left[\frac{1}{3} \tan^{-1} \left(\frac{2x}{3} \right) \right] + c$

$$= \frac{1}{6} \tan^{-1} \left(\frac{2x}{3} \right) + c$$

(ii) $\displaystyle \int \frac{1}{9x^2 - 4} dx = \int \frac{1}{(3x)^2 - (2)^2} dx = \frac{1}{3} \left[\frac{1}{(2)(2)} \log \left(\frac{3x - 2}{3x + 2} \right) \right] + c$

$$= \frac{1}{12} \log \left(\frac{3x - 2}{3x + 2} \right) + c$$

(iii) $\displaystyle\int \frac{1}{9-4x^2}dx = \int \frac{1}{(3)^2-(2x)^2}dx = \frac{1}{2}\left[\frac{1}{(2)\,(3)}\,\log\left(\frac{3+2x}{3-2x}\right)\right]+c$

$\displaystyle\qquad\qquad = \frac{1}{12}\log\left(\frac{3+2x}{3-2x}\right)+c$

[Hint 1: We will shortly show that when the final answer obtained by the above method, there is a possibility of committing mistake, unknowingly.]

[Hint 2: Besides, it will be seen from the integrals at (b) and (c) below that the student *may commit an even more serious mistake*, if he is *not careful* about the *standard integrals* in question. Now we will show, how the mistakes can creep in.]

Consider,

(a) $\displaystyle\int \frac{1}{4+x^2}dx = \int \frac{1}{(2)^2+x^2}dx = \frac{1}{2}\tan^{-1}\left(\frac{x}{2}\right)+c$

With this result, *one may be tempted to write,*

(b) $\displaystyle\int \frac{1}{1+\sin^2 x}dx \neq \tan^{-1}(\sin x)+c,\,\text{(why?)}$

Similarly, it will be wrong to write

(c) $\displaystyle\int \frac{1}{4+5\cos^2 x}dx \neq \frac{\sqrt{5}}{2}\tan^{-1}\left(\frac{\sqrt{5}}{2}\cos x\right)+c$

It is easy to see why the results of (b) and (c) above are wrong.

Recall that the *standard form* is $\int(1/(a^2+x^2))dx = (1/a)\tan^{-1}(x/a)+c$, wherein the function x^2 appears *(without any coefficient)*. Naturally x^2 *cannot be replaced by any other function* in this formula. (A similar statement is applicable for other standard results.)

This suggests that the given integral be first converted to standard form, before writing the final answer.

Now, we consider some *special cases of the integrals of Batch I*, and those that can be expressed in these forms. We classify such integrals in three types: Type (a), Type (b), and Type(c).[2]

Type (a): Integrals of the form(s) $\int \frac{dx}{ax^2\pm c}$ or $\int \frac{dx}{c-ax^2}$ and those that can be reduced to these forms.

For this purpose consider the following exercise.

Exercise (1)

(i) $\displaystyle\int \frac{dx}{3x^2-7}$

Ans. $\displaystyle\frac{1}{2\sqrt{21}}\log\left(\frac{\sqrt{3}x-\sqrt{7}}{\sqrt{3}x+\sqrt{7}}\right)+c$

---

[2] There are number of integrals involving algebraic, exponential, or trigonometric functions, which can be reduced to one of the three standard integrals of Batch (I). This will be clear from the problems listed in the given exercise.

(ii) $\int \dfrac{dx}{9-4x^2}$

Ans. $\dfrac{1}{12} \log \left(\dfrac{3+2x}{3-2x} \right) + c$

(iii) $\int \dfrac{dx}{9x^2+25}$

Ans. $\dfrac{1}{15} \tan^{-1} \left(\dfrac{3x}{5} \right) + c$

We emphasize that for computing such integrals, it is always necessary to make the coefficient of x^2 as 1 or (-1), *so that the given integral can be expressed in the standard form, that is,:*

$$\int \frac{dx}{x^2 \pm k^2} \quad \text{or} \quad \int \frac{dx}{k^2 - x^2}$$

Now we consider exercise (iii),

Let us evaluate $\int \dfrac{dx}{9x^2-25}$

Let $I = \int \dfrac{dx}{9x^2-25} = \int \dfrac{dx}{(3x)^2-(5)^2}$

Note that the integral on the right-hand side is not in standard form, but the reader is tempted to write the integral as $\frac{1}{(3)(5)} \log \left(\frac{3x-5}{3x+5} \right) + c$, which is wrong.

This mistake can be avoided if we substitute $3x = t$, so that $3dx = dt$, and we get $I = 1/3$ $\int dt/(t^2-5^2)$. Now, this is in the standard form, and its value is $1/30 \log ((3x-5)/(3x+5)) + c$.

Note that, $\left[I = \dfrac{1}{3} \int \dfrac{dt}{t^2-5^2} = \dfrac{1}{3} \left[\dfrac{1}{(2)(5)} \log \left(\dfrac{t-5}{t+5} \right) \right] \right]$.

Such mistake(s) can also be easily avoided by making the coefficient of x^2 as unity, as mentioned above.

Consider $9x^2-25 = 9 \left[x^2 - \dfrac{25}{9} \right] = 9 \left[x^2 - \left(\dfrac{5}{3} \right)^2 \right]$

$$\therefore \quad I = \frac{1}{9} \int \frac{dx}{x^2-(5/3)^2} = \frac{1}{9} \left[\frac{1}{2 \cdot (5/3)} \log \left(\frac{x-(5/3)}{x+(5/3)} \right) \right] + c$$

$$= \frac{1}{30} \log \left(\frac{3x-5}{3x+5} \right) + c \qquad \text{Ans.}$$

(Observe that method of substitution indicated above is more convenient.)
Now we consider the following problems of the Exercise (1) from (iv).

(iv) $\displaystyle\int \frac{dx}{e^x + 2e^{-x}}$

Ans. $\displaystyle\frac{1}{\sqrt{2}} \tan^{-1}\left(\frac{e^x}{\sqrt{2}}\right)$

(v) $\displaystyle\int \frac{\cos x}{9 - \sin^2 x}\,dx$

Ans. $\displaystyle\frac{1}{6}\log\left(\frac{3 + \sin x}{3 - \sin x}\right) + c$

(vi) $\displaystyle\int \frac{\cos x}{9 - \cos^2 x}\,dx$

Ans. $\displaystyle\frac{1}{\sqrt{8}} \tan^{-1}\left(\frac{\sin x}{\sqrt{8}}\right) + c$

[Hint : $\cos^2 x = 1 - \sin^2 x$.]

(vii) $\displaystyle\int \frac{3^x}{1 + 3^{2x}}\,dx$

Ans. $\displaystyle\frac{1}{2}\tan^{-1}(3^x) + c$

(viii) $\displaystyle\int \frac{x^2}{1 - x^6}\,dx$

Ans. $\displaystyle\frac{1}{6}\log\left(\frac{1 + x^2}{1 - x^2}\right) + c$

(ix) $\displaystyle\int \frac{\sec^2 x}{25 - 16\tan^2 x}\,dx$

Ans. $\displaystyle\frac{1}{40}\log\left(\frac{5 + 4\tan x}{5 - 4\tan x}\right) + c$

(x) $\displaystyle\int \frac{dx}{a^2 \sin^2 x + b^2 \cos^2 x}$

Ans. $\displaystyle\frac{1}{ab}\tan^{-1}\left(\frac{a \tan x}{b}\right) + c$

(xi) $\displaystyle\int \frac{dx}{1 + \sin^2 x}$

Ans. $\displaystyle\frac{1}{\sqrt{2}}\tan^{-1}\left[\sqrt{2}\tan x\right] + c$

(xii) $\int \dfrac{dx}{4 + 5 \cos^2 x}$

Ans. $\dfrac{1}{6} \tan^{-1} \left[\dfrac{2 \tan x}{3} \right] + c$

It is easy to show that the integrals at (ix)–(xii) *are of the same type and can be evaluated in the same way.* To understand this, it is proposed to evaluate *the last integral.*

Let $I = \int \dfrac{dx}{4 + 5 \cos^2 x}$

Write $4 = 4(\sin^2 x + \cos^2 x) = 4\sin^2 x + 4\cos^2 x$.

$$\therefore \quad I = \int \dfrac{dx}{4 \sin^2 x + 9 \cos^2 x} \tag{A}$$

Dividing N^r and D^r by $\cos^2 x$, we get

$$I = \int \dfrac{\sec^2 x \, dx}{4\tan^2 x + 9} = \int \dfrac{\sec^2 x \, dx}{(2 \tan x)^2 + (3)^2} \tag{B}$$

We put $2 \tan x = t \qquad \therefore \quad 2 \sec^2 x \, dx = dt$

$$\therefore \quad I = \dfrac{1}{2} \int \dfrac{dt}{t^2 + (3)^2} = \dfrac{1}{2} \left[\dfrac{1}{3} \tan^{-1} \dfrac{t}{3} \right] + c$$

$$= \dfrac{1}{6} \tan^{-1} \left(\dfrac{2 \tan x}{3} \right) + c \quad \text{Ans.}$$

Note: Looking at the integrals at (A) and (B), and comparing their forms with those at (ix)–(xii) of Exercise (1), it is easy to note *that all these integrals are of the same type and hence, they can be evaluated by the same method.*

Type (b): Integrals of the form $\int \frac{dx}{ax^2 + bx + c}$ *and those which can be reduced to this form by substitution.* [It is useful to be clear that all such integrals are finally reduced to the Type (a).]

Example: Let us evaluate the following integral:

$$I = \int \dfrac{dx}{8 - 6x - 9x^2}$$

Consider $\quad 8 - 6x - 9x^2 = -9 \left[x^2 + \dfrac{2}{3} x - \dfrac{8}{9} \right]$

$$= -9 \left[x^2 + 2 \left(\dfrac{1}{3} \right) x + \dfrac{1}{9} - \dfrac{1}{9} - \dfrac{8}{9} \right]$$

$$= -9 \left[\left(x + \dfrac{1}{3} \right)^2 - 1 \right] = 9 \left[1^2 - \left(x + \dfrac{1}{3} \right)^2 \right]$$

$$\therefore \quad I = \frac{1}{9} \int \frac{dx}{1^2 - \left(x + \frac{1}{3}\right)^2}$$

$$= \frac{1}{9}\left[\frac{1}{(2)(1)} \; \log\left\{\frac{1 + (x + (1/3))}{1 - (x + (1/3))}\right\}\right] + c$$

$$= \frac{1}{18} \log\left(\frac{4 + 3x}{2 - 3x}\right) + c \quad \text{Ans.}$$

Exercise (2)

Evaluate the following integrals:

(i) $\displaystyle\int \frac{dx}{3x^2 + 2x + 7}$

Ans. $\dfrac{1}{\sqrt{20}} \tan^{-1}\left(\dfrac{3x + 1}{\sqrt{20}}\right) + c$

(ii) $\displaystyle\int \frac{dx}{3 - 10\,x - 25\,x^2}$

Ans. $\dfrac{1}{20} \log\left(\dfrac{3 + 5x}{1 - 5x}\right) + c$

(iii) $\displaystyle\int \frac{e^x\,dx}{e^{2x} + e^x + 1}$

Ans. $\dfrac{1}{\sqrt{3}} \tan^{-1}\left(\dfrac{2e^x + 1}{\sqrt{3}}\right) + c$

(iv) $\displaystyle\int \frac{\cos x\,dx}{9\sin^2 x + 12\sin x + 5}$

Ans. $\dfrac{1}{3} \tan^{-1}\left(\dfrac{3\tan x}{2}\right) + c$

(v) $\displaystyle\int \frac{dx}{1 + 3(x - 5)^2}$

Ans. $\dfrac{1}{\sqrt{3}} \tan^{-1}\left[\sqrt{3}(x - 5)\right] + c$

(vi) $\displaystyle\int \frac{\sec^2 x\,dx}{2\tan^2 x + 6\tan x + 5}$

Ans. $\tan^{-1}(2\tan x + 3) + c$

(vii) $\displaystyle\int \frac{\sec^2 x \, dx}{\sec^2 x - 3 \tan x + 1}$ [Imp.]

Ans. $\log \left(\dfrac{\tan x - 2}{\tan x - 1} \right) + c$

(viii) $\displaystyle\int \frac{dx}{2 \cos^2 x + 2 \sin x \cdot \cos x + \sin^2 x}$

Ans. $\tan^{-1}(\tan x + 1) + c$

(ix) $\displaystyle\int \frac{dx}{4 \sin^2 x + 9 \cos^2 x}$

Ans. $\dfrac{1}{6} \tan^{-1} \left(\dfrac{2 \tan x}{3} \right) + c$

(x) $\displaystyle\int \frac{dx}{8 \sin^2 x + 12 \sin x + 1}$

Ans. $\dfrac{1}{6} \tan^{-1}(3 \sin x + 2) + c$

Let us evaluate the integral at (viii).

$$\text{Let} \quad I = \int \frac{dx}{2 \cos^2 x + 2 \sin x \cos x + \sin^2 x}$$

Dividing N^r and D^r by $\cos^2 x$, we get

$$I = \int \frac{\sec^2 x \, dx}{2 + 2 \tan x + \tan^2 x} = \int \frac{\sec^2 x \, dx}{\tan^2 x + 2 \tan x + 2}$$

Put $\tan x = t$ $\therefore$ $\sec^2 x \, dx = dt$

$$\therefore \quad I = \int \frac{dt}{t^2 + 2t + 2}$$

$$= \int \frac{dt}{(t+1)^2 + 1^2} = \tan^{-1}(t+1) + c$$

$$= \tan^{-1}(\tan x + 1) + c \qquad \text{Ans.}$$

It is useful to remember that an integral of the form, $I = \int dx/(a + b \sin 2x + c \cos 2x)$, where a, b, and c are *integers*, which can always be reduced to a standard form, by the method of substitution. Depending on the *integers a, b,* and *c,* the standard form of the integral are obtained and accordingly the final answer. As an example, consider the following integral:

$$I = \int \frac{dx}{3 \sin 2x + 2 \cos 2x + 3} \tag{P}$$

Put $\sin 2x = 2 \sin x \cos x$

$\quad \cos 2x = \cos^2 x - \sin^2 x$

and, $3 = 3(\sin^2 x + \cos^2 x)$

We get,

$$I = \int \frac{dx}{6 \sin x \cdot \cos x + 2(\cos^2 x - \sin^2 x) + 3 \cos^2 x + 3 \sin^2 x}$$

$$= \int \frac{dx}{\sin^2 x + 6 \sin x \cdot \cos x + 5 \cos^2 x}$$

Dividing N^r and D^r by $\cos^2 x$, we get

$$I = \int \frac{\sec^2 x}{\tan^2 x + 6 \tan x + 5} \qquad\qquad (Q)$$

Now, put $\tan x = t$ $\quad \therefore \quad \sec^2 x\, dx = dt$

We get,

$$I = \int \frac{dt}{t^2 + 6t + 5} = \int \frac{dt}{t^2 + 2(3)t + 9 - 4}$$

$$= \int \frac{dt}{(t+3)^2 - (2)^2}$$

$$= \frac{1}{2 \cdot 2} \log \left[\frac{(t+3) - 2}{(t+3) + 2} \right] + c = \frac{1}{4} \log \left(\frac{t+1}{t+5} \right) + c$$

$$= \frac{1}{4} \log \left(\frac{\tan x + 1}{\tan x + 5} \right) + c \qquad \text{Ans.}$$

Note: We have shown above that an integral $\int \frac{dx}{a + b \sin x + c \cos x}$, where a, b, and c are integers, can always be reduced to the form as listed in Batch I.

With a view to express any given integral of the form (P), to the standard form, we shall always use the same trigonometric identities (as shown above). Also, we shall use the substitution $\tan x = t$ to maintain uniformity in our approach.

Type (c): Now we will show that the integrals of the form $\int \frac{dx}{a + b \cos x}$ and $\int \frac{dx}{a + b \sin x}$ can also be *reduced to the forms of Batch (I)*.

Method: For integrating these functions of $\sin x$ and $\cos x$, we make use of the following identities:

$$\sin x = 2 \sin \frac{x}{2} \cdot \cos \frac{x}{2}$$

$$\cos x = \cos^2 \frac{x}{2} - \sin^2 \frac{x}{2}, \text{ and for constant } ''a'' \text{ we write,}$$

$$a = a \left(\cos^2 \frac{x}{2} + \sin^2 \frac{x}{2} \right)$$

to express the denominator in the desired form.

Finally, by dividing N^r and D^r by $\cos^2(x/2)$, we express the given integral in the form (See chapter 3a example on pg. 51) $\int f(\tan(x/2))\sec^2(x/2)dx$, and then put $\tan(x/2) = t$, to convert it to the form $\int (dt)/(At^2 + Bt + c)$, as in the solved examples above.

Remark: *This method is also applicable for evaluating integrals of the form* $\int (dx)/(a\sin x + b\cos x)$ *in which there is no separate constant term.* However, there is an alternate simpler method available, in which the expression $a\sin x + b\cos x$ is converted into a single trigonometric quantity $r\sin(x + \alpha)$, where $r = \sqrt{a^2 + b^2}$ and $\alpha = \tan^{-1}(b/a)$. We have introduced this alternate method in Chapter 2. Recall that, (in Chapter 2) we did not use the method of substitution for evaluating such integrals. Besides, that method is simpler than the method of substitution discussed here. Now, it is proposed to evaluate the integral $[\int dx/(4\sin x + 3\cos x)]$ by both the methods for comparison.

Solution: Let $I = \displaystyle\int \frac{dx}{4\sin x + 3\cos x}$

Method (I): Put $\quad \sin x = 2\sin\dfrac{x}{2}\cdot\cos\dfrac{x}{2},\quad$ and

$$\cos x = \cos^2\frac{x}{2} - \sin^2\frac{x}{2}$$

$$I = \int \frac{dx}{8\sin(x/2)\cos(x/2) + 3\cos^2(x/2) - 3\sin^2(x/2)}$$

Dividing N^r and D^r, by $\cos^2\dfrac{x}{2}$, we get

$$I = \int \frac{\sec^2(x/2)dx}{8\tan(x/2) + 3 - 3\tan^2(x/2)}$$

$$= \int \frac{\sec^2(x/2)dx}{3 + 8\tan(x/2) - 3\tan^2(x/2)}$$

Put $\qquad\qquad \tan\dfrac{x}{2} = t \quad \therefore \quad \dfrac{1}{2}\sec^2\dfrac{x}{2}dx = dt$

$$\therefore \quad I = \int \frac{2dt}{3 + 8t - 3t^2}$$

Consider $\qquad\qquad 3 + 8t - 3t^2 = -3\left[t^2 - \dfrac{8}{3}t - 3\right]$

$$= -3\left[t^2 - 2\left(\frac{4}{3}\right)t + \frac{16}{9} - \frac{16}{9} - 3\right]$$

$$= -3\left[\left(t - \frac{4}{3}\right)^2 - \left(\frac{\sqrt{43}}{3}\right)^2\right]$$

$$= 3\left[\left(\frac{\sqrt{43}}{3}\right)^2 - \left(t - \frac{4}{3}\right)^2\right]$$

$$\therefore I = \frac{2}{3}\int \frac{dt}{\left(\sqrt{43}/3\right)^2 - (t-(4/3))^2}$$

$$= \frac{2}{3}\left[\frac{1}{2}\cdot\frac{3}{\sqrt{43}}\log\left(\frac{(\sqrt{43}/3)+(t-(4/3))}{(\sqrt{43}/3)-(t-(4/3))}\right)\right] + c$$

$$= \frac{1}{\sqrt{43}}\log\left(\frac{\sqrt{43}-4+3t}{\sqrt{43}+4-3t}\right) + c \quad \text{Ans.}$$

Method (II): To evaluate $\displaystyle\int \frac{dx}{4\sin x + 3\cos x}$

Consider $\quad 4\sin x + 3\cos x = E$ (say)

$$\therefore \quad \text{Let} \quad 4 = r\cos\alpha \quad \text{and} \quad 3 = r\sin\alpha$$

$$\therefore \quad E = r\sin x\cos\alpha + r\cos x\sin\alpha$$

$$\therefore \quad = r\sin(x+\alpha), \text{ where } r = \sqrt{4^2+3^2} = 5$$

$$\therefore \quad I = \int \frac{dx}{r\sin(x+\alpha)} = \int\frac{1}{r}\text{cosec}\,(x+\alpha)$$

$$= \frac{1}{5}\log\left[\tan\left(\frac{x+\alpha}{2}\right)\right] + c$$

$$= \frac{1}{5}\log\left[\tan\left(\frac{x}{2}+\frac{1}{2}\tan^{-1}\frac{3}{4}\right)\right] + c \quad \text{Ans.}$$

Observe that Method (II) is simpler than Method (I), and less time consuming. Now, let us solve some problems of Type (c).

Example (1): Evaluate $I = \displaystyle\int \frac{dx}{1+2\sin x + 3\cos x}$

Solution: Consider $\quad 1 + 2\sin x + 3\cos x$

We write $\quad 1 = \cos^2\frac{x}{2} + \sin^2\frac{x}{2}$,

$$2\sin = x\,4\sin\frac{x}{2}\cdot\cos\frac{x}{2},$$

and $\quad 3\cos x = 3\cos^2\frac{x}{2} - 3\sin^2\frac{x}{2}$

$$\therefore \quad I = \int\frac{dx}{4\cos^2(x/2)+4\sin(x/2)\cdot\cos(x/2)-2\sin^2(x/2)}$$

Dividing N^r and D^r by $\cos^2\frac{x}{2}$, we get

$$I = \int\frac{\sec^2(x/2)dx}{4+4\tan(x/2)-2\tan^2(x/2)}$$

Put $\qquad\qquad \tan\frac{x}{2} = t, \qquad \therefore \frac{1}{2}\sec^2\frac{x}{2}dx = dt$

$$\therefore \quad I = \int\frac{2\,dt}{4+4t-2t^2} = \int\frac{dt}{2+2t-t^2}$$

Now consider,

$$2+2t-t^2 = -[t^2-2t+1-3]$$

$$= -\left[(t-1)^2-(\sqrt{3})^2\right] = \left[(\sqrt{3})^2-(t-1)^2\right]$$

$$\therefore \quad I = \int \frac{dt}{(\sqrt{3})^2-(t-1)^2} = \frac{1}{2\sqrt{3}} \log \left(\frac{\sqrt{3}+t-1}{\sqrt{3}-t+1}\right) + c$$

$$= \frac{1}{2\sqrt{3}} \log \left(\frac{\sqrt{3}-1+t}{\sqrt{3}+1-t}\right) + c$$

$$= \frac{1}{2\sqrt{3}} \log \left(\frac{(\sqrt{3}-1)+\tan(x/2)}{(\sqrt{3}+1)-\tan(x/2)}\right) + c \quad \text{Ans.}$$

Example (2): Evaluate $\int \frac{1}{\cos \alpha + \cos x} dx = I$ (say)

For convenience, let us put $\cos \alpha = a$,

$$\therefore \quad I = \int \frac{dx}{a + \cos x}$$

We write $\quad \cos x = \cos^2 \frac{x}{2} - \sin^2 \frac{x}{2}$

and $\quad a = a\cos^2 \frac{x}{2} + a\sin^2 \frac{x}{2}$

$$\therefore \quad I = \int \frac{dx}{(1+a)\cos^2(x/2)-(1-a)\sin^2(x/2)}$$

Dividing N^r and D^r by $\cos^2 \frac{x}{2}$, we get

$$I = \int \frac{\sec^2(x/2)dx}{(1+a)-(1-a)\tan^2(x/2)}$$

Put $\quad \tan \frac{x}{2} = t \quad \therefore \quad \frac{1}{2}\sec^2 \frac{x}{2} dx = dt \quad$ so that $\quad \sec^2 \frac{x}{2} dx = 2\,dt$

$$\therefore \quad I = \int \frac{2\,dt}{(1+a)-(1-a)t^2} = \frac{2}{(1-a)} \int \frac{dt}{((1+a)/(1-a))-t^2}$$

$$= \frac{2}{(1-a)} \int \frac{dt}{((\sqrt{1+a})/(\sqrt{1-a}))^2-t^2}$$

$$= \frac{2}{(\sqrt{1-a})^2} \left[\frac{1}{2}\frac{\sqrt{1-a}}{\sqrt{1+a}} \log \left(\frac{(\sqrt{1+a})/(\sqrt{1-a}+t)}{(\sqrt{1+a})/(\sqrt{1-a}-t)}\right)\right] + c$$

$$= \frac{1}{\sqrt{1-a^2}} \log \left(\frac{(\sqrt{1+a})/(\sqrt{1-a}+t)}{(\sqrt{1+a})/(\sqrt{1-a}-t)} \right) + c^{(3)}$$

$$= \frac{1}{\sqrt{1-\cos^2\alpha}} \log \left[\frac{\cot(\alpha/2) + \tan(x/2)}{\cot(\alpha/2) - \tan(x/2)} \right] + c$$

$$= \frac{1}{\sin\alpha} \log \left[\frac{\cot(\alpha/2) + \tan(x/2)}{\cot(\alpha/2) - \tan(x/2)} \right] + c \quad \text{Ans.}$$

Example (3): Evaluate $I = \int \dfrac{1 + \cos\alpha\cos x}{\cos\alpha + \cos x} dx$

Here, we write $1 = \sin^2\alpha + \cos^2\alpha$ [Imp. step]

$$\therefore \quad I = \int \frac{\sin^2\alpha + \cos^2\alpha + \cos\alpha\cos x}{\cos\alpha + \cos x} dx$$

$$= \int \frac{\sin^2\alpha + \cos\alpha(\cos\alpha + \cos x)}{\cos\alpha + \cos x} dx$$

$$= \sin^2\alpha \int \frac{1}{\cos\alpha + \cos x} dx + \int \cos\alpha\, dx$$

$$= \sin^2\alpha \int \frac{1}{\cos\alpha + \cos x} dx + x\cos\alpha$$

Note that we have already evaluated the first integral in the previous Example (2).

$$\therefore \quad I = \sin^2\alpha \left[\frac{1}{\sin\alpha} \log \left(\frac{\cot(\alpha/2) + \tan(\alpha/2)}{\cot(\alpha/2) - \tan(\alpha/2)} \right) + (\cos\alpha)x \right] + c$$

$$= \sin\alpha \log \left(\frac{\cot(\alpha/2) + \tan(\alpha/2)}{\cot(\alpha/2) - \tan(\alpha/2)} \right) + (\cos\alpha) \cdot x + c \quad \text{Ans.}$$

3b.3   SOME NEW INTEGRALS

(i) Now we are in a position to consider for evaluation, the integrals of the form $\int ((px+q)/(ax^2 + bx + c))dx$. Here, *the N is a linear polynomial and D is a quadratic polynomial.* In this case, we express the numerator in the form $[A((d/dx) \text{ of denominator}) + B]$, where A and B are constants. In other words, we express

$$\int \frac{px + q}{ax^2 + bx + c} dx = A \int \frac{2ax + b}{ax^2 + bx + c} dx + B \int \frac{dx}{ax^2 + bx + c}.$$

[3] Since $a = \cos\alpha$, we get $\sqrt{\frac{1+a}{1-a}} = \sqrt{\frac{1+\cos\alpha}{1-\cos\alpha}} = \left[\frac{1+(2\cos^2(\alpha/2)-1)}{1-(1-2\sin^2(\alpha/2))} \right]^{1/2} = \left(\frac{\cos^2\alpha}{\sin^2\alpha} \right)^{1/2} = \cot\frac{\alpha}{2}$

The first integral on the right-hand side is of the form $\int((f'(x))/(f(x)))dx = \log|f(x)| + c_1 = \log|ax^2 + bx + c| + c_1$ and *we know how to integrate the second integral.*

(ii) We can also evaluate the integrals of the form $\int((f(x))/(ax^2 + bx + c))dx$, where *f(x) is a polynomial of degree 2.*[4]

In this case, we divide the numerator by denominator, separate out the quotient and reduce the remaining to the following form

$$K\int \frac{px+q}{ax^2 + bx + x}dx \quad \text{or} \quad K\int \frac{1}{ax^2 + bx + c}dx.$$

For example, $\int \frac{x^3 + 4x^2 + x + 11}{x^2 - 5x + 6}dx = \int \left[x + 1 + \frac{2x+5}{x^2 - 5x + 6}\right]dx$, *which can be easily integrated, as we know.*

Remark: In the integral of the form $\int((px+q)/(ax^2 + bx + x))dx$, if the quadratic polynomial in the denominator (i.e., $ax^2 + bx + c$) has two *distinct linear factors*, then we can evaluate the integral by the *method of partial fractions.* For example, it can be shown that $(x+7)/(x^2 + 8x + 15) = (2/(x+3)) - (1/(x+5))$[5]

3b.4   FOUR MORE STANDARD INTEGRALS

(i) $\displaystyle\int \frac{1}{\sqrt{x^2 + a^2}}dx = \log\left(x + \sqrt{x^2 + a^2}\right) + c$

(ii) $\displaystyle\int \frac{1}{\sqrt{x^2 - a^2}}dx = \log\left(x + \sqrt{x^2 - a^2}\right) + c$

(iii) $\displaystyle\int \frac{1}{\sqrt{a^2 - x^2}}dx = \sin^{-1}\left(\frac{x}{a}\right) + c$

(iv) $\displaystyle\int \frac{1}{x\sqrt{x^2 - a^2}}dx = \frac{1}{a}\sec^{-1}\frac{x}{a} + c$

Now, we shall prove the above standard integrals, using the method of substitution.

(i) To prove that $\displaystyle\int \frac{dx}{\sqrt{x^2 + a^2}} = \log\left[x + \sqrt{x^2 + a^2}\right] + c$

Solution: Let $\displaystyle\int \frac{dx}{\sqrt{x^2 + a^2}} = I$

Put $x = a\tan t \qquad \therefore \quad dx = a\sec^2 t\, dt$

$$\therefore \quad I = \int \frac{a\sec^2 t\, dt}{\sqrt{a^2 \tan^2 t + a^2}} = \int \frac{\sec^2 t}{\sqrt{\tan^2 t + 1}}dt$$

---

[4] We shall restrict ourselves to the cases where the denominator is a quadratic polynomial, irrespective of whether it has linear factors or not.

[5] The (two) terms on the right-hand side are called the partial fractions of the given rational function. In algebra book, methods are discussed for finding the partial fractions provided they exist. [We have already used the method of partial fractions (earlier in this chapter) for evaluating the integrals $\int(1/(x^2 - a^2))dx$ and $\int(1/(a^2 - x^2))dx$.]

$$I = \int \frac{\sec^2 t}{\sec t}\, dt = \int \sec t\, dt \left[\because \quad \tan^2 t + 1 = \sec^2 t\right]$$

$$= \log[\tan t + \sec t] + k$$

$$= \log\left[\frac{x}{a} + \frac{\sqrt{x^2 + a^2}}{a}\right] + k \qquad \begin{cases} \because \quad x = a \tan t \quad \therefore \quad \tan t = \dfrac{x}{a} \quad \text{and} \\[2mm] \sec t = \dfrac{\sqrt{x^2 + a^2}}{a} \end{cases}$$

$$= \log\left[\frac{x + \sqrt{x^2 + a^2}}{a}\right] + k = \log\left[x + \sqrt{x^2 + a^2}\right] - \log a + k$$

$$= \log\left[x + \sqrt{x^2 + a^2}\right] + c, \quad \text{where } c \text{ is an arbitrary constant.}$$

[Here $(-\log a)$ and k both are absorbed in arbitrary constant c.]

Thus,

$$\int \frac{dx}{\sqrt{x^2 + a^2}} = \log\left[x + \sqrt{x^2 + a^2}\right] + c \quad \text{Ans.}$$

(ii) To prove that $\displaystyle\int \frac{dx}{\sqrt{x^2 - a^2}} = \log\left[x + \sqrt{x^2 - a^2}\right] + c$

Solution: Let $\displaystyle\int \frac{dx}{\sqrt{x^2 - a^2}} = I$

Put $x = a \sec t \qquad \therefore \quad dx = a \sec t \cdot \tan t \cdot dt$

$$\therefore \quad I = \int \frac{a \sec t \tan t}{\sqrt{a^2 \sec^2 t - a^2}}\, dt \qquad \left[\begin{array}{l} \because \tan^2 x + 1 = \sec^2 x \\[2mm] \therefore \sec^2 x - 1 = \tan^2 x \end{array}\right]$$

$$= \int \frac{a \sec t \cdot \tan t}{\sqrt{a^2 \tan^2 t}}\, dt = \int \sec t\, dt$$

$$I = \log[\sec t + \tan t] + c$$

$$= \log\left[\frac{x}{a} + \frac{\sqrt{x^2 - a^2}}{a}\right] + c \qquad \begin{cases} \because x = a \sec t \quad \therefore \quad \sec t = (x/a) \text{ and} \\[2mm] \tan t = \dfrac{\sqrt{x^2 - a^2}}{a} \end{cases}$$

$$= \log\left[\frac{x + \sqrt{x^2 - a^2}}{a}\right] + c$$

$$= \log\left[x + \sqrt{x^2 - a^2}\right] - \log a + c$$

$$= \log\left[x + \sqrt{x^2 - a^2}\right] + c$$

(Here again the constant "–log a" is absorbed in the arbitrary constant c.)

Thus, $\displaystyle\int \frac{dx}{\sqrt{x^2-a^2}} = \log\left[x+\sqrt{x^2-a^2}\right]+c$ Ans.

Remark: The above result can also be obtained from the result (i) by replacing a^2 by $-a^2$.

(iii) To prove that $\displaystyle\int \frac{dx}{\sqrt{a^2-x^2}} = \sin^{-1}\frac{x}{a}+c$

Solution: Let $\displaystyle\int \frac{dx}{\sqrt{a^2-x^2}} = I$

Put $x = a\sin t$ $\therefore$ $dx = a\cos t\, dt$

$\therefore\quad I = \displaystyle\int \frac{a\cos t\, dt}{\sqrt{a^2-a^2\sin^2 t}} = \int \frac{a\cos t}{a\cos t}dt = \int dt = t+c = \sin^{-1}\frac{x}{a}+c$

Thus, $\displaystyle\int \frac{dx}{\sqrt{a^2-x^2}} = \sin^{-1}\frac{x}{a}+c^{(6)}$

This result can also be obtained by using the substitution $x = a\cdot t$. We get $dx = a\, dt$.

$\therefore\quad \displaystyle\int \frac{dx}{\sqrt{a^2-x^2}} = \int \frac{a\, dt}{\sqrt{a^2-a^2 t^2}} = \int \frac{dt}{\sqrt{1-t^2}}$

$$= \sin^{-1} t+c = \sin^{-1}\frac{x}{a}+c\left[\because t = \sin^{-1}\frac{x}{a}\right]$$

[The reader may convince himself to note that the substitution $x = a\cdot t$ is not applicable in cases (i) and (ii).]

(iv) $I = \displaystyle\int \frac{1}{x\sqrt{x^2-a^2}}dx = \frac{1}{a}\sec^{-1}\frac{x}{a}+c^{(7)}$

Put $x = a\sec t$ $\therefore$ $dx = a\sec t\cdot\tan t\, dt$

$\therefore\quad I = \displaystyle\int \frac{a\sec t\cdot\tan t}{a\sec t\sqrt{a^2\sec^2 t-a^2}}dt$

$= \dfrac{1}{a}\displaystyle\int \frac{\tan t}{\sqrt{\sec^2 t-1}}dt = \frac{1}{a}\int dt = \frac{1}{a}\cdot t+c\qquad \left\{\begin{array}{l}\because\quad \sec t = x/a \\[2mm] \therefore\quad t = \sec^{-1}(x/a)\end{array}\right.$

$= \dfrac{1}{a}\sec^{-1}\left(\dfrac{x}{a}\right)+c$

$\therefore\quad \displaystyle\int \frac{1}{x\sqrt{x^2-a^2}}dx = \frac{1}{a}\sec^{-1}\left(\frac{x}{a}\right)+c$

---

[6] Note that, this formula is the more general form of the result $\int(1/(x\sqrt{x^2-1}))dx = \sec^{-1} x+c$, which follows from $(d/dx)(\sec^{-1} x) = (1/x\sqrt{x^2-1})$.

[7] Note that, this formula is the more general form of the result $\int(1/(x\sqrt{x^2-1}))dx = \sec^{-1} x+c$, which follows from $(d/dx)(\sec^{-1} x) = (1/x\sqrt{x^2-1})$.

Solved Examples

Evaluate the following:

Example (1): $\displaystyle \int \frac{dx}{\sqrt{5+4x-x^2}} = I$

Consider
$$5+4x-x^2 = -(x^2-4x-5)$$
$$= -[x^2-2(2)x+4-9]$$
$$= -[(x-2)^2-3^2] = [3^2-(x-2)^2]$$

$$\therefore \quad I = \int \frac{dx}{\sqrt{3^2-(x-2)^2}} = \sin^{-1}\left(\frac{x-2}{3}\right)+c \quad \text{Ans.}$$

Example (2): $\displaystyle \int \frac{2\cos x}{\sqrt{4-\sin^2 x}}\,dx$

Put $\sin x = t$ $\quad \therefore \quad \cos x\,dx = dt$

$$\therefore \quad I = \int \frac{2\,dt}{\sqrt{(2)^2-t^2}} = 2\int \frac{dt}{\sqrt{(2)^2-t^2}}$$

$$= 2\left[\sin^{-1}\left(\frac{t}{2}\right)\right]+c = 2\cdot\sin^{-1}\left(\frac{\sin x}{2}\right)+c \quad \text{Ans.}$$

Example (3): $\displaystyle \int \frac{dx}{\sqrt{2x-x^2}} = I$

Consider
$$2x-x^2 = -[x^2-2x]$$
$$= -[x^2-2x+1-1] = -[(x-1)^2-(1)^2]$$
$$= [(1)^2-(x-1)^2]$$

$$\therefore \quad I = \int \frac{dx}{\sqrt{1^2-(x-1)^2}} = \sin^{-1}\left(\frac{x-1}{1}\right)+c$$

$$= \sin^{-1}(x-1)+c \quad \text{Ans.}$$

Example (4): $\displaystyle \int \frac{\sqrt{e^x}}{\sqrt{e^x+4e^{-x}}}\,dx = I$

We have,

$$I = \int \frac{\sqrt{e^x}}{\sqrt{e^x+4e^{-x}}}\,dx = \int \sqrt{\frac{e^x}{e^x+4e^{-x}}}\,dx$$

$$= \int \sqrt{\frac{e^{2x}}{e^{2x}+4}}\,dx \qquad \left\{ \begin{array}{l} \because \quad e^x+4e^{-x} \\ = e^x+(4/e^x) \\ = (e^{2x}+4)/e^x \end{array} \right.$$

$$= \int \frac{e^x}{\sqrt{e^{2x}+4}}\,dx = \int \frac{e^x}{\sqrt{(e^x)^2+(2)^2}}\,dx$$

(It is clear how the numerator becomes free from square root.)

Now put $e^x = t$ $\therefore$ $e^x\,dx = dt$

$$\therefore\ \ I = \int \frac{dt}{\sqrt{t^2 + (2)^2}} = \log\left[t + \sqrt{t^2 + 2^2}\right] + c$$

$$= \log\left[e^x + \sqrt{e^{2x} + 4}\right] + c \quad \text{Ans.}$$

Example (5): $\displaystyle\int \frac{\sin x}{\sqrt{\cos 2x}}\,dx = I$

We have, $I = \displaystyle\int \frac{\sin x}{\sqrt{2\cos^2 x - 1}}\,dx$ $\left\{ \begin{array}{l} \text{Note that} \\ \cos 2x = 2\cos^2 x - 1 \\ = (\sqrt{2}\cos x)^2 - 1 \end{array} \right.$

$$\therefore\ \ I = \int \frac{\sin x}{\sqrt{(\sqrt{2}\cos x)^2 - 1^2}}\,dx$$

Put $\sqrt{2}\cos x = t$ $\therefore -\sqrt{2}\sin x\,dx = dt$

$$\therefore\ \ I = -\frac{1}{\sqrt{2}}\int \frac{dt}{\sqrt{t^2 - 1}} = -\frac{1}{\sqrt{2}}\log\left[t + \sqrt{t^2 - 1}\right] + c$$

$$I = -\frac{1}{\sqrt{2}}\log\left[\sqrt{2}\cos x + \sqrt{2\cos^2 x - 1}\right] + c$$

$$= -\frac{1}{\sqrt{2}}\log\left[\sqrt{2}\cos x + \sqrt{\cos(2x)}\right] + c \qquad \text{Ans.}$$

Example (6): $\displaystyle\int \frac{dx}{\sqrt{2x^2 + 3x + 4}} = I$

Consider $\quad 2x^2 + 3x + 4 = 2\left[x^2 + \frac{3}{2}x + 2\right]$

$$= 2\left[x^2 + 2\left(\frac{3}{4}\right)x + \left(\frac{3}{4}\right)^2 - \left(\frac{3}{4}\right)^2 + 2\right]$$

$$= 2\left[\left(x + \frac{3}{4}\right)^2 - \left(\frac{\sqrt{23}}{2}\right)^2\right]$$

$$\therefore\ \ I = \frac{1}{\sqrt{2}}\int \frac{dx}{\sqrt{(x + (3/4))^2 - (\sqrt{23}/2)^2}}$$

$$= \frac{1}{\sqrt{2}}\log\left[\left(x + \frac{3}{4}\right) + \sqrt{(x + (3/4))^2 - (\sqrt{23}/4)^2}\right] + c$$

$$= \frac{1}{\sqrt{2}}\log\left[\left(x + \frac{3}{4}\right) + \sqrt{x^2 + \frac{3}{2}x + 2}\right] + c \quad \text{Ans.}$$

(Recall that after making the coefficient of x^2 unity, we express the quadratic expression as the sum or difference of two squares and then apply the required standard result.)

Exercise (3)

Evaluate the following integrals:

(i) $\displaystyle\int \frac{\cos x}{\sqrt{1-4\sin^2 x}}\,dx$

Ans. $\dfrac{1}{2}\sin^{-1}(2\sin x)+c$

[Hint: Put $2\sin x = t$, so that $2\cos x\,dx = dt$]

(ii) $\displaystyle\int \frac{dx}{\sqrt{2ax-x^2}}$

Ans. $\sin^{-1}\left(\dfrac{x-a}{a}\right)+c$

(iii) $\displaystyle\int \frac{x^2}{\sqrt{x^6+2x^3+2}}\,dx$

Ans. $\dfrac{1}{3}\log\left[(x^3+1)+\sqrt{x^6+2x^3+2}\right]+c$

(iv) $\displaystyle\int \frac{x^2}{\sqrt{a^6-x^6}}\,dx$

Ans. $\dfrac{1}{3}\sin^{-1}\dfrac{x^3}{a^3}+c$

Solved Examples

Example (7) $\displaystyle\int \frac{1}{x\sqrt{3x^2-2}}\,dx = I$

$\therefore\quad I = \dfrac{1}{\sqrt{3}}\displaystyle\int \frac{1}{x\sqrt{x^2-(\sqrt{2/3})^2}}\,dx = \dfrac{1}{\sqrt{3}}\sec^{-1}\left(\dfrac{x}{\sqrt{2/3}}\right)+c$

$\qquad = \dfrac{1}{\sqrt{3}}\sec^{-1}\left(\dfrac{\sqrt{3}x}{\sqrt{2}}\right)+c$ Ans.

Example (8) $\displaystyle\int \frac{1}{x\cdot\sqrt{x^4-9}}\,dx$ [Imp.]

Let $I = \displaystyle\int \frac{1}{x\cdot\sqrt{(x^2)^2-(3)^2}}\,dx$

Here, if we put $x^2 = t$, we get $2x\,dx = dt$. But $2x$ is not there in the N^r. Therefore, we multiply N^r and D^r by $2x$ and get.

$$\therefore \qquad I = \frac{1}{2} \int \frac{2x}{x^2 \cdot \sqrt{(x^2)^2 - (3)^2}}\,dx$$

Now put $x^2 = t$ $\therefore$ $2x\ dx = dt$

$$= \frac{1}{2} \int \frac{dt}{t \cdot \sqrt{t^2 - (3)^2}} = \frac{1}{2} \left[\frac{1}{3} \sec^{-1} \frac{t}{3} \right] + c$$

$$= \frac{1}{6} \sec^{-1} \left(\frac{x^2}{3} \right) + c \quad \text{Ans.}$$

Exercise (4)

Evaluate the following integrals:

(a) $\displaystyle \int \frac{dx}{x \cdot \sqrt{16x^2 - 9}}$

Ans. $\dfrac{1}{4} \sec^{-1} \left(\dfrac{4x}{3} \right) + c$

(b) $\displaystyle \int \frac{1}{x^2 \sqrt{x^2 - 1}}\,dx$

Ans. $\dfrac{\sqrt{x^2 - 1}}{x} + c$

[Hint: Put $x = \sec t$]

(c) $\displaystyle \int \frac{1}{x^3 \cdot \sqrt{x^2 - 1}}\,dx$

Ans. $\dfrac{1}{2} \sec^{-1} x + \dfrac{1}{2} \dfrac{\sqrt{x^2 - 1}}{x^2} + c$

Now we shall consider integrals of the type $\int \left((px + q) / \sqrt{ax^2 + bx + c} \right) dx$.

Method: We find two constants A and B, such that

$$px + q = A \frac{d}{dx}(ax^2 + bx + c) + B$$

$$\text{i.e.,}\ px + q = A(2ax + b) + B$$

Now, by equating coefficients of x and the constants on both sides, we find the values of A and B. We then express the given integral as follows:

$$\int \frac{px + q}{\sqrt{ax^2 + bx + c}}\,dx = A \int \frac{2ax + b}{\sqrt{ax^2 + bx + c}}\,dx + B \int \frac{1}{\sqrt{ax^2 + bx + c}}\,dx$$

The first integral on the right-hand side is of the form $\int (f'(x)/\sqrt{f(x)})dx$, *which is easily integrated by putting* $f(x) = t$, *so that* $f'(x)\, dx = dt$, *and we get*

$$\int \frac{f'(x)}{\sqrt{f(x)}}dx = \int [f(x)]^{-1/2} f'(x)dx = \int t^{-1/2}\, dt = \frac{t^{1/2}}{1/2} + c = 2\sqrt{t} + c$$

and the second can be evaluated by expressing $(ax^2 + bx + c) = a\left[x^2 + \frac{b}{a}x + \frac{c}{a}\right]$ as a sum or difference of twoas a sum or difference of two squares and applying the relevant standard formula applicable.

Illustrative Examples

Example (9): Evaluate $\displaystyle\int \frac{4x+1}{\sqrt{x^2+3x+2}}dx = I$ (say)

Solution: Let $4x + 1 = A\left[\dfrac{d}{dx}(x^2 + 3x + 2)\right] + B$

Then $4x + 1 = A(2x + 3) + B$

$$= 2Ax + (3A + B) \qquad\qquad (4)$$

Equating the coefficients of x on both the sides of (4), we get

$$2A = 4 \quad \therefore \quad A = 2$$

Again, equating the constant terms on both sides of (4), we have

$$3A + B = 1 \quad \text{or} \quad 6 + B = 1$$

$$\therefore \quad B = -5$$

Substituting the values of A and B in (4), we get

$$(4x + 1) = 2(2x + 3) - 5$$

$$\therefore \quad \int \frac{4x+1}{\sqrt{x^2+3x+2}}dx = 2\int \frac{2x+3}{\sqrt{x^2+3x+2}}dx - 5\int \frac{dx}{\sqrt{x^2+3x+2}}$$

Now, $I_1 = 2\displaystyle\int \frac{(d/dx)(x^2+3x+2)}{\sqrt{x^2+3x+2}}dx = 2\int \frac{2x+3}{\sqrt{x^2+3x+2}}dx$

Put $x^2 + 3x + 2 = t$ $\therefore$ $(2x+3)dx = dt$

$$\therefore \quad I_1 = 2\int \frac{dt}{\sqrt{t}} = 2\int t^{-1/2}\, dt = 2\left[\frac{t^{1/2}}{1/2}\right] + c_1$$

$$= t\sqrt{t} + c_1$$

$$= 4\sqrt{x^2+3x+2} + c_1$$

Now, consider $x^2 + 3x + 2$

$$= x^2 + 2\left(\frac{3}{2}\right)x + \frac{9}{4} - \frac{9}{4} + 2$$

$$= \left(x + \frac{3}{2}\right)^2 - \left(\frac{1}{2}\right)^2$$

$$\therefore \quad I_2 = 5\int \frac{dx}{\sqrt{x^2 + 3x + 2}} = 5\int \frac{dx}{\sqrt{(x + (3/2))^2 - (1/2)^2}}$$

$$= 5\log\left[\left(x + \frac{3}{2}\right) + \sqrt{\left(x + \frac{3}{2}\right)^2 - \left(\frac{1}{2}\right)^2}\right] + c_2 \,^{(8)}$$

$$= 5\log\left[x + \frac{3}{2} + \sqrt{x^2 + 3x + 2}\right] + c_2$$

$$\therefore \quad I = 4\sqrt{x^2 + 3x + 2} - 5\log\left[x + \frac{3}{2} + \sqrt{x^2 + 3x + 2}\right] + c \quad \text{Ans.}$$

Note that $\left[\left(x + \dfrac{3}{2}\right)^2 - \left(\dfrac{1}{2}\right)^2 = x^2 + 3x + 2\right].$

Example (10): Evaluate $\displaystyle\int \frac{2x + 5}{\sqrt{2x^2 + 2x + 5}}dx = I \qquad \text{(say)}$

Solution: Let $2x + 5 = A\left[\dfrac{d}{dx}(2x^2 + 2x + 5)\right] + B$

$$= A(4x + 2) + B$$

$$= 4Ax + (2A + B)$$

$$\therefore \quad 4A = 2 \quad \therefore \quad A = {}^1\!/_2$$

and $\quad 2A + B = 5 \quad \therefore \quad 2(1/2) + B = 5 \quad \therefore \quad B = 4$

$$\therefore \quad I = \int \frac{(1/2)(4x + 2) + 4}{\sqrt{2x^2 + 2x + 5}}dx$$

$$\frac{1}{2}\int \frac{4x + 2}{\sqrt{2x^2 + 2x + 5}}dx + 4\int \frac{dx}{\sqrt{2x^2 + 2x + 5}}$$

Consider I_1 Put $\quad 2x^2 + 2x + 5 = t \qquad \therefore \; (4x + 2)\,dx = dt$

$$\therefore \quad I_1 = \frac{1}{2}\int \frac{dt}{\sqrt{t}} = \frac{1}{2}\int t^{-1/2}dt = \frac{1}{2}\left[\frac{t^{1/2}}{1/2}\right] + c_1$$

$$= \sqrt{t} + c_1 = \sqrt{2x^2 + 2x + 5} + c_1$$

Now $2x^2 + 2x + 5 = 2\left[x^2 + x + \dfrac{5}{2}\right]$

$$= 2\left[x^2 + 2\left(\dfrac{1}{2}\right)x + \left(\dfrac{1}{2}\right)^2 - \left(\dfrac{1}{2}\right)^2 + \dfrac{5}{2}\right]$$

$$= 2\left[\left(x + \dfrac{1}{2}\right)^2 + \left(\dfrac{3}{2}\right)^2\right]$$

$$\therefore \quad I_2 = \dfrac{4}{\sqrt{2}}\int \dfrac{dx}{\sqrt{(x+(1/2))^2+(3/2)^2}} = \dfrac{4}{\sqrt{2}} \cdot \dfrac{\sqrt{2}}{\sqrt{2}}\int \dfrac{dx}{\sqrt{(x+(1/2))^2+(3/2)^2}}$$

$$= 2\sqrt{2}\log\left[\left(x+\dfrac{1}{2}\right)+\sqrt{x^2+x+\dfrac{5}{2}}\right] + c_2$$

$$\therefore \quad I = \sqrt{2x^2+2x+5} + 2\sqrt{2}\log\left[x+\dfrac{1}{2}+\sqrt{x^2+x+\dfrac{5}{2}}\right] + c \qquad \text{Ans.}$$

Note that $(x+(1/2))^2+(3/2)^2 = x^2+x+(5/2)$ (and not $2x^2+2x+5$).

Now we give below some integrals which can be expressed in the following form $\int (px+q)/\sqrt{ax^2+bx+c}\,dx$.

Example (11): Evaluate $\displaystyle\int \sqrt{\dfrac{x+1}{x+3}}\,dx = I$ (say)

Consider $\quad \sqrt{\dfrac{x+1}{x+3}} = \sqrt{\dfrac{x+1}{x+3} \cdot \dfrac{x+1}{x+1}}$[9]

$$= \dfrac{x+1}{\sqrt{(x+3)(x+1)}} = \dfrac{x+1}{\sqrt{x^2+4x+3}}$$

$$\therefore \quad I = \int \dfrac{x+1}{\sqrt{x^2+4x+3}}\,dx = \dfrac{1}{2}\int \dfrac{2x+2}{\sqrt{x^2+4x+3}}\,dx$$

But $\quad \dfrac{d}{dx}(x^2+4x+3) = 2x+4$

$$\therefore \quad I = \dfrac{1}{2}\int \dfrac{(2x+4)-2}{\sqrt{x^2+4x+3}}\,dx$$

$$= \dfrac{1}{2}\left[\int \dfrac{2x+4}{\sqrt{x^2+4x+3}}\,dx - \int \dfrac{2}{\sqrt{x^2+4x+4-1}}\,dx\right]$$

$$= \dfrac{1}{2}\int \dfrac{2x+4}{\sqrt{x^2+4x+3}}\,dx - \int \dfrac{dx}{\sqrt{(x+2)^2-(1)^2}}$$

$$= \dfrac{1}{2}\left(2\sqrt{x^2+4x+3}\right) - \log\left[(x+2)+\sqrt{x^2+4x+3}\right] + c$$

$$= \sqrt{x^2+4x+3} - \log\left[(x+2)+\sqrt{x^2+4x+3}\right] + c \quad \text{Ans.}$$

[9] The basis behind the method is to release the numerator from the square-root sign.

Example (12): Evaluate $\int \sqrt{\dfrac{a-x}{a+x}}dx = I$ (say)

Consider $\sqrt{\dfrac{a-x}{a+x}} = \sqrt{\dfrac{a-x}{a+x} \cdot \dfrac{a-x}{a-x}}$

$\qquad\qquad = \dfrac{a-x}{\sqrt{a^2-x^2}}$

$\therefore \; I \;\; = \int \dfrac{a-x}{\sqrt{a^2-x^2}}dx = \int \dfrac{a}{\sqrt{a^2-x^2}}dx + \dfrac{1}{2}\int \dfrac{-2x}{\sqrt{a^2-x^2}}dx$

$\qquad\quad = a \cdot \sin^{-1}\dfrac{x}{a} + \dfrac{1}{2} \cdot 2\sqrt{a^2-x^2} + c$

$\qquad\quad = a \cdot \sin^{-1}\dfrac{x}{a} + \sqrt{a^2-x^2} + c$ Ans.

Now evaluate the following integrals:

(i) $\int \sqrt{\dfrac{x-2}{x}}dx$

Ans. $\sqrt{x^2-2x} - \log\left(x-1+\sqrt{x^2-2x}\right) + c$

(ii) $\int \sqrt{\dfrac{x-5}{x-7}}dx$

Ans. $\sqrt{x^2-12x+35} + \log\left[(x-6) + \sqrt{x^2-12x+35}\right] + c$

Observation: Evaluation of an integral of the type $\int ((px+q)/(\sqrt{ax^2+bx+c}))dx$, *is more time consuming.*

Note: With regards to the integrals of the functions $\sqrt{x^2+a^2}, \sqrt{x^2-a^2}$, and $\sqrt{a^2-x^2}$ [of Batch (III)], we can use the method of substitution, but as mentioned earlier, *there is a simpler method* (*known as integration by parts introduced in Chapter* 4b). *We shall obtain the integrals of these functions, by both the methods in Chapter* 4b.

4a Integration by Parts

4a.1 INTRODUCTION

As yet, we have no technique for evaluating integrals like $\int x \cos x \, dx$, which involve products of two functions. In this chapter, we give a method of integration that is useful in integrating *certain products* of two functions.[1]

We give below some more examples of the integrals (involving products of functions) to give an idea of the type of functions that we propose to handle in this chapter.

$$\int x \, e^x \, dx, \quad \int x^2 \cos x \, dx, \quad \int \sin^{-1} x \, dx, \quad \int \cos^{-1} x \, dx, \quad \int \log x \, dx,$$

$$\int (\log x)^2 \, dx, \quad \int \sec^3 x \, dx, \quad \int \left(\sin^{-1} x\right)^2 dx, \quad \int \sin\sqrt{x} \, dx, \text{ etc.}$$

The technique that we are going to discuss is one of the most widely used techniques of integration, known as *Integration by Parts*. It is obtained from the formula for derivative of the product of two functions just as the *sum rule of integration* is derived from the *sum rule for differentiation*. In fact, the operations of *differentiation* and *antidifferentiation* are closely related. Hence, it is natural that certain rules of *integral Calculus* follow from their counterparts in *differential Calculus*.

A most surprising and interesting fact comes to light when we study (the first and the second) *fundamental theorems of Calculus*, to be introduced later in Chapter 6a. The concept of the *Definite Integral* (discussed later in Chapter 5) clearly tells that in the development of the idea of the definite integral, the concept of derivatives does not come into play. On the other hand, *the fundamental theorems of Calculus* prove that computation of *Definite Integral(s)* can be done very easily using antiderivative(s). [The method of computing definite integrals is otherwise a very complicated process and it cannot be applied to many functions.] It is for this reason that the term '*integral*', (picked up from the definite integral) is also used to stand for antiderivative. Accordingly, the process of computing *both antiderivative(s) and definite integral(s)* is called *integration*.

4a-Method of integration by parts (When the integrand is in the form of product of two functions)

[1] Generally, such integrals arise in practical applications of integration, namely computation of areas, volumes, and other quantities, using the concept of definite integral.

Introduction to Integral Calculus: Systematic Studies with Engineering Applications for Beginners, First Edition. Ulrich L. Rohde, G. C. Jain, Ajay K. Poddar, and A. K. Ghosh.
© 2012 John Wiley & Sons, Inc. Published 2012 by John Wiley & Sons, Inc.

4a.2   OBTAINING THE RULE FOR INTEGRATION BY PARTS

Let u and v be functions of x possessing continuous derivatives. Then, we have

$$\frac{d}{dx}(u \cdot v) = u \cdot \frac{dv}{dx} + v \cdot \frac{du}{dx} \tag{1}$$

Also, we can *restate* (1) *in the language of indefinite integrals* as

$$u \cdot v = \int u \cdot \frac{dv}{dx} \cdot dx + \int v \cdot \frac{du}{dx} \cdot dx \tag{2}$$

or by rearranging, we get

$$\int u \cdot \frac{dv}{dx} \cdot dx = u \cdot v - \int v \cdot \frac{du}{dx} \cdot dx \tag{3(A)}$$

Or

$$\int u \cdot dv = u \cdot v - \int v \cdot du \tag{3(B)}$$

For computational purposes, a more convenient way of writing this formula is obtained, if we put, $u = f(x)$ and $v = g(x)$. Then, $du = f'(x)dx$ and $dv = g'(x)dx$. [These relations may also be visualized in (3(A)), in view of the definition of differential(s) discussed in Chapter 16 of Part I.]

Using the above expressions for u, v, du, and dv, we can write equation (3(B)) in the form

$$\int f(x) \cdot g'(x) \cdot dx = f(x) \cdot g(x) - \int g(x) \cdot f'(x) \, dx \tag{3(C)}[2]$$

The formula (3(C)) expresses the integral $\int f(x) \cdot g'(x) \cdot dx \ [=\int u \cdot dv]$ in terms of another integral $\int g(x) \cdot f'(x) \cdot dx \ [=\int v \cdot du]$.

4a.2.1   Important Notes for Proper Choice of First and Second Functions Needed for Applying the Rule of Integration by Parts

Now, *suppose we wish to evaluate* $\int h(x) \cdot dx$, *but cannot readily do so. If $h(x)$ can be rewritten as the product of $f(x) \cdot g'(x)$, then* (3(C)) *tells us that,*

$$\int h(x) \cdot dx = \int f(x) \cdot g'(x) \cdot dx$$
$$= f(x) \cdot g(x) - \int g(x) \cdot f'(x)dx \tag{4}$$

In addition, if $\int g(x) \cdot f'(x) \cdot dx$ *can be readily evaluated, then* $\int h(x) \cdot dx$ *can be evaluated by means of* (4).

It is useful to state the formula (3(C)) in words. For this purpose, we shall call the function $f(x)$ as *the first function* and the function $g'(x)dx$ as *the second function*.

---

[2] The formula (3(C)) is the most convenient statement of the rule for our purpose. It defines the intergral of the product of two functions namely $f(x)$ and $g'(x)$. Note that the function $g(x)$ [on the right-hand side of (3(C))] stands for $\int g'(x) \cdot dx$ and it occurs twice.

Then, the rule defined by equation (3(C)) can be remembered in words, as follows:

(Integral of the product of two functions) = (first function) × (integral of the second function) − $\int$[(derivative of first function) × (integral of the second function)]d$x^{(3)}$

The important point to note is to select from the product $f(x) \cdot g'(x)$ [in the integrand $h(x)$], *the first* and *the second* function, and the correct substitutions for the functions $f(x)$ and $g'(x)$ dx.

The selection of the functions has to be such that the integral $\int g(x) \cdot f'(x) dx$ appearing on the right-hand side is no more difficult and preferably less difficult to integrate then the integral $\int f(x) \cdot g(x) \cdot dx$.

The above discussion suggests that $f(x)$ should be a function that is easy to differentiate and $g'(x)dx$ should be chosen so that $g(x)$ can be readily found by integration [or $g(x)$ may be some standard integral, available in the table]. The method will be clearer once we see how it works in specific examples.$^{(4)}$

Note (1): The method of integration by parts, is applicable only if one of the two functions in the given integral $\int h(x) \cdot dx$ [=$\int f(x) \cdot g'(x) \cdot dx$] can be easily integrated. In fact, our ability to select such a function (i.e., the second function) correctly, will depend upon the integrand, and our experience. [This is so because in some problems, both the functions may be easily integrable whereas, in some others there may be only one function, and not a product of two]. These situations will become clearer shortly, as we proceed to solve problems. First, we list the *standard indefinite integration formulas* that will be needed frequently.

Standard Indefinite Integration Formulas

1. $\int k \cdot dx = kx + c$

2. $\int x \cdot dx = \frac{1}{2}x^2 + c$

3. $\int x^n \, dx = \frac{x^{n+1}}{n+1} + c, \quad n \neq -1, n \in R$

4. $\int e^x \, dx = e^x + c$

5. $\int a^x \, dx = \frac{a^x}{\log_e a} + c \quad (a > 0) \quad [\because \int a^x \cdot \log_e a \, dx = a^x + c \quad (a > 0)]$

6. $\int \sin x \, dx = -\cos x + c$ (Chapter 1)

7. $\int \cos x \, dx = \sin x + c$ (Chapter 1)

8. $\int \tan x \, dx = \log_e |\sec x| + c = \log|\sec x| + c$ (Chapter 3a)

9. $\int \cot x \, dx = \log_e |\sec x| + c = \log|\sec x| + c$ (Chapter 3a)

10. $\int \sec x \, dx = \log_e |\sec x + \tan x| + c = \log|\tan(\frac{x}{2} + \frac{\pi}{4})| + c$ (Chapter 3a)

11. $\int \operatorname{cosec} x \, dx = \log_e |\operatorname{cosec} x - \cot x| + c = \log|(\tan \frac{x}{2})| + c$ (Chapter 3a)

12. $\int \sec^2 x \, dx = \tan x + c$ (Chapter 1)

$^{(3)}$ This method of evaluating $h(x)$, by "splitting" the integrand $h(x)$ into two parts $f(x)$ and $g'(x)$, is known as "integration by parts".

$^{(4)}$ Remember that in the formula (3(C)), selection of the second function $g'(x)$ is very important since its integral [i.e., $g(x)$] appears twice on the right-hand side of the formula.

13. $\int \operatorname{cosec}^2 x \, dx = -\cot x + c$ (Chapter 1)

14. $\int \sec x \cdot \tan x \, dx = \sec x + c$ (Chapter 1)

15. $\int \operatorname{cosec} x \cdot \cot x \, dx = -\operatorname{cosec} x + c$ (Chapter 1)

16. $\int \frac{dx}{\sqrt{1-x^2}} = \sin^{-1} x + c$ or $-\cos^{-1} x + c$ (Chapter 1)

17. $\int \frac{dx}{1+x^2} = \tan^{-1} x + c$ or $-\cot^{-1} x + c$ (Chapter 1)

18. $\int \frac{dx}{x \cdot \sqrt{x^2-1}} = \sec^{-1} x + c$ or $-\operatorname{cosec}^{-1} x + c)$ (Chapter 1)

19. $\int \frac{1}{a^2+x^2} dx = \frac{1}{a} \tan^{-1}\left(\frac{x}{a}\right) + c, \quad (a \neq 0)$ (Chapter 3b)

20. $\int \frac{1}{a^2-x^2} dx = \frac{1}{2a} \log\left(\frac{a+x}{a-x}\right) + c, \quad x < a$ (Chapter 3b)

      $= \frac{1}{2a} \log\left|\frac{a+x}{a-x}\right| + c$

21. $\int \frac{dx}{\sqrt{a^2-x^2}} = \sin^{-1} \frac{x}{a} + c, \quad (a > 0)$ (Chapter 3b)

22. $I = \int \frac{1}{x\sqrt{x^2-a^2}} dx = \frac{1}{a} \sec^{-1} \frac{x}{a} + c, \quad (a > 0)$ (Chapter 3b)

23. $\int \frac{dx}{\sqrt{x^2+a^2}} = \log\left[x + \sqrt{x^2+a^2}\right] + c$ (Chapter 3b)

24. $\int \frac{dx}{\sqrt{x^2-a^2}} = \log\left[x + \sqrt{x^2-a^2}\right] + c, \quad (x > a > 0)$ (Chapter 3b)

25. $\int \sqrt{x^2+a^2} \, dx = \frac{x}{2}\sqrt{x^2+a^2} + \frac{a^2}{2}\log\left[x + \sqrt{x^2+a^2}\right] + c$ (Chapter 4b)

26. $\int \sqrt{x^2-a^2} \, dx = \frac{x}{2}\sqrt{x^2-a^2} - \frac{a^2}{2}\log\left[x + \sqrt{x^2-a^2}\right] + c$ (Chapter 4b)

27. $\int \sqrt{a^2-x^2} \, dx = \frac{x}{2}\sqrt{a^2-x^2} + \frac{a^2}{2}\sin^{-1}\frac{x}{a} + c$ (Chapter 4b)

Now, we give some important notes and supporting illustrative examples.

Note (2): Of the two functions in the integrand, if one function is a power function (i.e., x, x^2, x^3, ...) and *the other function is easy to integrate*, then we choose the other one as the second function. If power function is chosen as second function, then its index will keep on increasing when the rule of *integration by parts* is applied. As a result, the resulting integral so obtained will be more difficult to evaluate, than the given integral.

Example (1): Evaluate $I = \int e^x \cdot x^2 \, dx$

Solution: Observe that the above integral cannot be solved by any of our previous methods. Further, the integrals of both the functions (i.e., x^2 and e^x) are equally easy. However, since x^2 *is a power function*, in view of the Note (2) above, we choose e^x as the second function.

$$\therefore \quad I = \int (x^2)(e^x) \, dx \qquad \begin{cases} \int e^x \, dx = e^x \\[2mm] \dfrac{d}{dx}(x^2) = 2x \end{cases}$$

Integrating by parts, we get

$$I = x^2 \cdot e^x - \int (2x) \cdot e^x \, dx$$

$$= x^2 \cdot e^x - 2\int x \cdot e^x \, dx \quad \begin{cases} \int e^x \, dx = e^x \\ \dfrac{d}{dx}(x) = 1 \end{cases}$$

$$= x^2 \cdot e^x - 2\left[x\,e^x - \int 1 \cdot e^x \, dx \right]$$

$$= x^2 \cdot e^x - 2x\,e^x + 2\,e^x + c \qquad \text{Ans.}$$

Now *let us see what happens if we choose* x^2 *as the second function.*

Consider $\quad I = \int (e^x) \cdot (x^2) \, dx, \quad \begin{cases} \int x^2 \, dx = \dfrac{x^3}{3} \\ \dfrac{d}{dx} \, dx = e^x \end{cases}$

Integrating by parts, we get

$$I = e^x \cdot \frac{x^3}{3} - \int e^x \cdot \frac{x^3}{3} \, dx$$

$$I = \frac{1}{3} e^x \cdot x^3 - \frac{1}{3} \int e^x \cdot x^3 \, dx$$

Observe *that the resulting integral on right-hand side is more complicated than the given integral.* This is due to *our wrong choice of the second function.*

Example (2): Evaluate $I = \int x \sec^2 x \, dx$

Solution: $\quad \therefore \quad I = \int x \cdot \sec^2 x \, dx \quad \begin{cases} \dfrac{d}{dx}(x) = 1 \\ \int \sec^2 x \, dx = \tan x \end{cases}$

Integrating by parts, we get

$$I = x \cdot \tan x - \int (1) \cdot \tan x \, dx = x \cdot \tan x - \int \tan x \, dx$$

$$= x \cdot \tan x - \log(\sec x) + c \qquad \text{Ans.}$$

Example (3): Evaluate $\int x^3 \, e^x \, dx$

Solution: Let $I = \int x^3 \cdot e^x \, dx \quad \begin{cases} \dfrac{d}{dx} x^3 = 3x^2 \\ \int e^x \, dx = e^x \end{cases}$

Integrating by parts, we get

$$I = x^3 e^x - \int 3 x^2 e^x \, dx = x^3 e^x - 3 \int x^2 e^x \, dx$$

$$= x^3 e^x - 3 \left[x^2 \cdot e^x - 2 \int x e^x \, dx \right]$$

$$= x^3 e^x - 3x^2 e^x + 6 \left[x e^x - \int (1) e^x \, dx \right]$$

$$= x^3 e^x - 3x^2 e^x + 6x e^x - 6 e^x + c$$

$$= e^x [x^3 - 3x^2 + 6x - 6] + c \qquad \text{Ans.}$$

Note (3): In many cases, the formula for integrating by parts has to be applied more than once. Of course, the given integral is reduced to a simpler form but the new integral is such that it has to be evaluated by applying the rule repeatedly. Remember, that *the rule of integration by parts will be useful only when the resulting integral (after applying the rule) is simpler than the integral being evaluated.* This suggests that we make a proper choice of the second function every time. A wrong choice of the second function will complicate the situation [see Example (1)].

Example (4): Evaluate $I = \int x^2 \cos x \, dx$

Solution: Observe that

(i) The given integral cannot be evaluated by any of our previous methods.
(ii) The integrals of both the parts (i.e., x^2 and $\cos x$) are *equally simple.* But we should not choose x^2 as a second function. (Why?)

Therefore, we choose

x^2 as first function, and
$\cos x$ as second function.

$$\therefore \quad I = \int (x^2) \cdot \cos x \, dx \qquad \begin{cases} \dfrac{d}{dx}(x^2) = 2x \\[2mm] \int \cos x \, dx = \sin x \end{cases}$$

Now, integrating by parts, we get

$$I = x^2 \cdot \sin x - \int (2x)(\sin x) \, dx$$

$$= x^2 \cdot \sin x - 2 \int x \sin x \, dx \qquad \begin{cases} \dfrac{d}{dx}(x) = 1 \\[2mm] \int \sin x \, dx = -\cos x \end{cases}$$

Again, integrating by parts, the resulting integral, we get

$$I = x^2 \cdot \sin x - 2\left[x \cdot (-\cos x) - \int (1)(-\cos x)dx\right]$$

$$= x^2 \cdot \sin x + 2x \cos x - 2\int \cos x\, dx$$

$$= x^2 \cdot \sin x + 2x \cos x - 2(\sin x) + c$$

$$= x^2 \sin x + 2x \cos x - 2\sin x + c \qquad \text{Ans.}$$

Example (5): Evaluate $I = \int x^2\, e^{-2x}\, dx$

Solution: $I = \int x^2 e^{-2x}\, dx$ $\begin{cases} \dfrac{d}{dx}x^2 = 2x \\[2mm] \int e^{-2x}\, dx = \dfrac{e^{-2x}}{-2} \end{cases}$

Integrating by parts, we get

$$I = x^2\left(-\frac{e^{-2x}}{2}\right) - \int 2x \cdot \left(-\frac{e^{-2x}}{2}\right)dx = -\frac{1}{2}x^2\, e^{-2x} + \int x e^{-2x}\, dx$$

Now consider

$$\int x e^{-2x}\, dx \qquad \begin{cases} \dfrac{d}{dx}x = 1 \\[2mm] \int e^{-2x}\, dx = \dfrac{e^{-2x}}{-2} \end{cases}$$

Integrating by parts, we get

$$\int x e^{-2x}\, dx = x\left(\frac{e^{-2x}}{-2}\right) - \int 1 \cdot \frac{e^{-2x}}{-2}\, dx$$

$$= -\frac{1}{2}x e^{-2x} + \frac{1}{2}\int e^{-2x}\, dx$$

$$\therefore \qquad I = -\frac{1}{2}x^2\, e^{-2x} - \frac{1}{2}x e^{-2x} - \frac{1}{4}e^{-2x} + c \qquad \text{Ans.}$$

Note (4): To evaluate integrals like $\int x \cdot \tan^{-1} x\, dx$, $\int x^2 \log x\, dx$, wherein the integrals of $\tan^{-1}x$ and $\log x$, and so on, are *not known*, and the *other function is a power function*, then we choose the power function as the second function. This choice (of the second function) *helps in evaluating such integrals, which otherwise cannot be evaluated by any other method.*

Example (6): Evaluate $I = \int x \log x \, dx$

Solution: Observe that

(i) The given integral cannot be evaluated by any of our previous methods.

(ii) The integral of the function x is known (i.e., $= x^2/2$) but integral of the other part, that is, $\log x$ is *not known*. Hence, we choose

$\log x$ as first function, and

x as second function.

$$\therefore \quad I = \int (\log x) \cdot (x) \, dx \qquad \left. \begin{array}{l} \dfrac{d}{dx}(\log x) = \dfrac{1}{x} \\[2ex] \displaystyle\int x \, dx = \dfrac{x^2}{2} \end{array} \right\} {}^{(5)}$$

Now, integrating by parts, we get

$$I = (\log x) \cdot \frac{x^2}{2} - \int \frac{1}{x} \cdot \frac{x^2}{2} dx = \frac{x^2}{2} \log x - \frac{1}{2} \int x \, dx$$

$$= \frac{x^2}{2} \log x - \frac{1}{2} \left(\frac{x^2}{2} \right) + c = \frac{x^2}{2} \log x - \frac{x^2}{4} + c \qquad \text{Ans.}$$

Example (7): Evaluate $I = \int x^2 \sin^{-1} x \, dx$

Solution: Observe that

(i) The given integral cannot be evaluated by any of our previous methods.

(ii) The integral of the function $\sin^{-1} x$ is not known but integral of the other function x^2 is known (i.e., $= x^3/3$). Hence, we choose,

$\sin^{-1} x$ as first function, and

x^2 as second function.

$$\therefore \quad I = \int (\sin^{-1} x) \cdot (x^2) \, dx \qquad \left\{ \begin{array}{l} \dfrac{d}{dx}(\sin^{-1} x) = \dfrac{1}{\sqrt{1 - x^2}} \\[2ex] \displaystyle\int x^2 \, dx = \dfrac{x^3}{3} \end{array} \right.$$

---

[5] For applying the rule of Integration by Parts it is useful to remember the rule in words. It adds to our further convenience if we write down d/dx of first function and the integral of the second function. This will be clearer when we solve problems.

Now, integrating by parts, we get

$$I = (\sin^{-1}x) \cdot \left(\frac{x^3}{3}\right) - \int \frac{1}{\sqrt{1-x^2}} \cdot \left(\frac{x^3}{3}\right) dx$$

$$= \frac{x^3}{3} \cdot \sin^{-1} x - \frac{1}{3} \int \frac{x^2 \cdot x}{\sqrt{1-x^2}} dx \qquad \begin{cases} \text{Put } 1 - x^2 = t^2 \\ \therefore \quad x^2 = 1 - t^2 \\ \therefore \quad 2x\,dx = -2t\,dt \\ \therefore \quad x\,dx = -t\,dt \end{cases}$$

$$= \frac{x^3}{3} \cdot \sin^{-1} x - \frac{1}{3} \int \frac{(1-t^2)\,(-t\,dt)}{t} = \frac{x^3}{3} \cdot \sin^{-1} x + \int (1-t^2)\,dt$$

$$= \frac{x^3}{3} \cdot \sin^{-1} x + \frac{1}{3}\left(t - \frac{t^3}{3}\right) + c, \quad \text{where } t = (1-x^2)^{1/2}$$

$$= \frac{x^3}{3} \cdot \sin^{-1} x + \frac{1}{3}\sqrt{1-x^2} - \frac{1}{9}(1-x^2)^{3/2} + c \qquad \text{Ans.}$$

Example (8): Evaluate $\int x \tan^{-1} x \, dx$

Solution: $I = \int x \tan^{-1} x \, dx \qquad \begin{cases} \dfrac{d}{dx}\tan^{-1} = \dfrac{1}{1+x^2} \\ \\ \int x\,dx = \dfrac{x^2}{2} \end{cases}$

$$\therefore \quad I = \int (\tan^{-1} x)(x)\, dx$$

Integrating by parts, we get

$$I = (\tan^{-1}x) \cdot \frac{x^2}{2} - \int \left(\frac{1}{1+x^2}\right) \cdot \frac{x^2}{2} dx$$

$$= \frac{x^2}{2} \tan^{-1}x - \frac{1}{2}\int \frac{x^2}{1+x^2}dx = \frac{x^2}{2}\tan^{-1}x - \frac{1}{2}\int \left[\frac{(1+x^2)-1}{1+x^2}dx\right]$$

$$= \frac{x^2}{2}\tan^{-1}x - \frac{1}{2}\int \left[1 - \frac{1}{1+x^2}\right] dx$$

$$= \frac{x^2}{2}\tan^{-1}x - \frac{1}{2}\left[x - \tan^{-1}x\right] + c$$

$$= \frac{x^2}{2}\tan^{-1}x - \frac{x}{2} + \frac{1}{2}\tan^{-1}x + c \qquad \text{Ans.}$$

Note (5): To evaluate the integrals $\int \log x \, dx$, $\int \sin^{-1} x \, dx$, $\int \tan^{-1} x \, dx$, and so on, *whose integrals are not known as standard results and that cannot be evaluated by any other method*, we choose "1" (unity) as the second function. [Also, there are functions $\sqrt{x^2 + a^2}$, $\sqrt{x^2 - a^2}$, and $\sqrt{a^2 - x^2}$, which are easily integrated by parts, taking "1" as the second function, but we will not be considering these functions in this chapter. Integrals of these functions are discussed in Chapter 4b, by two methods: by parts and by trigonometric substitutions].

Example (9): Evaluate $\int \sin^{-1} x \, dx$

Solution:

Method (I):

Let $I = \int \sin^{-1} x \, dx$

$$\therefore \quad I = \int \left(\sin^{-1} x \right)(1) dx \qquad \begin{cases} \dfrac{d}{dx} \left(\sin^{-1} x \right) = \dfrac{1}{\sqrt{1-x^2}} \\[3mm] \displaystyle\int 1 \, dx = x \end{cases}$$

Integrating by parts, we get

$$I = \left(\sin^{-1} x \right)(x) - \int \frac{1}{\sqrt{1-x^2}} \cdot x \, dx$$

Consider $\displaystyle\int \frac{x}{\sqrt{1-x^2}} dx$

Put $1 - x^2 = t^2 \qquad \therefore \quad -2x \, dx = 2t \, dt \qquad \therefore \quad x \, dx = -t \, dt$

$$\therefore \quad \int \frac{x \, dx}{\sqrt{1-x^2}} = \int \frac{-t \, dt}{t} = \int -dt = -t = -\sqrt{1-x^2} + c$$

$$\therefore \quad I = x \cdot \sin^{-1} x + \sqrt{1-x^2} + c \qquad \text{Ans.}$$

Method (II): $\int \sin^{-1} x \, dx = I$ (say)

Put $\sin^{-1} x = t \quad \therefore \quad x = \sin t \quad$ and $\quad dx = \cos t \, dt$

$$\therefore \quad I = \int t \cdot \cos t \, dt \qquad \begin{cases} \dfrac{d}{dt}(t) = 1 \\[3mm] \displaystyle\int \cos t \, dt = \sin t \end{cases}$$

$$I = t \cdot \sin t - \int (1) \cdot \sin t \, dt = t \cdot \sin t - (-\cos t) + c$$

$$= t \cdot \sin t + \cos t = \left(\sin^{-1} x \right) x + \sqrt{1-x^2} + c$$

$$= x \sin^{-1} x + \sqrt{1-x^2} + c \qquad \text{Ans.}$$

Example (10): Evaluate $\int \log x \, dx$

Solution: Let $I = \int \log x \, dx$

$$= \int \log x \cdot (1) dx \qquad \begin{cases} \dfrac{d}{dx}(\log x) = \dfrac{1}{x} \\[3mm] \displaystyle\int 1 \cdot dx = x \end{cases}$$

On integrating by parts, we get

$$I = (\log x) \cdot (x) - \int \left(\frac{1}{x}\right) \cdot (x)dx = x \cdot \log x - \int 1 \cdot dx$$

$$= x \cdot \log x - x + c \qquad \text{Ans.}$$

Example (11): Evaluate $\int \tan^{-1} x \, dx$

Solution:

Method (I): Let $I = \int \tan^{-1}x \, dx = \int (\tan^{-1}x)(1)dx$
Integrating by parts, we get

$$\therefore \quad I = (\tan^{-1} x)(x) - \int \frac{d}{dx}(\tan^{-1}x) \cdot (x)dx = x \cdot \tan^{-1}x - \int \frac{x}{1 + x^2}dx$$

Consider $\int \dfrac{x}{1 + x^2}dx$

Put $1 + x^2 = t \quad \therefore \quad 2x \, dx = dt \quad \therefore \quad x \, dx = \dfrac{1}{2}dt$

$$\therefore \quad \int \frac{x}{1 + x^2}dx = \frac{1}{2}\int \frac{dt}{t} = \frac{1}{2}\log t + c$$

$$= \log\sqrt{t} + c = \log\sqrt{1 + x^2} + c$$

$$\therefore \quad I = x \cdot \tan^{-1} x - \log\sqrt{1 + x^2} + c \qquad \text{Ans.}$$

Method (II): $\int \tan^{-1} x \, dx = I \quad$ (say)

Put $\tan^{-1}x = t \quad \therefore \quad \tan t = x$

$$\therefore \quad \frac{1}{1 + x^2}dx = dt \quad \therefore \quad dx = (1 + x^2) \, dt = (1 + \tan^2 t)dt$$

$$I = \int t \cdot (1 + \tan^2 t)dt = \int t(1 + \sec^2 t - 1)dt = \int t \cdot \sec^2 t \, dt$$

Integrating by parts, we get

$$\begin{cases} \dfrac{d}{dt}(t) = 1 \\ \int \sec^2 t dt = \tan t \end{cases}$$

$$I = t \cdot \tan t - \int (1) \cdot \tan t dt = t \cdot \tan t - \log(\sec t) + c$$

$$= (\tan^{-1}x) \cdot x - \log\left(\sqrt{1 + x^2}\right) + c$$

$$= x\tan^{-1}x - \log\left(\sqrt{1 + x^2}\right) + c \qquad \text{Ans.}$$

Note (6): Sometimes a simple substitution reduces the given integral into a form that can be easily integrated by parts.

Example (12): Evaluate $\int \left(\sin^{-1} x\right)^2 dx = I$ (say)

Solution:

$$\sin^{-1}x = t \quad \therefore \quad x = \sin t \quad \therefore \quad dx = \cos t\, dt$$

$$\therefore \quad I = \int t^2 \cdot \cos t\, dt \qquad \begin{cases} \dfrac{d}{dt}\left(t^2\right) = 2t \\[2mm] \int \cos t\, dt = \sin t \end{cases}$$

$$= t^2 \cdot \sin t - \int (2t)\sin t\, dt$$

Consider $\int 2t \sin t\, dt$

$$= 2\int t \cdot \sin t\, dt \qquad \begin{cases} \dfrac{d}{dt}(t) = 1 \\[2mm] \int \sin t\, dt = -\cos t \end{cases}$$

$$= 2\left[t \cdot (-\cos t) - \int (1)(-\cos t)dt\right]$$

$$= -2t\cos t + 2\int \cos t\, dt$$

$$= -2t \cos t + 2 \sin t + c$$

$$\therefore \quad I = t^2 \cdot \sin t + 2t\cos t - 2 \sin t + c$$

$$= \left(\sin^{-1}x\right)^2 \cdot x + 2\sin^{-1}x\left(\sqrt{1-x^2}\right) - 2x + c$$

$$= x \cdot \left(\sin^{-1}x\right)^2 + 2\sqrt{1-x^2}\,\sin^{-1}x - 2x + c \qquad \text{Ans.}$$

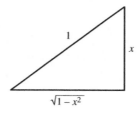

Example (13): Evaluate $\int \sin\sqrt{x}\, dx = I$ (say)

Solution: Put $x = t^2 \quad \therefore \quad dx = 2t\, dt \quad \therefore \quad \sqrt{x} = t$

$$\therefore \quad I = \int (\sin t)(2t\, dt) = 2\int t \cdot \sin t\, dt$$

$$= 2\left[t \cdot (-\cos t) - \int (1)(-\cos t)dt\right] \qquad \begin{cases} \dfrac{d}{dt}(t) = 1 \\[2mm] \int \sin t\, dt = -\cos t \end{cases}$$

$$= -2t \cos t + 2\int \cos t\, dt = -2t\cos t + 2 \sin t + c$$

$$= -2\sqrt{x}\cos\sqrt{x} + 2\sin\sqrt{x} + c \qquad \text{Ans.}$$

Example (14): Evaluate $\int \tan^{-1}\sqrt{x}\,dx = I$ (say)

Solution: Put $x = t^2$ $\therefore$ $dx = 2t\,dt$ $\therefore$ $\sqrt{x} = t$

$$\therefore \quad I = \int (\tan^{-1}t)(2t\,dt) = 2\int (\tan^{-1}t)(t)dt$$

$$I = 2\left[(\tan^{-1}t)\left(\frac{t^2}{2}\right) - \int\left(\frac{1}{1+t^2}\right)\frac{t^2}{2}dt\right] \qquad \begin{cases} \dfrac{d}{dt}(\tan^{-1}t) = \dfrac{1}{1+t^2} \\[2mm] \int t\,dt = \dfrac{t^2}{2} \end{cases}$$

$$= t^2 \cdot \tan^{-1}t - \int \frac{t^2}{1+t^2}dt = t^2 \cdot \tan^{-1}t - \int \frac{1+t^2-1}{1+t^2}dt$$

$$= t^2 \cdot \tan^{-1}t - \int dt + \int \frac{1}{1+t^2}dt = t^2 \cdot \tan^{-1}t - t + \tan^{-1}t + c$$

$$= (t^2+1)\tan^{-1}t - t + c = (x+1)\tan^{-1}\sqrt{x} - \sqrt{x} + c \qquad \text{Ans.}$$

Miscellaneous Solved Examples

Example (15): Evaluate $\int x^2\tan^{-1}(x^3)dx$

Solution: $I = \int x^2\tan^{-1}(x^3)dx$

Put $x^3 = t$ $\therefore$ $3x^2\,dx = dt$

$$\therefore \quad I = \int \tan^{-1}(x^3) \cdot x^2\,dx = \frac{1}{3}\int \tan^{-1}t\,dt$$

Now, in order to integrate $\tan^{-1}t$, we take "1" as the second function, and proceed as follows:

$$\therefore \quad I = \frac{1}{3}\int (\tan^{-1}t)(1)dt \qquad \begin{cases} \dfrac{d}{dt}\tan^{-1}t = \dfrac{1}{1+t^2} \\[2mm] \int 1 \cdot dt = t \end{cases}$$

Integrating by parts, we get

$$I = \frac{1}{3}\left[(\tan^{-1}t)(t) - \int\left(\frac{1}{1+t^2}\right)t\,dt\right] = \frac{1}{3}t \cdot \tan^{-1}t - \frac{1}{3}\int \frac{t}{1+t^2}dt$$

Now, consider $\dfrac{1}{3}\int \dfrac{1}{1+t^2}dt$

Put $1 + t^2 = u$ $\quad \therefore \quad 2t\,dt = du$ $\quad \therefore \quad t\,dt = \dfrac{1}{2}\,du$

$$\therefore \quad \frac{1}{3}\int \frac{t}{1+t^2}\,dt = \frac{1}{6}\int \frac{du}{u} = \frac{1}{6}\log u + c = \frac{1}{6}\log(1+t^2) + c$$

$$\therefore \quad I = \frac{1}{3}t \cdot \tan^{-1}t - \frac{1}{6}\log(1+t^2) + c \quad \text{where } t = x^3$$

$$\therefore \quad I = \frac{1}{3}x^3 \cdot \tan^{-1}(x^3) - \frac{1}{6}\log(1+x^6) + c \qquad \text{Ans.}$$

Example (16): Evaluate $\int e^{\sin x} \cdot \sin 2x\,dx$

Solution: Let $\quad I = \displaystyle\int e^{\sin x} \cdot \sin 2x\,dx$

$$= 2\int e^{\sin x} \cdot \sin x \cdot \cos x\,dx$$

Put $\sin x = t$ $\quad \therefore \quad \cos x\,dx = dt$

$$\therefore \quad I = 2\int e^t \cdot t\,dt = 2\int t \cdot e^t\,dt$$

$$= 2\left[t \cdot e^t - \int (1)e^t\,dt\right] = 2t \cdot e^t - 2e^t + c$$

$$= 2\sin x \cdot e^{\sin x} - 2e^{\sin x} + c = 2e^{\sin x}(\sin x - 1) + c \qquad \text{Ans.}$$

Example (17): Evaluate $\displaystyle\int \frac{x}{1 + \sin x}\,dx = I \quad \text{(say)}$

Solution: Consider $\quad \dfrac{x}{1 + \sin x} = \dfrac{x}{1 + \sin x} \cdot \dfrac{1 - \sin x}{1 - \sin x}$

$$= \frac{x - x\sin x}{1 - \sin^2 x} = \frac{x - x\sin x}{\cos^2 x}$$

$$= x\sec^2 x - x\sec x \cdot \tan x$$

$$\therefore \quad I = \int x\sec^2 x\,dx - \int x\sec x \cdot \tan x\,dx$$

Consider $\quad I_1 = \displaystyle\int x\sec^2 x\,dx \qquad \begin{cases} \dfrac{d}{dx}(x) = 1 \\ \int \sec^2 x\,dx = \tan x \end{cases}$

$$\therefore \quad I = x \cdot \tan x - \int \tan x\,dx$$

$$= x \cdot \tan x - \log(\sec x) + c_1 \tag{5}$$

Now consider $I_2 = \int x \sec x \cdot \tan x \, dx$

$\therefore \quad I_2 = x \cdot \sec x - \int (1) \cdot \sec x \, dx, \quad \left[\because \int \sec x \tan x \, dx = \sec x \right]$

$\qquad = x \cdot \sec x - \log(\sec x + \tan x) + c_2 \qquad\qquad (6)$

$\qquad I = I_1 + I_2$

$\qquad\qquad [x \cdot \tan x - \log(\sec x)] - [x \cdot \sec x - \log(\sec x + \tan x)] + c$

$\qquad = x(\tan x - \sec x) + \log(\sec x + \tan x) - \log(\sec x) + c$

$\qquad = x(\tan x - \sec x) + \log \left[\dfrac{\sec x + \tan x}{\sec x} \right] + c$

$\qquad = x(\tan x - \sec x) + \log(1 + \sin x) + c \qquad$ Ans.

Example (18): Evaluate $\int (\log x)^2 \, dx$

Solution:

Method (I):

Let $\int (\log x)^2 \, dx = I = \int (\log x)^2 \cdot (1) \, dx$

Put $\log x = t \qquad \therefore \quad \dfrac{1}{x} dx = dt$

$\therefore \quad x = e^t \qquad\qquad \therefore \quad dx = x \cdot dt = e^t \, dt$

$\therefore \quad I = \int t^2 \cdot e^t dt = t^2 \cdot e^t - \int 2t \cdot e^t dt = t^2 \cdot e^t - 2 \left[t \cdot e^t - \int 1 \cdot e^t dt \right]$

$\qquad = t^2 \cdot e^t - 2t \, e^t + 2e^t + c = (\log x)^2 \cdot (x) - 2 \log x (x) + 2x + c$

$\qquad = x(\log x)^2 - 2x \log x + 2x + c \qquad$ Ans.

Method (II):

$I = \int (\log x)^2 \cdot (1) \, dx \qquad \begin{cases} \dfrac{d}{dx}(\log x)^2 = 2 \log x \cdot \dfrac{1}{x} \\[2mm] \int 1 \cdot dx = x \end{cases}$

$\therefore \quad I = (\log x)^2 \cdot x - \int 2 \dfrac{\log x}{x} \cdot x dx$

$\qquad = x(\log x)^2 - 2 \int \log x dx = x(\log x)^2 - 2 \left[\int (\log x) \cdot (1) dx \right]$

$\qquad = x(\log x)^2 - 2 \left[\log x(x) - \int \dfrac{1}{x} \cdot x dx \right]$

$\qquad = x(\log x)^2 - 2x \log x + 2 \int dx$

$\qquad = x(\log x)^2 - 2x \log x + 2x + c \qquad$ Ans.

Example (19): Evaluate $\int x^3\, e^{x^2}\, dx$

Solution: Let $I = \int x^3\, e^{x^2} dx = \int x^2 \cdot e^{x^2} \cdot x\, dx$

Put $x^2 = t$ $\therefore$ $2x\, dx = dt$

$$\therefore \quad I = \int t \cdot e^t \cdot \frac{1}{2}\, dt = \frac{1}{2}\int t \cdot e^t dt$$

$$= \frac{1}{2}\left[t \cdot e^t - \int (1) \cdot e^t\, dt \right] = \frac{1}{2}[t \cdot e^t - e^t] + c$$

$$= \frac{1}{2}\left[x^2 \cdot e^{x^2} - e^{x^2} \right] + c \qquad \text{Ans.}$$

Remark: We should not forget about basic integration forms. For example, to evaluate the integral $\int x\, e^{x^2}\, dx$, *the method of integration by parts is not needed.* (Why).[6]

Example (20): Evaluate $\int x^n \log x\, dx = I$ (say)

Solution: $I = \int (\log x)\, (x^n) dx$
$$\begin{cases} \dfrac{d}{dx}(\log x) = \dfrac{1}{x} \\[2mm] \int x^n = \dfrac{x^{n+1}}{n+1} \end{cases}$$

$$\therefore \quad I = (\log x) \cdot \frac{x^{n+1}}{n+1} - \int \frac{1}{x} \cdot \frac{x^{n+1}}{n+1} dx$$

$$= \frac{x^{n+1}}{n+1} \cdot \log x - \frac{1}{n+1}\int x^n\, dx$$

$$= \frac{x^{n+1}}{n+1} \cdot \log x - \frac{1}{n+1} \cdot \frac{x^{n+1}}{n+1} + c$$

$$= \frac{x^{n+1}}{n+1} \cdot \log x - \frac{x^{n+1}}{(n+1)^2} + c \qquad \text{Ans.}$$

Example (21): Evaluate $\int 2x \sin 4x \cdot \cos \cdot 2x\, dx$

Solution: Let $I = \int 2x \sin 4x \cdot \cos 2x\, dx$

$$= \int 2x \cdot \frac{1}{2}[\sin 6x + \sin 2x] dx$$

$$\therefore \quad I = \int x \sin 6x\, dx + \int x \sin 2x\, dx$$

Now, we can apply the method of *Integration by Parts* to each integral on right-hand side.

---

[6] Check if this integral can be evaluated by simple substitution.

Remark: If the integrand were $2x \cdot \sin 2x \cdot \cos 2x$, then it could be written as $x[2 \sin 2x \cdot \cos 2x] = x \cdot \sin 4x$, which can be easily integrated by parts.

Example (22): Evaluate $\int x \cos^2 x \, dx = I$ (say)

Here again we must use trigonometric identities to convert the integrand into some convenient form.

We have $\cos 2x = 2 \cos^2 x - 1$

$$\therefore \quad \cos^2 x = \frac{1 + \cos 2x}{2} \quad \therefore \quad I = \int x \left[\frac{1 + \cos 2x}{2} \right] dx = \frac{x^2}{4} + \frac{1}{2} \int x \cos 2x \, dx$$

Note (7): Integrals like $\int \sin^{-1} x \cdot \log x \, dx$ (where in both the functions cannot be integrated easily) *cannot be evaluated by the method of integration by parts.*

4a.3   HELPFUL PICTURES CONNECTING INVERSE TRIGONOMETRIC FUNCTIONS WITH ORDINARY TRIGONOMETRIC FUNCTIONS

In practice, inverse trigonometric functions are often combined with ordinary functions. Trigonometric substitutions, if successful, help in converting a troublesome integral in x to a simpler integral in t. *However, the problem of translating back to an expression in x always remains.* Such difficulties are easily overcome by drawing a suitable triangle based on the equation like $x = \sin t$. In this case we can write $\sin t = (x/1)$, and draw the required picture as done above. Similarly, the following pictures will be useful in other situations.

Trigonometric substitutions: pictorial aids

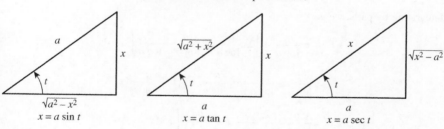

Note (8): In Chapter 4b, we shall evaluate a class of integrals (involving a product of two functions) in which the given integral repeats itself on right-hand side. A repetition is also observed in the following example.

Example (23): Evaluate $\int \sec^3 x \, dx$

Solution: Let $\quad I = \int \sec^3 x \, dx$

$$\therefore \quad I = \int \sec x \cdot \sec^2 x \, dx \qquad \begin{cases} \dfrac{d}{dx}(\sec x) = \sec x \tan x \\[2mm] \int \sec^2 x \, dx = \tan x \end{cases}$$

$$I = \sec x \cdot \int \sec^2 x \, dx - \int \left[\frac{d}{dx}(\sec x) \cdot \left(\int \sec^2 x \, dx \right) \right] dx$$

$$= \sec x \cdot \tan x - \int (\sec x \tan x) \cdot \tan x \, dx$$

$$= \sec x \cdot \tan x - \int \sec x \cdot \tan^2 x \, dx = \sec x \cdot \tan x - \int \sec x (\sec^2 x - 1) \, dx$$

$$= \sec x \cdot \tan x - \int \sec^3 x \, dx + \int \sec x \, dx$$

$$= \sec x \cdot \tan x - I + \log(\sec x + \tan x)$$

$$\therefore \quad 2I = \sec x \cdot \tan x + \log(\sec x + \tan x)$$

$$\therefore \quad I = \frac{1}{2} \sec x \cdot \tan x + \frac{1}{2} \log(\sec x + \tan x) + c \quad \text{Ans.}$$

Example (24): Evaluate $\int x^2 \, a^x \, dx$

Solution: Let $\quad I = \int x^2 \, a^x \, dx \qquad \begin{cases} \dfrac{d}{dx}(x^2) = 2x \\[2mm] \int a^x \, dx = \dfrac{a^x}{\log a} \end{cases}$

Integrating by parts, we get

$$I = x^2 \frac{a^x}{\log a} - \int 2x \frac{a^x}{\log a} \, dx \frac{x^2 a^x}{\log a} - \frac{2}{\log a} \int x a^x \, dx$$

$$= \frac{x^2 a^x}{\log a} - \frac{2}{\log a} \left[x \cdot \frac{a^x}{\log a} - \int (1) \cdot \frac{a^x}{\log a} \, dx \right]$$

$$= \frac{x^2 a^x}{\log a} - \frac{2x \cdot a^x}{(\log a)^2} + \frac{2}{(\log a)^2} \int a^x \, dx$$

$$= \frac{x^2 a^x}{\log a} - \frac{2x \cdot a^x}{(\log a)^2} + \frac{2}{(\log a)^2} \left[\frac{a^x}{\log a} \right] + c$$

$$= \frac{x^2 a^x}{\log a} - \frac{2x \cdot a^x}{(\log a)^2} + \frac{2a^x}{(\log a)^3} + c \qquad \text{Ans.}$$

4a.4   RULE FOR PROPER CHOICE OF FIRST FUNCTION

We know that the method of *Integration By Parts* is necessary when the integrand consists of *a product of two different types of functions*. If the integrand cannot be reduced to standard form by using method of substitution, trigonometric identities, or by algebraic/trigonometric simplification methods, then the simplest approach for integrating such a function is to select the second function (as discussed all throughout the chapter) and apply the method of integration by parts.

However, there is another way for selecting the first and second function, which some authors suggest to be convenient for the students. If we denote Logarithmic, Inverse trigonometric, Algebraic, Trigonometric, and Exponential functions by their first alphabet respectively, then the first function is selected according to the letters of the group LIATE.

Exercise

Evaluate the following integrals:

(1) $\int e^{\sqrt{x}}\,dx$

Ans. $2\,e^{\sqrt{x}}\left[\sqrt{x}-1\right]+c$

(2) $\int \cos x \cdot \log(\sin x)dx$

Ans. $\sin x[\log(\sin x)-1]+c$

(3) $\int \dfrac{x}{1+\cos 2x}\,dx$

Ans. $\dfrac{1}{2}[x\tan x - \log(\sec x)]+c$

(4) $\int x\sin x\cos 2x\,dx$

Ans. $\dfrac{1}{2}\left[x\left(\cos x - \dfrac{\cos 3x}{3}\right)+\dfrac{\sin 3x}{9}-\sin x\right]+c$

(5) $\int x\sin^3 x\,dx$

Ans. $\dfrac{x\cos 3x}{12}-\dfrac{\sin 3x}{36}-\dfrac{3x\cos x}{4}+\dfrac{3}{4}\sin x + c$

(6) $\int \log(x^2+1)dx$

Ans. $x\log(x^2+1)-2x+2\tan^{-1}x+c$

(7) $\displaystyle\int \frac{\log x}{x^2}\,dx$

Ans. $-\dfrac{\log x}{x} - \dfrac{1}{x} + c$

(8) $\displaystyle\int \cos^{-1} x\,dx$

Ans. $x\cos^{-1} x - \sqrt{1 - x^2} + c$

(9) $\displaystyle\int \cos\sqrt{x}\,dx$

Ans. $2\left[\sqrt{x}\sin\sqrt{x} + \cos\sqrt{x}\right] + c$

(10) $\displaystyle\int x^2 \sin x\,dx$

Ans. $-x^2 \cos x + 2x \sin x + 2\cos x + c$

(11) $\displaystyle\int x \sin x \cos x\,dx$

Ans. $1/4[-x\cos 2x^2 + (1/2)\sin 2x] + c$

(12) $\displaystyle\int x^2 \tan^{-1} x^3\,dx$

Ans. $(x^3/3)\tan^{-1} x^3 - (1/6)\log(1 + x^6) + c$

4b Further Integration by Parts: Where the Given Integral Reappears on the Right-Hand Side

4b.1 INTRODUCTION

There are *certain integrals*, which are *slightly special* in the sense that they reappear on the right-hand side (along with other terms) when the formula of *Integration by Parts* is applied. Some such examples are as follows: $\int e^{ax} \sin bx \, dx$, $\int e^{ax} \cos bx$, $\int a^x \sin bx \, dx$, $\int a^x \cos bx \, dx$, $\int \sqrt{x^2 + a^2} \, dx$, $\int \sqrt{x^2 - a^2} \, dx$, $\int \sqrt{a^2 - x^2} \, dx$, and many more.

For the purpose of applying the formula of Integration by Parts, these functions are similar to those considered in the previous Chapter 4a.

Note: For evaluating $\int e^{ax} \sin bx \, dx$ or $\int e^{ax} \cos bx \, dx$ (by parts), observe that, the functions e^{ax} and $\sin bx$ (or $\cos bx$) both, can be easily integrated, and so *any of them can be chosen as the second function*. However, experience suggests that the computation becomes somewhat simpler, if the trigonometric function [i.e., $\sin bx$ or $\cos bx$] in the integrand, is chosen as the *second function*. This suggestion proves to be more useful when the numbers "a" and "b" appear as they are, and not given particular integral values. Now consider the following solved examples.

Illustrative Examples

Example (1): Evaluate $\int e^{3x} \cdot \cos 2x \, dx$

Solution: Let $I = \int e^{3x} \cdot \cos 2x \, dx$, $\qquad \begin{cases} \dfrac{d}{dx} e^{3x} = 3\, e^{3x} \\ \int \cos 2x \, dx = \dfrac{\sin 2x}{2} \end{cases}$

$$\therefore \quad I = e^{3x} \cdot \frac{\sin 2x}{2} - \int 3\, e^{3x} \cdot \frac{\sin 2x}{2} dx$$

$$= \frac{1}{2} e^{3x} \cdot \sin 2x - \frac{3}{2} \int e^{3x} \sin 2x \, dx \quad \left\{ \int \sin 2x \, dx = -\frac{\cos 2x}{2} \right.$$

$$= \frac{1}{2} e^{3x} \cdot \sin 2x - \frac{3}{2} \left[e^{3x} \left(-\frac{\cos 2x}{2} \right) - \int 3\, e^{3x} \frac{(-\cos 2x)}{2} dx \right]$$

4b-Further integration by parts [Cases in which the given integral reappears as a resulting integral on the right-hand side (along with other terms), when the rule of integration by parts is applied]

Introduction to Integral Calculus: Systematic Studies with Engineering Applications for Beginners, First Edition. Ulrich L. Rohde, G. C. Jain, Ajay K. Poddar, and A. K. Ghosh.
© 2012 John Wiley & Sons, Inc. Published 2012 by John Wiley & Sons, Inc.

$$\therefore \quad I = \frac{1}{2}e^{3x}\sin 2x + \frac{3}{4}e^{3x}\cdot\cos 2x - \frac{9}{4}\int e^{3x}\cdot\cos 2x\,dx$$

$$\therefore \quad I = \frac{1}{2}e^{3x}\cdot\sin 2x + \frac{3}{4}e^{3x}\cos 2x - \frac{9}{4}I + c$$

$$\therefore \quad I + \frac{9}{4}\,I = \frac{2}{4}e^{3x}\cdot\sin 2x + \frac{3}{4}e^{3x}\cos 2x + c$$

$$\therefore \quad \frac{13}{4}\,I = \frac{1}{4}e^{3x}[2\sin 2x + 3\cos 2x] + c$$

$$\therefore \quad I = \frac{1}{13}e^{3x}[2\sin 2x + 3\cos 2x] + c \qquad \text{Ans.}$$

Example (2): Evaluate $\int 2^x\cdot\sin 6x\,dx$

Solution: Let $\quad I = \int 2^x\cdot\sin 6x\,dx,$
$$\begin{cases} \dfrac{d}{dx}2^x = 2^x\cdot\log 2 \\[2mm] \displaystyle\int \sin 6\,dx = -\dfrac{\cos 6x}{6} \end{cases}$$

$$\therefore \quad I = 2^x\cdot\left(-\frac{\cos 6x}{6}\right) - \int 2^x\log 2\left(-\frac{\cos 6x}{6}\right)dx$$

$$= -\frac{2^x}{6}\cos 6x + \frac{\log 2}{6}\int 2^x\cdot\cos 6x\,dx, \quad \left\{\int\cos 6x\,dx = \frac{\sin 6x}{6}\right.$$

$$I = -\frac{2^x}{6}\cos 6x + \frac{\log 2}{6}\left[2^x\cdot\frac{\sin 6x}{6} - \int 2^x\log 2\cdot\frac{\sin 6x}{6}dx\right]$$

$$= -\frac{2^x}{6}\cos 6x + \frac{\log 2}{6^2}\cdot 2^x\cdot\sin 6x - \frac{(\log 2)^2}{6^2}\,I$$

$$\therefore \quad I + \frac{(\log 2)^2}{6^2}\,I = \frac{\log 2}{6^2}2^x\sin 6x - \frac{1}{6}2^x\cos 6x$$

$$\therefore \quad \frac{6^2 + (\log 2)^2}{6^2}\,I = \frac{1}{6^2}2^x\log 2\sin 6x - \frac{1}{6}2^x\cos 6x$$

$$\therefore \quad I = \frac{2^x}{6^2 + (\log 2)^2}\,[\log 2\sin 6x - 6\cos 6x] + c \qquad \text{Ans.}$$

Example (3): Evaluate $\int a^x\cdot\cos bx\,dx$

Solution: Let $\quad I = \int a^x\cdot\cos bx\,dx$
$$\begin{cases} \dfrac{d}{dx}a^x = a^x\cdot\log a \\[2mm] \displaystyle\int \cos bx\,dx = \dfrac{\sin bx}{b} \end{cases}$$

$$\therefore \quad I = a^x\cdot\frac{\sin bx}{b} - \int a^x\log a\cdot\frac{\sin bx}{b}dx$$

$$= \frac{1}{b}a^x\sin bx - \frac{\log a}{b}\int a^x\cdot\sin bx\,dx, \quad \left\{\int\sin bx\,dx = -\frac{\cos bx}{b}\right.$$

$$= \frac{1}{b} a^x \sin bx - \frac{\log a}{b} \left[a^x \cdot \left(-\frac{\cos bx}{b} \right) - \int a^x \log a \cdot \frac{(-\cos bx)}{b} dx \right]$$

$$I = \frac{1}{b} a^x \sin bx + \frac{\log a}{b^2} a^x \cos 6x - \frac{(\log a)^2}{b^2} \int a^x \cos bx \, dx$$

$$= \frac{1}{b} a^x \sin bx + \frac{\log a}{b^2} a^x \cos bx - \frac{(\log a)^2}{b^2} \cdot I$$

$$\therefore \quad I + \frac{(\log a)^2}{b^2} \quad I = \frac{1}{b} a^x \sin bx + \frac{\log a}{b^2} a^x \cos bx$$

$$\therefore \quad \frac{b^2 + (\log a)^2}{b^2} \quad I = \frac{1}{b} a^x \sin bx + \frac{\log a}{b^2} a^x \cos bx$$

$$\therefore \quad I = \frac{a^x}{b^2 + (\log a)^2} [b \sin bx + \log a \cos bx] + c \qquad \text{Ans.}$$

Example (4): Evaluate $\int e^{2x} \cos^2 x \, dx$

Solution: Let $\quad I = \int e^{2x} \cos^2 x \, dx$

Note: Here we must use the identity

$$\cos 2x = 2\cos^2 x - 1 \qquad \therefore \quad \cos^2 x = \frac{1 + \cos 2x}{2}$$

$$\therefore \quad I = \int e^{2x} \cdot \left[\frac{1 + \cos 2x}{2} \right] dx = \frac{1}{2} \int e^{2x} \, dx + \frac{1}{2} \int e^{2x} \cos 2x \, dx$$

$$= \frac{1}{2} \cdot \frac{e^{2x}}{2} + \frac{1}{2} \int e^{2x} \cos 2x \, dx \qquad \therefore \quad I = \frac{1}{4} e^{2x} + \frac{1}{2} \int e^{2x} \cos 2x \, dx$$

Now, it is quite easy to find the integral $(1/2) \int e^{2x} \cos 2x \, dx$, as in Example (1).

$$I = \frac{1}{8} e^{2x} [2 + \sin 2x + \cos 2x] + c \qquad \text{Ans.}$$

Example (5): Evaluate $\int \sin(\log x) \, dx = I \quad$ (say)

Solution: Put $\log x = t$ so that $x = e^t$

$$\therefore \quad \frac{1}{x} dx = dt \quad \text{or} \quad dx = x \, dt = e^t \, dt$$

$$\therefore \quad I = \int (\sin t) \cdot e^t \, dt = \int e^t \cdot \sin t \, dt \quad \begin{cases} \dfrac{d}{dt} e^t = e^t \\ \\ \displaystyle\int \sin t \, dt = -\cos t \end{cases}$$

$$\therefore \quad I = e^t \cdot (-\cos t) - \int e^t \cdot (-\cos t) \, dt$$

$$= -e^t \cdot \cos t + \int e^t \cdot \cos t \, dt \qquad \left\{ \int \cos t \, dt = \sin t \right.$$

$$= -e^t \cdot \cos t + \left[e^t \cdot \sin t - \int e^t \cdot \sin t \, dt \right] = -e^t \cdot \cos t + e^t \cdot \sin t - I$$

$$\therefore \quad 2I = e^t \sin t - e^t \cos t \quad \therefore \quad I = \frac{e^t}{2} [\sin t - \cos t] + c$$

$$\therefore \quad I = \frac{x}{2} [\sin(\log x) - \cos(\log x)] + c \qquad \text{Ans.}$$

Example (6): Evaluate $\int e^{-x} \cos x \, dx$

Solution: Let $I = \int e^{-x} \cdot \cos x \, dx,$ $\qquad \begin{cases} \dfrac{d}{dx}(e^{-x}) = -e^{-x} \\[2mm] \displaystyle\int \cos x \, dx = \sin x \end{cases}$

$$\therefore \quad I = e^{-x} \cdot \sin x - \int (-e^{-x}) \sin x \, dx$$

$$= e^{-x} \cdot \sin x + \int e^{-x} \cdot \sin x \, dx \quad \left\{ \int \sin x \, dx = -\cos x \right.$$

$$\therefore \quad I = e^{-x} \cdot \sin x + e^{-x} \cdot (-\cos x) - \int (-e^{-x}) \cdot (-\cos x) dx$$

$$= e^{-x} \cdot \sin x - e^{-x} \cos x - \int e^{-x} \cos x \, dx$$

$$= e^{-x} \cdot \sin x - e^{-x} \cos x - I$$

$$\therefore \quad 2I = e^{-x}(\sin x - \cos x) \qquad \therefore \quad I = \frac{1}{2} e^{-x}(\sin x - \cos x) + c \qquad \text{Ans.}$$

Example (7): Evaluate $\int e^{2x} \sin x \cos x \, dx$

Solution: Let $I = \int e^{2x} \sin x \cos x \, dx$

$$= \frac{1}{2} \int e^{2x} (2 \sin x \cdot \cos x) dx$$

$$= \frac{1}{2} \int e^{2x} \cdot \sin 2x \, dx$$

Now, it is very simple to evaluate the above integral.

$$\frac{1}{8} e^{2x}(\sin 2x - \cos 2x) + c \qquad \text{Ans.}$$

4b.2   AN IMPORTANT RESULT: A COROLLARY TO INTEGRATION BY PARTS

Statement: If $f(x)$ is a *differentiable function* of x, then

$$\int e^x [f(x) + f'(x)] dx = e^x \cdot f(x) + c \tag{1}$$

This result is treated as a standard formula.

Remark: The above result suggests that in an integral of the form $\int e^x F(x) dx$, *if the function F(x) can be expressed in the form* $[f(x) + f'(x)]$, then we can directly write its integral using (1).

Therefore, if we have to evaluate integrals of the form $\int e^x F(x) \, dx$ [where $F(x)$ is a combination of functions], then we must try to express $F(x)$ in the form $[f(x) + f'(x)]$, (if possible), and that is all.

Now we shall prove the above result.
To prove

$$\int e^x [f(x) + f'(x)] dx = e^x \cdot f(x) + c$$

where $f(x)$ is a differentiable function of x.

Proof: Consider left-hand side of the Equation (1)

$$\text{We have, LHS} = \int e^x [f(x) + f'(x)] \, dx$$

$$= \int e^x f(x) \, dx + \int e^x f'(x) \, dx$$

$$= \int f(x) \cdot e^x \, dx + \int f'(x) \cdot e^x \, dx$$

Applying the rule of *Integrating by Parts* to the integral $\int f(x) \cdot e^x \, dx$.
We get,

$$I = f(x) \cdot e^x - \int f'(x) \cdot e^x \, dx + \int f'(x) e^x \, dx + c \qquad \left\{ \begin{array}{l} \dfrac{d}{dx}(f(x)) = f'(x) dx \\[2mm] \int e^x dx = e^x \end{array} \right.$$

$$= f(x) \cdot e^x + c = \text{RHS}$$

Method (2): The above result can also be proved by *differentiating* right-hand side of Equation (1).
We have RHS $= e^x \cdot f(x) + c$.

$$\text{Now} \quad \frac{d}{dx}[e^x \cdot f(x) + c] = e^x \cdot f'(x) + f(x) \cdot e^x$$

$$= e^x [f'(x) + f(x)] = e^x [f(x) + f'(x)] = \text{LHS}$$

$\therefore$ Thus, we have proved the results,

$$\int e^x [f(x) + f'(x)] dx = e^x f(x) + c$$

Now we shall evaluate some integrals of this type.

Example (8):   Evaluate $\int e^x \left(\dfrac{1 + \sin x}{1 + \cos x} \right) dx = I$ (say)

Solution: Consider $\dfrac{1 + \sin x}{1 + \cos x}$

$$= \dfrac{1 + 2\sin(x/2) \cdot \cos(x/2)}{1 + 2\cos^2(x/2) - 1} = \dfrac{1}{2\cos^2(x/2)} + \dfrac{\sin(x/2)}{\cos(x/2)}$$

$$= \dfrac{1}{2}\sec^2\dfrac{x}{2} + \tan\dfrac{x}{2} = \tan\dfrac{x}{2} + \dfrac{1}{2}\sec^2\dfrac{x}{2}$$

Let $f(x) = \tan\dfrac{x}{2}$. Then $f'(x) = \dfrac{1}{2}\sec^2\dfrac{x}{2}$

$$\therefore \quad I = \int e^x \left[\tan\dfrac{x}{2} + \dfrac{1}{2}\sec^2\dfrac{x}{2} \right] dx$$

$$= e^x \tan\dfrac{x}{2} + c \qquad \text{Ans.} \qquad \left\{ \begin{array}{l} \because \displaystyle\int e^x [f(x) + f'(x)]dx \\ \qquad = e^x f(x) + c \end{array} \right.$$

Example (9): Evaluate $\displaystyle\int \dfrac{x\,e^x}{(x+1)^2}\,dx = I$ (say)

Solution: Consider $\dfrac{x}{(x+1)^2} = \dfrac{x+1-1}{(x+1)^2}$

$$= \dfrac{1}{x+1} - \dfrac{1}{(x+1)^2}$$

Let $f(x) = \dfrac{1}{x+1}$. Then $f'(x) = -\dfrac{1}{(x+1)^2}$

$$\therefore \quad I = \int e^x \left[\dfrac{1}{x+1} + \dfrac{-1}{(x+1)^2} \right] dx$$

$$= \dfrac{e^x}{x+1} + c \qquad \text{Ans.} \qquad \left\{ \because \quad \displaystyle\int e^x [f(x) + f'(x)]dx = e^x \cdot f(x) + c \right.$$

Example (10): Evaluate $\displaystyle\int (5 + \tan x + \sec^2 x)e^x\,dx = I$ (say)

Solution: $I = \displaystyle\int e^x \big[(5 + \tan x) + \sec^2 x\big]dx$

Let $f(x) = 5 + \tan x$ $\therefore f'(x) = \sec^2 x$

$$\therefore \quad I = \int e^x [f(x) + f'(x)]dx$$

$$= e^x \cdot f(x) + c = e^x(5 + \tan x) + c \qquad \text{Ans.}$$

Example (11): Evaluate $\displaystyle\int \dfrac{x^2 + 1}{(x+1)^2}e^x\,dx = I$ (say)

Solution: Consider $\dfrac{x^2 + 1}{(x+1)^2}$

$$= \frac{(x^2 - 1) + 2}{(x+1)^2} = \frac{x^2 - 1}{(x+1)^2} + \frac{2}{(x+1)^2} = \frac{x-1}{x+1} + \frac{2}{(x+1)^2}$$

Let $f(x) = \dfrac{x-1}{x+1}$ $\therefore$ $f'(x) = \dfrac{(x+1)(1) - (x-1)(1)}{(x+1)^2} = \dfrac{2}{(x+1)^2}$

$$\therefore \quad I = \int e^x \left[\left(\frac{x-1}{x+1} \right) + \frac{2}{(x+1)^2} \right] dx$$

$$= \int e^x [f(x) + f'(x)] dx$$

$$= e^x f(x) + c = e^x \left(\frac{x-1}{x+1} \right) + c \qquad \text{Ans.}$$

Example (12): $\displaystyle\int e^{\sin x} (\sin x \cdot \cos x + \cos x) dx = I$ (say)

Solution: $I = \displaystyle\int e^{\sin x} \cos x (\sin + 1) dx$

$$= \int e^{\sin x} (\sin x + 1) \cos x \, dx$$

Put $\sin x = t$ $\therefore$ $\cos x \, dx = dt$

$$\therefore \quad I = \int e^t (t + 1) dt$$

Let $f(t) = t$ $\therefore$ $f'(t) = 1$

$$\therefore \quad I = \int e^t [f(t) + f'(t)] dt = e^t \cdot f(t) + c$$

$$= e^t \cdot t + c = e^{\sin t} \cdot \sin t + c \qquad \text{Ans.}$$

Example (13): $\displaystyle\int e^x [\cot x + \log(\sin x)] dx = I$ (say)

Solution: $I = \displaystyle\int e^x [\log(\sin x) + \cot x] dx$

Let $f(x) = \log(\sin x)$

$$\therefore \quad f'(x) = \frac{1}{\sin x} \cdot \cos x = \cot x$$

$$\therefore \quad I = \int e^x [f(x) + f'(x)] dx$$

$$= e^x f(x) + c = e^x \log(\sin x) + c \qquad \text{Ans.}$$

Remark: The result: $\int e^x[f(x) + f'(x)]dx = e^x f(x) + c$ where $f(x)$ is any derivable function of x, *is very important. In this form of integral, the part $\int f'(x)e^x\, dx$ cancels out when the part $\int f(x)e^x dx$ is integrated by parts.* In all the examples given above, the same thing happens.

4b.3 APPLICATION OF THE COROLLARY TO INTEGRATION BY PARTS TO INTEGRALS THAT CANNOT BE SOLVED OTHERWISE

Now, we give below, *some integrals of a different form*, where also the same phenomenon occurs.

Example (14): Evaluate $\displaystyle\int \frac{x + \sin x}{1 + \cos x}dx$

Solution: Let $I = \displaystyle\int \frac{x + \sin x}{1 + \cos x}dx$

$$\therefore \quad I = \int \frac{x + 2\sin(x/2) \cdot \cos(x/2)}{2\cos^2(x/2)}dx$$

$$= \int \frac{x}{2\cos^2(x/2)}dx + \int \tan\frac{x}{2}dx$$

$$= \frac{1}{2}\int x\sec^2\frac{x}{2}dx + \int \tan\frac{x}{2}dx \qquad (2)$$

Now, we may integrate $x\sec^2(x/2)$, *taking $\sec^2(x/2)$ as the second function or integrate tan $(x/2)$ taking 1 as the second function.*

$$\text{Thus,} \quad \int \tan\frac{x}{2} \cdot (1)\, dx = \tan\frac{x}{2}.x - \int \frac{1}{2}\sec^2\frac{x}{2}.x\, dx$$

$$= x\tan\frac{x}{2} - \frac{1}{2}\int x\sec^2\frac{x}{2}dx \qquad (3)$$

Substituting this value of

$$\int \tan\frac{x}{2}dx \text{ in Equation (2), we get} \quad I = x\tan\frac{x}{2} + c \qquad \text{Ans.}$$

Example (15): Evaluate $\displaystyle\int \frac{\log x - 1}{(\log x)^2}dx$

Solution: Let $I = \displaystyle\int \frac{\log x - 1}{(\log x)^2}dx$

$$\therefore \quad I = \int \frac{\log x}{(\log x)^2} dx - \int \frac{1}{(\log x)^2} dx$$

$$I = \int \frac{1}{\log x} dx - \int \frac{1}{(\log x)^2} dx \qquad\qquad (4)$$

Consider $\displaystyle\int \frac{1}{\log x} dx = \int (\log x)^{-1} \cdot (1) dx$

$$= (\log x)^{-1} \cdot x - \int \left\{ -1 (\log x)^{-2} \cdot \frac{1}{x} \right\} \cdot x \, dx$$

$$= \frac{x}{\log x} + \int \frac{1}{(\log x)^2} dx$$

Putting this value of $\displaystyle\int \frac{1}{\log x} dx$ in Equation (4), we get

$$I = \frac{x}{\log x} + \int \frac{1}{(\log x)^2} dx - \int \frac{1}{(\log x)^2} + c$$

$$\therefore \quad I = \frac{x}{\log x} + c \qquad \text{Ans.}$$

Now, we shall evaluate this integral by the method of substitution.

Method of Substitution:

Solution: Let $I = \displaystyle\int \frac{\log x - 1}{(\log x)^2} dx$

Put $\log x - 1 = t \qquad \therefore \quad \log x = t + 1$

$\therefore \quad x = e^{t+1} + 1 = e \cdot e^t \qquad \therefore \quad dx = e \cdot e^t \, dt, \quad [\because \ e \ \text{is a constan } t]$

$\therefore \quad I = e \displaystyle\int \frac{t}{(t+1)^2} e^t \, dt = e \int \frac{t+1-1}{(t+1)^2} e^t \, dt$

$$= e \int \left[\frac{1}{t+1} - \frac{1}{(t+1)^2} \right] e^t \, dt$$

Let $f(t) = \dfrac{1}{t+1} \qquad \therefore \quad f'(t) = -\dfrac{1}{(t+1)^2},$

$\therefore \quad I = e \cdot \displaystyle\int e^t [f(t) + f'(t)] dt = e \cdot e^t \cdot f(t) + c$

$\therefore \quad I = e \left[e^t \cdot \dfrac{1}{t+1} \right] + c = e^{t+1} \cdot \dfrac{1}{t+1} + c$

$$= e^{\log_e x} \cdot \frac{1}{\log_e x} + c \quad [\because \ t + 1 = \log_e x]$$

$$= \frac{x}{\log_e x} + c \qquad \text{Ans.}$$

Example (16): Evaluate $\displaystyle\int \frac{\log x}{[1 + \log x]^2}\, dx = I$ (say)

Solution: $\displaystyle I = \int \frac{1 + \log x - 1}{[1 + \log x]^2}\, dx$

$$I = \int \frac{1}{[1 + \log x]}\, dx - \int \frac{1}{[1 + \log x]^2}\, dx \qquad (5)$$

Consider $\displaystyle\int \frac{1}{[1 + \log x]}\, dx = \int (1 + \log x)^{-1} \cdot (1)dx$

$$= (1 + \log x)^{-1}(x) - \int \frac{d}{dx}[1 + \log x]^{-1} \cdot x\, dx$$

$$= \frac{x}{1 + \log x} - \int \left\{-1 \cdot (1 + \log x)^{-2} \cdot \frac{1}{x}\right\} \cdot x\, dx$$

$$= \frac{x}{1 + \log x} + \int \frac{1}{(1 + \log x)^2}\, dx$$

Putting this value in Equation (5), we get

$$I = \frac{x}{1 + \log x} + \int \frac{1}{(1 + \log x)^2}\, dx - \int \frac{1}{[1 + \log x]^2}\, dx$$

$$\therefore \quad I = \frac{x}{1 + \log x} + c \qquad \text{Ans.}$$

4b.4   SIMPLER METHOD(S) FOR EVALUATING STANDARD INTEGRALS

Now, we shall prove the following *three standard results*, by using the rule of integration by parts

Identity (1) $\displaystyle\int \sqrt{x^2 + a^2}\, dx = \frac{x}{2}\sqrt{x^2 + a^2} + \frac{a^2}{2}\log\left[x + \sqrt{x^2 + a^2}\right] + c$

Identity (2) $\displaystyle\int \sqrt{x^2 - a^2}\, dx = \frac{x}{2}\sqrt{x^2 - a^2} - \frac{a^2}{2}\log\left[x + \sqrt{x^2 - a^2}\right] + c$

Identity (3) $\displaystyle\int \sqrt{a^2 - x^2}\, dx = \frac{x}{2}\sqrt{a^2 - x^2} + \frac{a^2}{2}\sin^{-1}\frac{x}{a} + c$

In the process of computing the above integrals, the following integrals are used.

(i) $\displaystyle\int \frac{dx}{\sqrt{x^2 + a^2}}\, dx = \log\left[x + \sqrt{x^2 + a^2}\right] + c$

(ii) $\displaystyle\int \frac{dx}{\sqrt{x^2 - a^2}}\, dx = \log\left[x + \sqrt{x^2 - a^2}\right] + c$

(iii) $\displaystyle\int \frac{dx}{\sqrt{a^2 - x^2}}\, dx = \sin^{-1}\left(\frac{x}{a}\right) + c$

Recall that these results were obtained in Chapter 3b and that they are also treated as *standard results*. Now we proceed to prove Identity (1), Identity (2), and Identity (3) by two methods: (a) by parts and (b) by substitution to compare and show that *the method of integration by parts is comparatively simpler and less time consuming.*

Example (17): $\int \sqrt{x^2 + a^2}\, dx = \dfrac{x}{2}\sqrt{x^2 + a^2} + \dfrac{a^2}{2}\log\left[x + \sqrt{x^2 + a^2}\right] + c$

Solution:

$$\text{Let}\quad I = \int \sqrt{x^2 + a^2}\, dx$$

$$= \int \sqrt{x^2 + a^2}(1)\, dx \qquad \begin{cases} \dfrac{d}{dx}\sqrt{x^2 + a^2} \\[2mm] = \dfrac{1}{2\sqrt{x^2 + a^2}} \cdot 2x = \dfrac{x}{\sqrt{x^2 + a^2}} \end{cases}$$

Integrating by parts, we get

$$I = \sqrt{x^2 + a^2} \cdot x - \int \frac{x}{\sqrt{x^2 + a^2}} \cdot x\, dx = x \cdot \sqrt{x^2 + a^2} - \int \frac{x^2}{\sqrt{x^2 + a^2}}\, dx$$

$$= x \cdot \sqrt{x^2 + a^2} - \int \frac{(x^2 + a^2) - a^2}{\sqrt{x^2 + a^2}}\, dx$$

$$I = x \cdot \sqrt{x^2 + a^2} - \int \sqrt{x^2 + a^2}\, dx + a^2 \int \frac{1}{\sqrt{x^2 + a^2}}\, dx$$

$$= x \cdot \sqrt{x^2 + a^2} - I + a^2 \cdot \log\left[x + \sqrt{x^2 + a^2}\right] + c_1$$

$$\therefore\quad 2I = x \cdot \sqrt{x^2 + a^2} + a^2 \cdot \log\left[x + \sqrt{x^2 + a^2}\right] + c_1$$

$$\therefore\quad I = \frac{x}{2} \cdot \sqrt{x^2 + a^2} + \frac{a^2}{2}\log\left[x + \sqrt{x^2 + a^2}\right] + c$$

Now, let us evaluate $\int \sqrt{x^2 + a^2}\, dx$, using trigonometric substitution.

Solution: Put $x = a \tan t$ $\therefore$ $dx = a \sec^2 t\, dt$

$$\therefore\quad \sqrt{x^2 + a^2} = \sqrt{a^2 \tan^2 t + a^2} = \sqrt{a^2 \sec^2 t} = a \sec t$$

$$\therefore\quad I = \int \sqrt{x^2 + a^2}\, dx = \int a \sec t \cdot a \sec^2 t\, dt$$

$$= a^2 \int \sec^3 t\, dt \qquad\qquad (6)$$

To find $\int \sec^3 t\, dt$, *we resort to integration by parts,* taking sec t as the first function and $\sec^2 t$ as the second function.

Now, $\displaystyle\int \sec^3 t\, dt = \int \sec t \cdot \sec^2 t\, dt$ $\left\{\begin{array}{l}\dfrac{d}{dt}(\sec t) = \sec t \cdot \tan t \\[2mm] \displaystyle\int \sec^2 t\, dt = \tan t\end{array}\right.$

$$= \sec t \cdot \tan t - \int (\sec t \cdot \tan t) \cdot \tan t\, dt$$

$$= \sec t \cdot \tan t - \int \sec t \cdot \tan^2 t\, dt$$

$$= \sec t \cdot \tan t - \int \sec t (\sec^2 t - 1)\, dt$$

$$= \sec t \cdot \tan t - \int \sec^3 t\, dt + \int \sec t\, dt$$

Shifting the term $\int \sec^3 t\, dt$ to left-hand side, we get

$$2 \int \sec^3 t\, dt = \sec t \cdot \tan t + \log(\sec t + \tan t) + c_1$$

$$\int \sec^3 t\, dt = \frac{1}{2}[\sec t \cdot \tan t + \log(\sec t + \tan t)] + c_1$$

$$\therefore \quad a^2 \int \sec^3 t\, dt = \frac{a^2}{2}[\sec t \cdot \tan t + \log(\sec t + \tan t)] + c_1$$

(7)

where, $\tan t = \dfrac{x}{a}$, $[\because \quad x = a \tan t]$

and $\displaystyle a^2 \int \sec^3 t\, dt = \int \sqrt{x^2 + a^2}\, dx$ (7A)

[see Equation (6) above]

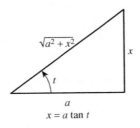

$x = a \tan t$

Now, consider the right-hand side of Equation (7), we get

$$\sec t \cdot \tan t = \frac{\sqrt{x^2 + a^2}}{a} \cdot \frac{x}{a} = \frac{x\sqrt{x^2 + a^2}}{a^2} \qquad (7B)$$

and $\log[\sec t + \tan t] = \log\left[\dfrac{\sqrt{x^2 + a^2}}{a} + \dfrac{x}{a}\right]$

$$= \log\left[\frac{x + \sqrt{x^2 + a^2}}{a}\right] = \log[x + \sqrt{x^2 + a^2}] - \log a \qquad (7C)$$

Using (7A), (7B), and (7C) in Equation (7), we get

$$\int \sqrt{x^2 + a^2} \, dx = \frac{a^2}{2} \left[\frac{x}{a^2} \sqrt{x^2 + a^2} + \log\left(x + \sqrt{x^2 + a^2}\right) - \log a \right] + c_1$$

$$= \frac{x}{2} \sqrt{x^2 + a^2} + \frac{a^2}{2} \log\left[x + \sqrt{x^2 + a^2} \right] - \frac{a^2}{2} \log a + c_1$$

$$= \frac{x}{2} \sqrt{x^2 + a^2} + \frac{a^2}{2} \log\left[x + \sqrt{x^2 + a^2} \right] + c$$

Note that for evaluating $\int \sqrt{x^2 + a^2} \, dx$, by the above method (of substitution) is quite lengthy and time consuming. *The simplest approach to integrate the above function is to use the rule of integration by parts, taking unity as the second function.*

Example (18): $\int \sqrt{x^2 - a^2} \, dx = \frac{x}{2} \sqrt{x^2 - a^2} - \frac{a^2}{2} \log\left[x + \sqrt{x^2 - a^2} \right] + c$

Solution:

Let $\quad I = \int \sqrt{x^2 - a^2} \, dx$

$$= \int \sqrt{x^2 - a^2}\,(1) \, dx \qquad \begin{cases} \dfrac{d}{dx} \sqrt{x^2 - a^2} \\[2mm] = \dfrac{1}{2\sqrt{x^2 - a^2}} \cdot 2x = \dfrac{x}{\sqrt{x^2 - a^2}} \end{cases}$$

Integrating by parts, we get

$$I = \sqrt{x^2 - a^2} \cdot x - \int \frac{x}{\sqrt{x^2 - a^2}} \cdot x \, dx = x \cdot \sqrt{x^2 - a^2} - \int \frac{x^2}{\sqrt{x^2 - a^2}} \, dx$$

$$= x \cdot \sqrt{x^2 - a^2} - \int \frac{(x^2 - a^2) + a^2}{\sqrt{x^2 - a^2}} \, dx$$

$$I = x \cdot \sqrt{x^2 - a^2} - \int \sqrt{x^2 - a^2} \, dx - a^2 \int \frac{1}{\sqrt{x^2 - a^2}} \, dx$$

$$= x \cdot \sqrt{x^2 - a^2} - I - a^2 \int \frac{1}{\sqrt{x^2 - a^2}} \, dx$$

$$\therefore \quad 2I = x \cdot \sqrt{x^2 - a^2} - a^2 \int \frac{1}{\sqrt{x^2 - a^2}} \, dx$$

$$\therefore \quad I = \frac{x}{2} \cdot \sqrt{x^2 - a^2} - \frac{a^2}{2} \log\left[x + \sqrt{x^2 - a^2} \right] + c$$

Now, let us evaluate $\int \sqrt{x^2 - a^2} \, dx$, using trigonometric substitution.

$$\int \sqrt{x^2 - a^2} \, dx = \frac{x}{2} \sqrt{x^2 - a^2} - \frac{a^2}{2} \log\left[x + \sqrt{x^2 - a^2} \right] + c$$

Solution: Let $\int \sqrt{x^2 - a^2}\, dx = I$

Put $x = a \sec t$ $\therefore dx = a \sec t \tan t \cdot dt$

$$\therefore \quad I = \int \sqrt{a^2 \sec^2 t - a^2}\, a \sec t \cdot \tan t\, dt$$

$$= \int \sqrt{a^2(\sec^2 t - 1)} \cdot a \sec t \cdot \tan t\, dt$$

$$= \int a \tan t \cdot a \sec t \cdot \tan t\, dt$$

$$= a^2 \int \sec t \cdot \tan^2 t\, dt = a^2 \int \sec t \cdot (\sec^2 t - 1) dt$$

$$= a^2 \int \sec^3 t\, dt - a^2 \int \sec t\, dt \tag{8}$$

Now, to find $\int \sec^3 t\, dt$ we resort to integration by parts, taking $\sec t$ as the first function and $\sec^2 t$ as the second function.

$$\int \sec^3 t\, dt = \int \sec t \cdot \sec^2 t\, dt \qquad \begin{cases} \dfrac{d}{dt}(\sec t) = \sec t \cdot \tan t \\[2mm] \displaystyle\int \sec^2 t\, dt = \tan t \end{cases}$$

$$= \sec t \cdot \tan t - \int \sec t \tan t \cdot \tan t\, dt$$

$$= \sec t \cdot \tan t - \int \sec t \tan^2 t \cdot dt$$

$$= \sec t \cdot \tan t - \int \sec t \, (\sec^2 t - 1) dt$$

$$= \sec t \cdot \tan t - \int \sec^3 t\, dt + \int \sec t\, dt$$

$$\therefore \quad 2 \int \sec^3 t\, dt = \sec t \tan t + \int \sec t\, dt$$

$$\therefore \quad \int \sec^3 t\, dt = \frac{1}{2} \sec t \cdot \tan + \frac{1}{2} \int \sec t\, dt$$

$$\therefore \quad a^2 \int \sec^3 t\, dt = \frac{a^2}{2} \sec t \cdot \tan t + \frac{a^2}{2} \int \sec t\, dt$$

Using this result in Equation (8), we get

$$I = \frac{a^2}{2} \sec t \cdot \tan t + \frac{a^2}{2} \int \sec t\, dt - a^2 \int \sec t\, dt$$

$$= \frac{a^2}{2} \sec t \cdot \tan t - \frac{a^2}{2} \int \sec t\, dt$$

$$= \frac{a^2}{2} \sec t \cdot \tan t - \frac{a^2}{2} \log(\sec t + \tan t) + c_1 \tag{9}$$

Now, consider $x = a \sec t$

$$x = a \sec t$$

$\therefore \quad \sec t = \dfrac{x}{a}$

$\therefore \quad \dfrac{a^2}{2} \sec t \cdot \tan t = \dfrac{a^2}{2} \left[\dfrac{x}{a} \cdot \dfrac{\sqrt{x^2 - a^2}}{a} \right]$

$$= \dfrac{x}{2} \cdot \sqrt{x^2 - a^2} \tag{9A}$$

and $\dfrac{a^2}{2} \log[\sec t + \tan t] = \dfrac{a^2}{2} \log \left[\dfrac{x}{a} + \dfrac{\sqrt{x^2 - a^2}}{a} \right]$

$$= \dfrac{a^2}{2} \log \left[\dfrac{x + \sqrt{x^2 - a^2}}{a} \right] = \dfrac{a^2}{2} \left[\log \left(x + \sqrt{x^2 - a^2} \right) - \log a \right]$$

$$= \dfrac{a^2}{2} \log \left[x + \sqrt{x^2 - a^2} \right] - \dfrac{a^2}{2} \log a \tag{9B}$$

Using (9A) and (9B) in (9), we get

$$I = \dfrac{x}{2} \sqrt{x^2 - a^2} - \dfrac{a^2}{2} \log \left[x + \sqrt{x^2 - a^2} \right] + c \qquad \text{Ans.}$$

Example (19): To prove $\displaystyle\int \sqrt{a^2 - x^2} \, dx = \dfrac{x}{2} \sqrt{a^2 - x^2} + \dfrac{a^2}{2} \sin^{-1} \dfrac{x}{a} + c$

Solution:

Let $I = \displaystyle\int \sqrt{a^2 - x^2} \, dx$

$\therefore I = \displaystyle\int \sqrt{a^2 - x^2}(1) dx$
$\qquad \begin{cases} \dfrac{d}{dx} \sqrt{a^2 - x^2} \\[2mm] = \dfrac{1}{2\sqrt{a^2 - x^2}} \cdot (-2x) = -\dfrac{x}{\sqrt{a^2 - x^2}} \end{cases}$

Integrating by parts, we get

$$I = \sqrt{a^2 - x^2} \cdot x - \int -\frac{x}{\sqrt{a^2 - x^2}} \cdot (x)dx$$

$$= x \cdot \sqrt{a^2 - x^2} - \int \frac{-x^2}{\sqrt{a^2 - x^2}} dx = x \cdot \sqrt{a^2 - x^2} - \int \frac{(a^2 - x^2) - a^2}{\sqrt{a^2 - x^2}} dx$$

$$= x \cdot \sqrt{a^2 - x^2} - \int \sqrt{a^2 - x^2} dx + a^2 \int \frac{1}{\sqrt{a^2 - x^2}} dx$$

$$= x \cdot \sqrt{a^2 - x^2} - I + a^2 \sin^{-1}\left(\frac{x}{a}\right) + c_1$$

$$\therefore \quad 2I = x \cdot \sqrt{a^2 - x^2} + a^2 \cdot \sin^{-1}\left(\frac{x}{a}\right) + c_1$$

$$\therefore \quad I = \frac{x}{2} \cdot \sqrt{a^2 - x^2} + \frac{a^2}{2} \sin^{-1}\left(\frac{x}{a}\right) + c$$

$$\text{or} \int \sqrt{a^2 - x^2} \, dx = \frac{x}{2} \cdot \sqrt{a^2 - x^2} + \frac{a^2}{2} \sin^{-1}\left(\frac{x}{a}\right) + c$$

Observation

Here, it may be mentioned that *this particular integral can also be easily evaluated by the method of substitution, without resorting to the method of integration by parts as will be clear from the following proof.*

Method of Substitution:

$$\int \sqrt{a^2 - x^2} \, dx = \frac{x}{2} \sqrt{a^2 - x^2} + \frac{a^2}{2} \sin^{-1} \frac{x}{a} + c$$

Solution: Let $\quad I = \int \sqrt{a^2 - x^2} dx$

Put $\quad x = a \sin t \quad \therefore \quad dx = a \cos t \, dt$

$$\therefore \quad I = \int \sqrt{a^2 - a^2 \sin^2 t} \, a \cos t \, dt = \int a \cos t \cdot a \cos t \, dt$$

$$= a^2 \int \cos^2 t \, dt \qquad \begin{cases} \cos 2t = 2\cos^2 t - 1 \\ \\ \therefore \quad \cos^2 t = \frac{1}{2}[\cos 2t + 1] \end{cases}$$

$$= a^2 \int \frac{\cos 2t + 1}{2} \, dt = \frac{a^2}{2} \int \cos 2t \, dt + \frac{a^2}{2} \int dt$$

$$= \frac{a^2}{2} \left[\frac{\sin 2t}{2}\right] + \frac{a^2}{2} t + c = \frac{a^2}{2} \left[\frac{2 \sin t \cdot \cos t}{2}\right] + \frac{a^2}{2} t + c$$

$$= \frac{a^2}{2} [\sin t \cdot \cos t] + \frac{a^2}{2} t + c \tag{10}$$

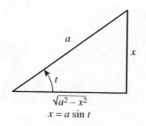

$$x = a \sin t$$

Now $x = a \sin t$ $\therefore$ $\sin t = \dfrac{x}{a}$ $\therefore$ $t = \sin^{-1}\dfrac{x}{a}$

$\therefore$ $\cos t = \dfrac{\sqrt{a^2 - x^2}}{a}$

$\therefore$ $I = \dfrac{a^2}{2}\left[\dfrac{x}{a} \cdot \dfrac{\sqrt{a^2 - x^2}}{a}\right] + \dfrac{a^2}{2}\sin^{-1}\left(\dfrac{x}{a}\right) + c$

or $\displaystyle\int \sqrt{a^2 - x^2}\,dx = \dfrac{x}{2}\sqrt{a^2 - x^2} + \dfrac{a^2}{2}\sin^{-1}\left(\dfrac{x}{a}\right) + c$ Ans.

4b.5   TO EVALUATE $\int \sqrt{ax^2 + bx + c}\,dx$

Now, we are in a position to evaluate integrals of the form $\int \sqrt{ax^2 + bx + c}\,dx$. For this purpose we express the quadratic expression $ax^2 + bx + c$ as the sum or difference of two squares. Besides, we can also evaluate those integrals that can be reduced to this form.

Illustrative Examples

Example (20):   Evaluate $\displaystyle\int \sqrt{16 - 9x^2}\,dx$

Solution: Let $I = \displaystyle\int \sqrt{16 - 9x^2}\,dx$

Consider $16 - 9x^2 = 9\left(\dfrac{16}{9} - x^2\right) = 3^2\left[\left(\dfrac{4}{3}\right)^2 - x^2\right]$ [1]

$\therefore$ $I = \displaystyle\int \sqrt{3^2\left[(4/3)^2 - x^2\right]}\,dx = 3\int \sqrt{(4/3)^2 - x^2}\,dx$

$= 3\left[\dfrac{x}{2}\sqrt{(4/3)^2 - x^2} + \dfrac{1}{2}\left(\dfrac{4}{3}\right)^2 \sin^{-1}\left(\dfrac{x}{4/3}\right)\right] + c$

$= 3\left[\dfrac{x}{2}\sqrt{\dfrac{16}{9} - x^2} + \dfrac{16}{18}\sin^{-1}\left(\dfrac{3x}{4}\right)\right] + c$

$= 3\left[\dfrac{x}{2}\dfrac{\sqrt{16 - 9x^2}}{3} + \dfrac{16}{18}\sin^{-1}\left(\dfrac{3x}{4}\right)\right] + c$

$= \dfrac{x}{2}\sqrt{16 - 9x^2} + \dfrac{8}{3}\sin^{-1}\left(\dfrac{3x}{4}\right) + c$ Ans.

[1] $\int \sqrt{a^2 - x^2}\,dx = \frac{x}{2}\sqrt{a^2 - x^2} + \frac{a^2}{2}\sin^{-1}\left(\frac{x}{a}\right) + c$

Example (21): Evaluate $\int x^2 \sqrt{a^6 - x^6}\, dx$

Solution: Let $\quad I = \int x^2 \sqrt{a^6 - x^6}\, dx = \int x^2 \sqrt{(a^3)^2 - (x^3)^2}\, dx$

Put $\quad x^3 = t \qquad \therefore \quad 3x^2\, dx = dt \quad \therefore \quad x^2\, dx = \dfrac{dt}{3}$

$$\therefore \quad I = \frac{1}{3} \int \sqrt{(a^3)^2 - t^2}\, dt = \frac{1}{3}\left[\frac{t}{2}\sqrt{(a^3)^2 - t^2} + \frac{(a^3)^2}{2}\cdot \sin^{-1}\left(\frac{t}{a^3}\right)\right] + c$$

$$= \frac{1}{3}\left[\frac{t}{2}\sqrt{a^6 - t^2} + \frac{a^6}{2}\sin^{-1}\left(\frac{t}{a^3}\right)\right] + c$$

$$= \frac{1}{3}\left[\frac{x^3}{2}\sqrt{a^6 - x^6} + \frac{a^6}{2}\sin^{-1}\frac{x^3}{a^3}\right] + c \qquad \text{Ans.}$$

Example (22): Evaluate $\int \cos x \sqrt{4\sin^2 x + 9}\, dx$

Solution: Let $I = \int \cos x \sqrt{4\sin^2 x + 9}\, dx$

Put $\quad \sin x = t \qquad \therefore \quad \cos x\, dt$

$$\therefore \quad I = \int \sqrt{4t^2 + 9}\, dt = 2 \int \sqrt{t^2 + (3/2)^2}\, dt$$

$$= 2\left[\frac{t}{2}\sqrt{t^2 + (3/2)^2} + \frac{1}{2}\left(\frac{3}{2}\right)^2 \log\left(t + \sqrt{t^2 + (3/2)^2}\right)\right] + c_1$$

$$= 2\left[\frac{t}{2}\frac{\sqrt{4t^2 + 9}}{2} + \frac{1}{2}\left(\frac{9}{4}\right)\log\left(t + \frac{\sqrt{4t^2 + 9}}{2}\right)\right] + c_1$$

$$= \frac{t\sqrt{4t^2 + 9}}{2} + \frac{9}{4}\log\left(\frac{2t + \sqrt{4t^2 + 9}}{2}\right) + c_1$$

$$= \frac{\sin x \sqrt{4\sin^2 x + 9}}{2} + \frac{9}{4}\log\left(2\sin x + \sqrt{4\sin^2 x + 9}\right) + c \qquad \text{Ans.}$$

where $c = c_1 - \log 2$

Example (23): Evaluate $\int \sqrt{2x^2 + 2x + 5}\, dx$

Solution: Let $I = \int \sqrt{2x^2 + 2x + 5}\, dx$

Consider $2x^2 + 2x + 5$

$$= 2\left(x^2 + x + \frac{5}{2}\right) = 2\left[x^2 + 2\left(\frac{1}{2}\right)x + \frac{1}{4} + \frac{9}{4}\right]$$

$$= 2\left[\left(x + \frac{1}{2}\right)^2 + \left(\frac{3}{2}\right)^2\right]$$

$$\therefore \quad I = \sqrt{2}\int \sqrt{\left(x + \frac{1}{2}\right)^2 + \left(\frac{3}{2}\right)^2}\,dx^{(2)}$$

$$= \sqrt{2}\left[\frac{(x + 1/2)}{2}\sqrt{\left(x + \frac{1}{2}\right)^2 + \left(\frac{3}{2}\right)^2} + \frac{1}{2}\left(\frac{3}{2}\right)^2 \log\left\{\left(x + \frac{1}{2}\right) + \sqrt{\left(x + \frac{1}{2}\right)^2 + \left(\frac{3}{2}\right)^2}\right\}\right] + c$$

Note: This expression can be further simplified as given below. However, it can be left at this stage also, thus avoiding possible mistakes in the process of simplification.

$$I = \sqrt{2}\left[\frac{(2x + 1)}{4} \cdot \frac{\sqrt{2x^2 + 2x + 5}}{\sqrt{2}} + \frac{9}{8}\log\left\{\left(x + \frac{1}{2}\right) + \sqrt{x^2 + x + \frac{5}{2}}\right\}\right] + c$$

$$= \frac{(2x + 1)\sqrt{2x^2 + 2x + 5}}{4} + \frac{9\sqrt{2}}{8}\log\left[\left(x + \frac{1}{2}\right) + \sqrt{x^2 + x + \frac{5}{2}}\right] + c \qquad \text{Ans.}$$

Example (24):   To evaluate $\int (px + q)\sqrt{ax^2 + bx + c}\,dx = I$ (say)

Method: We find two constants A and B such that

$$px + q = A\frac{d}{dx}(ax^2 + bx + c) + B$$

$$\text{i.e.,} \quad px + q = (2aAx + Ab) + B$$

$$\therefore \quad px + q = 2aAx + (Ab + B) \qquad (11)$$

Equating coefficients of x and constants on both sides, we find A and B. Then, we use Equation (11) to substitute for $(px + q)$ in the given integral, and get

$$I = A\int (2ax + b)\sqrt{ax^2 + bx + c}\,dx + B\int \sqrt{ax^2 + bx + c}\,dx$$

The second integral on right-hand side is of the type, already considered above. The first integral is of the type $\int f'(x)\sqrt{f(x)}\,dx$ that can be easily computed by substitution.

(2) $\int \sqrt{x^2 + a^2}\,dx = \frac{x}{2}\sqrt{x^2 + a^2} + \frac{a^2}{2}\log\left(x + \sqrt{x^2 + a^2}\right)c$

We put $f(x) = t$ $\qquad \therefore \quad f'(x)dx = dt$

$$\therefore \quad \int f'(x)\sqrt{f(x)}dx = \int \sqrt{t}\, dt = \int t^{1/2}\, dt$$

$$= \frac{t^{3/2}}{3/2} + c = \frac{2}{3}[f(x)]^{3/2} + c \qquad \text{Ans.}$$

Exercise

Evaluate the following integrals:

(1) $\displaystyle\int e^{3x} \sin 4x\, dx$

Ans. $\dfrac{1}{25} e^{3x}(3 \sin 4x - 4 \cos 4x) + c$

(2) $\displaystyle\int 5^x \cos 3x\, dx$

Ans. $\dfrac{5^x}{9 + (\log 5)^2}[\log 5 \cos 3x + 3 \sin 3x] + c$

(3) $\displaystyle\int e^{2x} \sin^2 3x\, dx$

Ans. $\dfrac{1}{4} e^{2x}\left[1 - \dfrac{1}{20}(2 \cos 6x + 6 \sin 6x)\right] + c$

(4) $\displaystyle\int \dfrac{2 - \sin 2x}{1 - \cos 2x} e^x$

Ans. $-e^x \cot x + c$

(5) $\displaystyle\int \dfrac{1 + x + x^2}{1 + x^2} e^{\tan^{-1} x}$

Ans. $x\, e^{\tan^{-1} x} + c$

(6) $\displaystyle\int e^x\left[\dfrac{x + 2}{(x + 3)^2}\right] dx$

Ans. $\dfrac{e^x}{x + 3} + c$

(7) $\displaystyle\int e^{\sin^{-1} x}\left[\dfrac{x + \sqrt{1 - x^2}}{\sqrt{1 - x^2}}\right] dx$

Ans. $x \cdot e^{\sin^{-1} x} + c$

(8) $\int \sqrt{3x^2 + 4x + 1}\,dx$

Ans. $\left(\dfrac{3x+2}{6}\right)\sqrt{3x^2+4x+1} - \dfrac{\sqrt{3}}{18}\log\left[x + \dfrac{2}{3} + \sqrt{x^2 + \dfrac{4x}{3} + \dfrac{1}{3}}\right] + c$

(9) $\int \sqrt{8 + 2x - x^2}\,dx$

Ans. $\left(\dfrac{x-1}{2}\right)\sqrt{8 + 2x - x^2} + \dfrac{9}{2}\sin^{-1}\left(\dfrac{x-1}{3}\right) + c$

(10) $\int (x+3)\sqrt{5 - 4x - x^2}\,dx$

Ans. $-\dfrac{1}{3}(5 - 4x - x^2)^{3/2} + \dfrac{x+2}{2}\sqrt{5 - 4x - x^2} + \dfrac{9}{2}\sin^{-1}\left(\dfrac{x+2}{3}\right) + c$

5 Preparation for the Definite Integral: The Concept of Area

5.1 INTRODUCTION

We have an intuitive idea of area. It is a measure that tells us about the size of a region which is "the part of a plane" enclosed by a closed curve. Since the time of the ancient Greeks, mathematicians have attempted to calculate areas of *plane regions. The most basic plane region is the rectangle whose area is the product: base × height* (Figure 5.1a).

The ancient Greeks used Euclidean geometry to compute the areas of parallelograms and triangles. They also knew how to compute the *area of any polygon by partitioning it into triangles* (Figure 5.1b). We know that area of a triangle is given by $A = \frac{1}{2}bh$.[1] In this section, *we define the area of a region in a plane, if the region is bounded by a curve.* For this purpose, it must be realized that the area of a polygon can be defined *as the sum of the areas of triangles into which it is decomposed, and it can be proved that the area thus obtained is independent of how the polygon is decomposed into triangles.*[2]

When we consider *a region with a curved boundary, the problem of assigning the area is more difficult.* It was Archimedes (about 287–212 BC), who provided the key to a solution by ingenious use of the *"method of exhaustion". With this method, he found the area of certain complex regions by inscribing larger and larger polygons of known area in such a region* so that it would eventually be "exhausted."

Archimedes went further, considering circumscribed polygons as well. He showed that you get *the same value for the area of the circle of radius 1* (≈3.14159), whether you use *inscribed* or *circumscribed* polygons. The fact that the modern definition of area stems from Archimedes' method of exhaustion is a *tribute to his genius.*

In this chapter, *we will use the problem of computing area to motivate the definition of what we will call the definite integral of a continuous function.* Then, we will use *the definite integral to define the area of a region.* Finally, the *fundamental theorem of integral Calculus* will provide a simple method of computing many definite integrals, in terms of numbers that may represent various quantities (Figure 5.2).

5-The definite integral (The concept of area, definite integral as an area, definite integral as limit of a sum, Riemann sums, and analytical definition of definite integral)

[1] Now, we also know that if a, b, c are the lengths of sides of a triangle, then its area A is given by $A = \sqrt{s(s-a)(s-b)(s-c)}$, where $s = (a+b+c)/2$, the semiperimeter.

[2] These considerations are useful since we are laying the foundation that is necessary to motivate geometrically the definition of the definite integral.

Introduction to Integral Calculus: Systematic Studies with Engineering Applications for Beginners, First Edition.
Ulrich L. Rohde, G. C. Jain, Ajay K. Poddar, and A. K. Ghosh.
© 2012 John Wiley & Sons, Inc. Published 2012 by John Wiley & Sons, Inc.

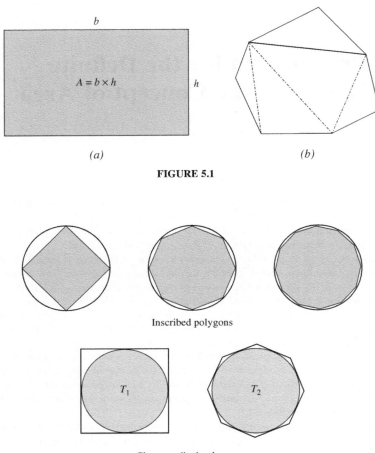

FIGURE 5.1

Inscribed polygons

Circumscribed polygons

FIGURE 5.2

Note: Before discussing the area of a plane region, we indicate why we use the terminology *"measure of the area"*. The word *measure* refers to a number (no units are included). For example, if the area of a triangle is $20\,\text{cm}^2$, we say that *the square-centimeter measure* of the area of the triangle is 20. When the word *measurement* is applied, the units are included. Thus, the measurement of the area of the triangle is $20\,\text{cm}^2$.

5.2   PREPARATION FOR THE DEFINITE INTEGRAL

Consider now a region R in the plane as shown in Figure 5.3. It is bounded by the x-axis, the lines $x = a$ and $x = b$, and the *curve having the equation $y = f(x)$*, where f is a function continuous on the closed interval $[a, b]$.

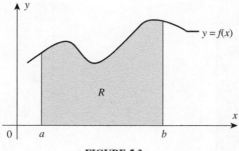

FIGURE 5.3

For simplicity, *assume that all values taken by the function f are non-negative, so that the graph of the function is a curve above the x-axis.*

We wish to assign *a number A to the measure of the area of R,* and we use a limiting process similar to the one used in defining the area of a circle. The area of a circle is defined as *the limit of the areas of inscribed regular polygons as the number of sides increases without bound.*[3]

Now, let us consider the region R bounded by parabola $f(x) = x^2$, the x-axis and the line $x = 2$ (Figure 5.4). (*Since R and R_0 together comprise a triangle whose area is 4, finding the area of R is equivalent to finding the area of R_0.*)

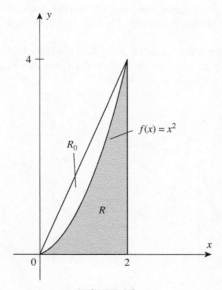

FIGURE 5.4

---

[3] We realize intuitively that, whatever number is chosen to represent A, that number must be at least as great as the measure of area of any polygonal region contained in R, and it must be no greater than the measure of the area of any polygonal region containing R.

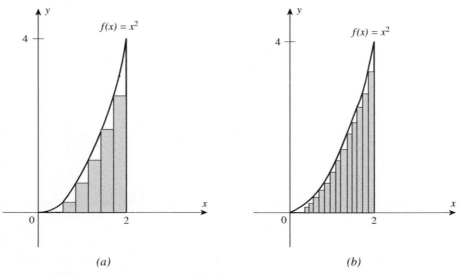

FIGURE 5.5

We can define a *polygonal region contained in R*. For this purpose, we *inscribe rectangles in the region R*, as shown in Figure 5.5a and b. Then the sum of the areas of the rectangles is less than the area of R.

Similarly, if *we circumscribe rectangles about R, as in Figure 5.6a and b*, then the sum of the areas of the rectangles is greater than the area of R. *Of course, we can find the area of each rectangle as the product of its base and height.*

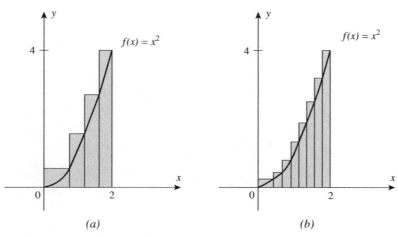

FIGURE 5.6

A *crucial observation to make about this process* is that as the bases of the rectangles become smaller and smaller, *the sum of the areas of the rectangles appears to approach the area of R.* This suggests that the area of R should be defined as the limit (in a sense to be clarified later) *of the sum of the areas of inscribed or circumscribed rectangles.* Our definition of area will be based on this idea.

Our assertion thus far about the area of R has rested on the following three basic properties we expect area to possess:

(1) *The Rectangle Properties*: The area of a rectangle is the product of its base and height. (This property is treated as the definition of area of a rectangle.)

(2) *The Addition Properties*: The area of a region composed of several smaller regions that *overlap in at most a line segment* is the sum of the areas of the smaller regions.[4]

(3) *The Comparison Property*: The area of a region that contains a second region is at least as large as the area of the second region.[5]

It is important to understand where each of these properties was employed in the preceding discussion. *They will play a major role in the definition of area to be discussed.*

5.3   THE DEFINITE INTEGRAL AS AN AREA

Consider a function $y = f(x)$ which is *continuous and positive in a closed interval* $[a, b]$. We think of the function as represented by a curve, and consider the area of the region which is *bounded above by the curve, at the sides by the straight lines* $x = a$ *and* $x = b$, and *below by the portion of the x-axis* between the points a and b (Figure 5.7).

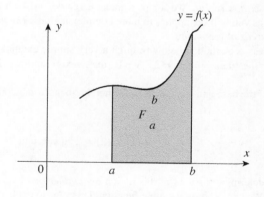

FIGURE 5.7

---

[4] This is so because a line has only one dimension namely "length", and it has no width. Thus, a line has no area.

[5] In other words, if two regions "A" and "B" are such that A contains B, then *area of A is at least as large as the area of B*.

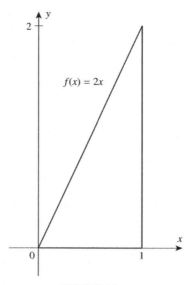

FIGURE 5.8

That there is a definite meaning in speaking of the area of this region is an assumption inspired by intuition. We denote the area of this region by F_a^b and call it the *definite integral of the function f(x) between the limits a and b*. When we actually seek to assign a numerical value to this area, we find that we are, in general, unable to measure areas of such regions with curved boundaries.

However, there is a way out. We adopt a method (based on Archimedes' method of exhaustion), which as you will see applies to more complex regions. *The method involves the summation of the areas of rectangles.*

The whole process is explained below through a very simple example. Consider a right triangle formed by the lines $y = f(x) = 2x$, $y = 0$ (the x-axis), and $x = 1$, as shown in the Figure 5.8.

Let $b =$ length of the base and $h =$ length of the height, then, from geometry, the area A of the triangle is

$$A = \frac{1}{2} \times b \times h = \frac{1}{2} \times (1) \times (2) = 1 \text{ square unit}$$

Next, we can also determine the area of this region *by another method*, as suggested in the discussion above. [We have chosen a simple function $f(x) = 2x$, to explain the method easily and simplify calculations for checking the results.]

Let us divide the interval $[0,1]$ on the x-axis into *four subintervals of equal length* Δx. This is done by equally spaced points $x_0 = 0$, $x_1 = 1/4$, $x_2 = 2/4$, $x_3 = 3/4$, and $x_4 = 4/4 = 1$ (see Figure 5.9). Each subinterval has length $\Delta x = 1/4$.

These subintervals determine four subregions: R_1, R_2, R_3, and R_4. With each subregion, we can associate a *circumscribed* rectangle (Figure 5.10); that is, a rectangle whose base

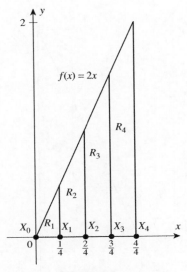

FIGURE 5.9

is the corresponding subinterval and whose *height is the maximum value of f* on that subinterval.

In this case, since *f* is *an increasing function*, the maximum value of *f* on each subinterval will occur when *x* is the right-hand end point of the subinterval.

The areas of the *circumscribed* rectangles (Figure 5.10) associated with regions R_1, R_2, R_3, and R_4 are $1/4 \, f \, (1/4)$, $1/4 \, f \, (2/4)$, $1/4 \, f \, (3/4)$, and $1/4 \, f \, (4/4)$, respectively. The area of each rectangle is *an approximation* to the area of its corresponding *subregion*.

Thus, the sum of the areas of the circumscribed rectangles, denoted by $\overline{F}_4$ (*upper sum*), is an approximation to the area *A* of the triangle.

$$\overline{F}_4 = \frac{1}{4}f\left(\frac{1}{4}\right) + \frac{1}{4}f\left(\frac{2}{4}\right) + \frac{1}{4}f\left(\frac{3}{4}\right) + \frac{1}{4}f\left(\frac{4}{4}\right)$$

$$= \frac{1}{4}\left[2\left(\frac{1}{4}\right) + 2\left(\frac{2}{4}\right) + 2\left(\frac{3}{4}\right) + 2\left(\frac{4}{4}\right)\right], \quad (\because \ f(x) = 2x)$$

$$= \frac{1}{4}\left[\frac{1}{2} + \frac{2}{2} + \frac{3}{2} + \frac{4}{2}\right] = \frac{10}{8} = \frac{5}{4}$$

Using sigma notation, we can write

$$\overline{F}_4 = \sum_{i=1}^{4} f(x_i)\Delta x$$

Obviously, $\overline{F}_4$ is *greater than the actual area* of the triangle, since *it includes areas of shaded regions* that are not in the triangle (Figure 5.10).

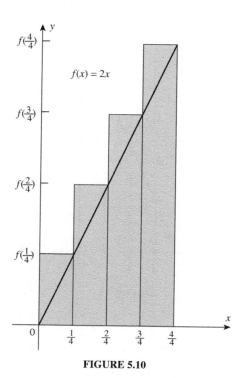

FIGURE 5.10

Similarly, the areas of the four inscribed rectangles (Figure 5.11) associated with R_1, R_2, R_3, and R_4 are $\frac{1}{4}f(0), \frac{1}{4}f\left(\frac{1}{4}\right), \frac{1}{4}f\left(\frac{2}{4}\right)$, and $\frac{1}{4}f\left(\frac{3}{4}\right)$, respectively. Their sum, denoted by $\underline{F}_4$ (*lower sum*), is also an approximation to the area A of the triangle.[6]

$$\underline{F}_4 = \frac{1}{4}f(0) + \frac{1}{4}f\left(\frac{1}{4}\right) + \frac{1}{4}f\left(\frac{2}{4}\right) + \frac{1}{4}f\left(\frac{3}{4}\right)$$

$$= \frac{1}{4}\left[2(0) + 2\left(\frac{1}{4}\right) + 2\left(\frac{2}{4}\right) + 2\left(\frac{3}{4}\right)\right], \quad (\because \ f(x) = 2x)$$

$$= \frac{1}{4}\left[0 + \frac{1}{2} + \frac{2}{2} + \frac{3}{2}\right] = \frac{6}{8} = \frac{3}{4}$$

Using sigma notation, we can write

$$\underline{F}_4 = \sum_{i=1}^{4} f(x_{i-1})\Delta x$$

---

[6] Since f is an increasing function, the minimum value of f on each subinterval will occur when x is the left-hand end point. In general, maximum or minimum values of a function, on each subinterval, may occur at any point in the subinterval.

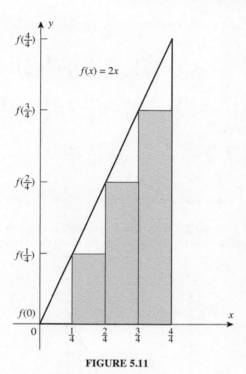

FIGURE 5.11

Clearly, $\underline{F}_4$ is less than the area of the triangle because the rectangles do not account for that portion of the triangle, which is not shaded. Note that

$$\frac{3}{4} = \underline{F}_4 \leq A \leq \overline{F}_4 = \frac{5}{4}$$

We say that $\underline{F}_4$ is an approximation to A *from below* and $\overline{F}_4$ is an approximation to A *from above*.

If $[0,1]$ is divided into more subintervals, better approximations to A will occur. For example, let us use *six subintervals of equal length* $\Delta x = 1/6$. Then, the total area of *six circumscribed rectangles* (i.e., *the upper sum*) is given by

$$\overline{F}_6 = \frac{1}{6}f\left(\frac{1}{6}\right) + \frac{1}{6}f\left(\frac{2}{6}\right) + \frac{1}{6}f\left(\frac{3}{6}\right) + \frac{1}{6}f\left(\frac{4}{6}\right) + \frac{1}{6}f\left(\frac{5}{6}\right) + \frac{1}{6}f\left(\frac{6}{6}\right)$$

$$= \frac{1}{6}\left[2\left(\frac{1}{6}\right) + 2\left(\frac{2}{6}\right) + 2\left(\frac{3}{6}\right) + 2\left(\frac{4}{6}\right) + 2\left(\frac{5}{6}\right) + 2\left(\frac{6}{6}\right)\right]$$

$$= \frac{1}{6}\left[\frac{1}{3} + \frac{2}{3} + \frac{3}{3} + \frac{4}{3} + \frac{5}{3} + \frac{6}{3}\right] = \frac{21}{18} = \frac{7}{6}$$

and, the total area of *six inscribed rectangles* (i.e., *the lower sum*) is given by

$$\underline{F} = \frac{1}{6}f(0) + \frac{1}{6}f\left(\frac{1}{6}\right) + \frac{1}{6}f\left(\frac{2}{6}\right) + \frac{1}{6}f\left(\frac{3}{6}\right) + \frac{1}{6}f\left(\frac{4}{6}\right) + \frac{1}{6}f\left(\frac{5}{6}\right)$$

$$= \frac{1}{6}\left[2(0) + 2\left(\frac{1}{6}\right) + 2\left(\frac{2}{6}\right) + 2\left(\frac{3}{6}\right) + 2\left(\frac{4}{6}\right) + 2\left(\frac{5}{6}\right)\right]$$

$$= \frac{1}{6}\left[0 + \frac{1}{3} + \frac{2}{3} + \frac{3}{3} + \frac{4}{3} + \frac{5}{3}\right] = \frac{15}{18} = \frac{5}{6}$$

Note that $\underline{F}_6 \le A \le \overline{F}_6$ and, *with appropriate labeling*, both $\overline{F}_6$ and $\underline{F}_6$ will be of the form $\sum f(x)\Delta x$.[7]

More generally, if we divide $[0,1]$ into *n subintervals of equal length* Δx, then $\Delta x = 1/n$ and the end points of the subintervals are $x = 0, 1/n, 2/n, \ldots, (n-1)/n$ and $n/n = 1$ (see Figure 5.12).

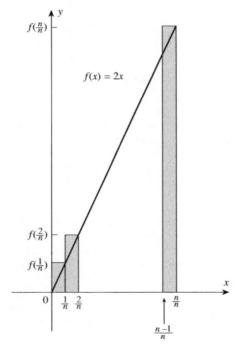

FIGURE 5.12

---

[7] $\overline{F}_6 = \sum_{i=1}^{6} f(x_i)\Delta x, \ \underline{F}_6 = \sum_{i=0}^{5} f(x_i)\Delta x.$

The total area of *n circumscribed rectangles* is

$$
\begin{aligned}
\overline{F}_n &= \frac{1}{n}f\left(\frac{1}{n}\right) + \frac{1}{n}f\left(\frac{2}{n}\right) + \frac{1}{n}f\left(\frac{3}{n}\right) + \ldots + \frac{1}{n}f\left(\frac{n}{n}\right) \\
&= \frac{1}{n}\left[2\left(\frac{1}{n}\right) + 2\left(\frac{2}{n}\right) + 2\left(\frac{3}{n}\right) + \ldots + 2\left(\frac{n}{n}\right)\right], \quad [\because f(x) = 2x] \\
&= \frac{2}{n^2}[1 + 2 + 3 + \ldots + n], \quad \text{(by taking out } 2/n \text{ from each term)} \\
&= \left(\frac{2}{n^2}\right)\frac{n(n+1)}{2}, \quad \left[\text{since } 1 + 2 + \ldots + n = \frac{n(n+1)}{2}\right] = \frac{n+1}{n}
\end{aligned}
\tag{1}
$$

And for *n inscribed rectangles*, the total area determined by the subintervals is (see Figure 5.13)

$$
\begin{aligned}
\underline{F}_n &= \frac{1}{n}f(0) + \frac{1}{n}f\left(\frac{1}{n}\right) + \ldots + \frac{1}{n}f\left(\frac{2}{n}\right)\ldots + \frac{1}{n}f\left(\frac{n-1}{n}\right) \\
&= \frac{1}{n}\left[2(0) + 2\left(\frac{1}{n}\right) + 2\left(\frac{2}{n}\right) + \ldots + 2\left(\frac{n-1}{n}\right)\right] \\
&= \frac{2}{n^2}[0 + 1 + 2 + 3 + \ldots + (n-1)] \\
&= \frac{2}{n^2}\frac{(n-1)n}{2} = \frac{n-1}{n}, \quad \left[\text{since } 1 + 2 + \ldots + (n-1) = \frac{(n-1)(n)}{2}\right]
\end{aligned}
\tag{2}
$$

From equations (1) and (2), we observe that both $\overline{F}_n$ and $\underline{F}_n$ are sums of the form $\sum f(x)\Delta x$. From the nature of $\underline{F}_n$ and $\overline{F}_n$, it is reasonable and indeed true to write $\underline{F}_n \leq A \leq \overline{F}_n$. As *n* becomes larger, $\underline{F}_n$ and $\overline{F}_n$ become better approximations to *A from below* and *from above*, respectively. If we take the limit of $\underline{F}_n$ and $\overline{F}_n$ as $n \to , \infty$ through positive integral values, we get

$$
\lim_{n\to\infty} \underline{F}_n = \lim_{n\to\infty}\frac{n-1}{n} = \lim_{n\to\infty}\left(1 - \frac{1}{n}\right) = 1, \text{ and}
$$

$$
\lim_{n\to\infty} \overline{F}_n = \lim_{n\to\infty}\frac{n+1}{n} = \lim_{n\to\infty}\left(1 + \frac{1}{n}\right) = 1
$$

Since $\underline{F}_n$ and $\overline{F}_n$ both have the *same common limit*, we write

$$
\lim_{n\to\infty} \underline{F}_n = \lim_{n\to\infty} \overline{F}_n = 1
\tag{3}
$$

and since $\underline{F}_n \leq A \leq \overline{F}_n$, *we take this common limit to be the area of the triangle.* Thus we get, the area $A = 1$ square unit. *This also agrees with our earlier finding.*

Mathematically, *the sums $\overline{F}_n$ and $\underline{F}_n$, as well as their common limit have a meaning, which is independent of area.* For the function $f(x) = 2x$, over the interval [0, 1], *we define the common limit of $\overline{F}_n$ and $\underline{F}_n$ to be the definite integral of $f(x) = 2x$, from $x = 0$ to $x = 1$.* Symbolically we write this as

$$
\int_0^1 f(x)dx = \int_0^1 2x\, dx = 1
\tag{4}
$$

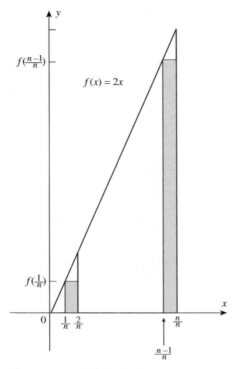

FIGURE 5.13

The numbers 0 and 1 appearing with the integral sign $\int$ in equation (4) are called *the limits of integration*; 0 is the *lower limit* and 1 is the *upper limit*.

Two points must be mentioned about the *definite integral:*

(i) Aside from any geometrical interpretation (such as area) *it is nothing more than a real number.*

(ii) The definite integral is the limit of a sum of the form $\sum f(x)\Delta x$.

The definite integral of a function f(x) over an interval from x = a to x = b, where a ≤ b, is the common limit of the upper sum (i.e., $\overline{F}_n$) and the lower sum (i.e., $\underline{F}_n$), if it exists, and is written as $\int_a^b f(x)\,dx$. In terms of the limiting process, we have $\sum f(x)\,\Delta x \to \int_b^b f(x)\,d(x)$. We take this limiting value as the *definition of the definite integral*. In particular, the *definite integral also stands for the area under a curve*, as discussed above. From a subdivision of the interval [a, b] into finite portions of the form Δx, *the process of passage to the limit (as $\Delta x \to 0$) is suggested by the use of the letter d in place of Δ.*

Note: It will be wrong to think that dx is an infinitely small quantity or an infinitesimal (i.e., a variable whose limit is 0) or that the definite integral $\int_a^b f(x)\,dx$ is the sum of an infinite number of infinitely small quantities. *This type of thinking is quite misleading and it is a sign of being in the state of confusion.* Hence, care must be taken to protect and preserve what we have carried out with precision. We now formally define the area in terms of the definite integral.

5.4 DEFINITION OF AREA IN TERMS OF THE DEFINITE INTEGRAL

Definition: Let f be *continuous* and *non-negative* on $[a, b]$, and let R be the region bounded above by the graph of f, below by the x-axis and on the left and right by the lines $x = a$ and $x = b$. Then, we call R *the region between the graph of f and the x-axis on $[a, b]$*, and the area of R is defined by $\int_a^b f(x)\mathrm{d}x$.

We emphasize that in the notation for the definite integral $\int_a^b f(x)\mathrm{d}x$, dx *has no independent meaning*. It arose originally in connection with the concept of the differential (see Chapter 16 of Part I). This expression will play a role later, when we develop special methods for computing definite integrals. The symbol $\int$ is an *integral sign*. The *integral sign* resembles the capital S, which is appropriate because *the definite integral is the limit of a sum*. Note that, it is the *same symbol* we have been using to indicate *the operation of antidifferentiation. The reason for the common symbol is that a theorem called the second fundamental theorem of the Calculus enables us to evaluate a definite integral by finding an antiderivative (which is also called an indefinite integral).*

We have seen in equation (3), that $\lim\limits_{n \to \infty} \underline{F}_n = \lim\limits_{n \to \infty} \overline{F}_n$.

For an arbitrary function, this is not always true. The statement "the function f is integrable on the closed interval $[a, b]$" is synonymous with the statement "the definite integral of f from a to b exists". The functions for which this is true are called *integrable* functions.

We now go for *more refined considerations, which permit us to separate the notion of definite integral from the simple intuitive idea of area.* This is done in the analytical definition of the definite integral. It expresses the *definite integral analytically* in terms of the notion of a number only. We shall find that this definition is of great significance not only because it alone enables us to attain complete clarity in our concepts but also because its applications extend far beyond the calculation of areas.

5.5 RIEMANN SUMS AND THE ANALYTICAL DEFINITION OF THE DEFINITE INTEGRAL

Both Newton and Leibniz introduced early versions of this concept. However, it was Riemann who gave us the *modern definition*. In formulating this definition, we are guided by the ideas that we have discussed earlier in this section. First, we describe certain *terms* that will be used in the *analytical definition* of definite integral.

(i) *Partition*: Let $[a, b]$ be a closed interval. Let $x_0, x_1, x_2, x_3, \ldots, x_n$ be *any* $(n + 1)$ points such that $a = x_0 < x_1 < x_2 < x_3 < \ldots < x_{n-1} < x_n = b$.

Then, the set $P = \{x_0, x_1, x_2, x_3, \ldots, x_n\}$ is called a partition of $[a, b]$.

Remarks:

(a) Any partition P of $[a, b]$ must contain the *end points a and b so that it is a nonempty set*.

(b) A partition P of $[a, b]$ containing $(n + 1)$ points $a = x_0, x_1, x_2, \ldots, x_n$ such that $a = x_0 < x_1 < x_2, \ldots, < x_n = b$ divides the interval $[a, b]$ into n *parts or subintervals*, $[x_0, x_1], [x_1, x_2], [x_2, x_3], \ldots, [x_{n-1}, x_n]$, and $x_1 - x_0, x_2 - x_1, x_3 - x_2, \ldots, x_n - x_{n-1}$, are called

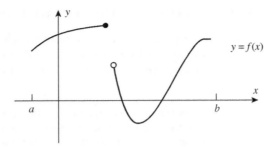

FIGURE 5.14

the lengths of the respective subintervals, given in order. Obviously, *these lengths of subintervals need not be equal.*

(c) The simplest partition of $[a, b]$ is $\{a, b\}$. If we go on adding more and *more points* to this partition then *the lengths of the subintervals will go on decreasing.*

(ii) Norm of a Partition: Let $P = \{x_0, x_1, x_2, \ldots, x_n \mid a = x_0 < x_1 < x_2 \ldots < x_n = b\}$ *be a partition of* $[a, b]$. *The greatest of the lengths* $x_1 - x_0, x_2 - x_1, \ldots, x_3 - x_2, \ldots, x_n - x_{n-1}$, of the subintervals formed by the partitions, is called *the norm of the partition* and is denoted by $||P||$ (read as the norm of the partition P). Now, we describe the *notion of a Riemann sum.*

5.5.1 Riemann Sums

Consider a function f defined on a *closed interval* $[a, b]$. It may have both *positive* and *negative* values on the interval and *it does not even need to be continuous.* Its graph might look something like the one in Figure 5.14.

Consider a partition P of the interval $[a, b]$ into n subintervals (*not necessarily of equal length*) by means of points $a = x_0 < x_1 < x_2 < \ldots < x_{n-1} < x_n = b$, and let $\Delta x_i = x_i - x_{i-1}$ ($i \varepsilon N$).

On each subinterval $[x_{i-1}, x_i]$, *choose a perfectly arbitrary point* $\overline{x}_i$ (which could even be an end point of the subinterval). We call it a *sample point* for the ith subinterval. An example of this construction is shown in Figure 5.15, for $n = 6$.

Now, form the sum $R_p = \sum_{i=1}^{6} f(\overline{x}_i) \Delta x_i$.

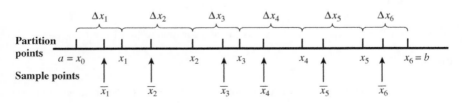

A partition of $[a, b]$ with sample points X_i

FIGURE 5.15

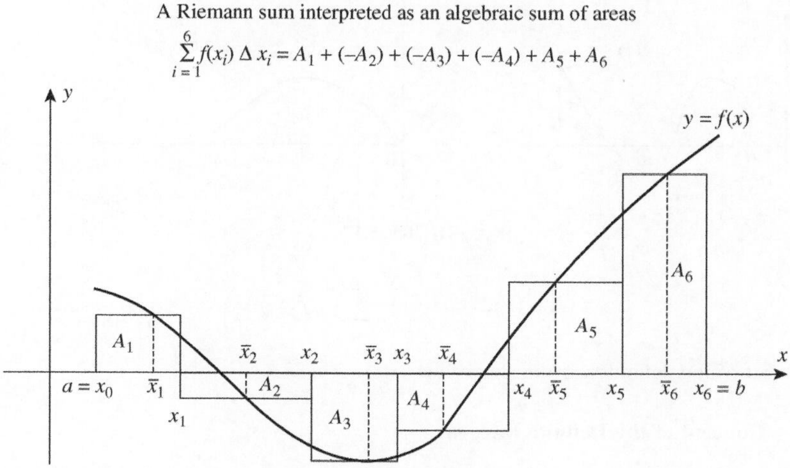

A Riemann sum interpreted as an algebraic sum of areas

$$\sum_{i=1}^{6} f(x_i)\,\Delta\,x_i = A_1 + (-A_2) + (-A_3) + (-A_4) + A_5 + A_6$$

FIGURE 5.16

We call R_p a *Riemann sum* for f, corresponding to the partition P. Its geometric interpretation is shown in Figure 5.16.

Note that, *the contributions from the rectangles below the x-axis are the negatives of their areas.* Thus, a *Riemann sum* is interpreted as *an algebraic sum of areas.*[8]

$$\sum_{i=1}^{6} f(\overline{x}_i)\Delta x_i = A_1 + (-A_2) + (-A_3) + (-A_4) + A_5 + A_6$$

Riemann sums corresponding to various choices of $\overline{x}_1, \overline{x}_2, \ldots, \overline{x}_n$ can be different from one another. However, all Riemann sums must lie between the lower sum and the upper sum. An important feature of a Riemann sum $\sum_{i=1}^{n} f(\overline{x}_i)\Delta x_i$ is that it approximates the definite integral $\int_a^b f(x)\mathrm{d}x$.

Therefore, we write $\int_a^b f(x)\mathrm{d}x \approx \sum_{i=1}^{n} f(\overline{x}_i)\Delta x_i.$[9]

Suppose now that P, Δx_i and $\overline{x}_i$ have the meanings discussed above. Also, let $\|P\|$ be the *norm* (i.e., the length of the longest subinterval of the partition P). Then, we give the following definition.

5.5.2   Definition: The Definite Integral

Let f be a function that is defined on the *closed interval* $[a, b]$. If $\lim_{\|P\| \to 0} \sum_{i=1}^{n} f(\overline{x}_i)\Delta x_i$ exists, then we say *that f is integrable* on $[a, b]$. Moreover, $\int_a^b f(x)\mathrm{d}x$, [called the *definite integral (or Riemann integral)* of f from a to b], is then given by [10]

---

[8] Riemann sums are named after the nineteenth century mathematician Georg Bernhard Riemann (1826–1866), who clarified the concept of the integral while employing such sums. The first formal definition of the integral is attributed to him.

[9] For detailed discussion, see *Calculus with Analytic Geometry* (Alternate Edition) by Robert Ellis and Denny Gulick, Chapter 5.

[10] *Calculus with Analytic Geometry* (Fifth Edition) by Edwin J. Purcell and Dale Varberg (pp. 234–235), Prentice-Hall, Inc, New Jersey.

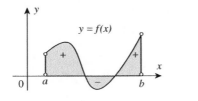
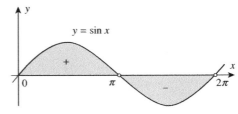

FIGURE 5.17

$$\int_a^b f(x)\mathrm{d}x = \lim_{\|P\| \to 0} \sum_{i=1}^n f(\overline{x}_i)\Delta x_i$$

The heart of the definition is the above line.

5.5.3   Concept of the Definite Integral

The concept captured in the above equation grows out of our area of discussion. However, we have considerably modified the notion presented here. For example,

(i) We now allow f to be *negative on part or all of* [a, b],

(ii) We use partitions with *subintervals* that may be of *unequal lengths*, and

(iii) We allow $\overline{x}_i$ to be *any point on the ith subinterval*.

Since we have made these changes, it is important *to state precisely how the definite integral relates to area.*

Note (1): In general, $\int_a^b f(x)\mathrm{d}x$ gives the *signed area of the region trapped between the curve* $y = f(x)$ *and the x-axis, on the interval* [a, b], meaning that a plus sign is attached to areas of parts above the x-axis and a minus sign is attached to areas of parts below the x-axis. In symbols, $\int_a^b f(x)\mathrm{d}x = A_{\mathrm{up}} - A_{\mathrm{down}}$, where A_{up} and A_{down} are the areas corresponding to the $+$ and $-$ regions as shown in Figure 5.17.

Note (2): The meaning of limit in the definition of the definite integral is more general than in earlier usage, and should be explained.

The equality $\lim_{\|P\| \to 0} \sum_{i=1}^n f(\overline{x}_i)\Delta x_i = L$ means that, corresponding to each $\varepsilon > 0$, there is a $\delta > 0$, such that $\left| \sum_{i=1}^n f(\overline{x}_i)\Delta x_i - L \right| < \varepsilon$, for *all Riemann sums* $\sum_{i=1}^n f(\overline{x}_i)\Delta x_i$ for f on [a, b], for which the norm $\|P\|$ of the associated partition is less than δ. In this case, we say that the indicated limit exists and has the value L.

Note (3): In the symbol $\int_a^b f(x)\mathrm{d}x$, most authors use the terminology "a", as the *lower limit* of integration and "b", as the *upper limit* of integration, which is fine provided we realize that this usage of the word *limit* has nothing to do with its more technical meaning.

5.5.4   Further Modification in the Notion of Definite Integral: Removal of One More Restriction

In our definition of $\int_a^b f(x)\mathrm{d}x$, since $f(x)$ is defined on the interval [a, b], we *implicitly assumed* that $a < b$. If we omit the condition $a < b$, and assume that $a > b$, *we can still retain our*

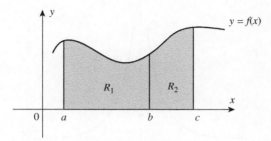

FIGURE 5.18

arithmetical definition of definite integral; the only change is that *when we traverse the interval from a to b, the differences* Δx, *are negative.* Thus, we get the following relation

$$\int\limits_{a}^{b} f(x)\mathrm{d}x = -\int\limits_{b}^{a} f(x)\mathrm{d}x \qquad \text{(I)}$$

which holds for all values of a and b ($a \neq b$). From the above relation, it follows that

$$\int\limits_{a}^{a} f(x)\mathrm{d}x = 0 \qquad \text{(II)}$$

This must be treated as a definition. Our definition (of the definite integral) immediately gives the basic relation

$$\int\limits_{a}^{b} f(x)\mathrm{d}x + \int\limits_{b}^{c} f(x)\mathrm{d}x = \int\limits_{a}^{c} f(x)\mathrm{d}x \qquad \text{(III)}$$

for $a < b < c$ (see Figure 5.18).

Remark: By means of the preceding relations, we at once find that the equation (III) is also true for any position of the point a, b, and c relative to one another. Also, we obtain a simple but important fundamental rule by considering the function $c\,f(x)$, *where c is a constant.* From the definition of the definite integral, we immediately obtain,

$$\int\limits_{a}^{b} c \cdot f(x)\mathrm{d}x = c \int\limits_{a}^{b} f(x)\mathrm{d}x \qquad \text{(IV)}$$

Further, we assert the following addition rule:

If $f(x) = \phi(x) + \psi(x)$ then,

$$\int\limits_{a}^{b} f(x)\mathrm{d}x = \int\limits_{a}^{b} \phi(x)\mathrm{d}x + \int\limits_{a}^{b} \psi(x)\mathrm{d}x \qquad \text{(V)}$$

This can be easily proved from the definition of definite integral using Riemann sums.

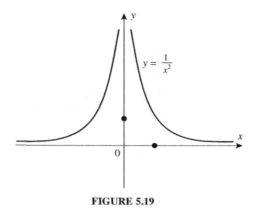

FIGURE 5.19

5.5.5   An Important Remark About the Variable of Integration

We have written the definite integral in the form $\int_a^b f(x)\mathrm{d}x$. For evaluating the integral, it does not matter whether we use the letter x or any other letter to denote the independent variable. The particular symbol we use for the variable of integration is therefore not important; instead of $\int_a^b f(x)\mathrm{d}x$, we could as well write $\int_a^b f(t)\mathrm{d}t$ or $\int_a^b f(u)\mathrm{d}u$ or any other expression. The importance of this remark will be realized shortly in applications, when we prove the *second fundamental theorem of Calculus*, in the next chapter.

5.5.6   What Functions Are Integrable?

Not every function is integrable.

For example, *the unbounded function* $f(x) = \begin{cases} 1/x^2 & \text{if } x \neq 0 \\ 1 & \text{if } x = 0 \end{cases}$ in Figure 5.19, is *not integrable on* $[-2, 2]$.

$$y = f(x) = \begin{cases} 1/x^2 & \text{if } x \neq 0 \\ 1 & \text{if } x = 0 \end{cases}$$

This is because the contribution of any Riemann sum for the subinterval containing $x = 0$ can be made arbitrarily large by choosing the corresponding sample point $\overline{x}_i$, sufficiently close to zero. In fact, this reasoning shows that *"any function that is integrable on* $[a, b]$*, must be bounded in the interval* $[a, b]$*"*. In other words, there must exist a constant M such that $|f(x)| \leq M$ for all x in $[a, b]$.

Remark: *Even some bounded functions can fail to be integrable.* For example, the function

$$f(x) = \begin{cases} 1 & \text{if } x \text{ is rational} \\ 0 & \text{if } x \text{ is irrational} \end{cases}$$

is *not integrable* on $[0,1]$.[11]

---

[11] This is so because between *any* two real numbers, there are infinite number of rationals and also infinite number of irrationals. Hence, no matter how small the norm of partition (i.e., $||P||$), the Riemann sum $\sum_{i=1}^{n} f(\overline{x}_i)\Delta x_i$ cannot have a unique value. Of course, it can have the value *either* 0 *or* 1.

By all odds, the theorem given in Section 5.5.7 is *the most important theorem* about integrability of a function.

5.5.7 Integrability Theorem

If *f* is *bounded* on [*a*, *b*] and if it is *continuous* there, *except at a finite number of points*, then *f* is integrable on [*a*, *b*]. We do not prove it here.

In particular, *if f is continuous on the whole interval* [*a*, *b*], *it is integrable on* [*a*, *b*].[12]

As a consequence of this theorem, the following functions are integrable on every closed interval [*a*, *b*]

(1) Polynomial functions.

(2) Sine and cosine functions.

(3) Rational functions (provided the interval [*a*, *b*] contains no points where denominator is 0).

From the definition of the definite integral, $\int_a^b f(x)\mathrm{d}x$, *as the limit of a sum*, we have

$$\int_b^b f(x)\mathrm{d}x = \lim_{\|P\| \to 0} \sum_{i=1}^{n} f(\overline{x}_i)\Delta x_i \tag{A}^{[13]}$$

We know that, if *f* is integrable on [*a*, *b*], then the limit on the right-hand side of equation (A) must exist. For *the functions that we shall consider here, this limit will always exist*. We now suggest the following so that the method for evaluating a definite integral, as the limit of a sum, is simplified to some extent.

(i) We shall consider a *convenient partition* of [*a*, *b*], that is, choose $\overline{x}_1, \overline{x}_2, \overline{x}_3, \ldots, \overline{x}_n$ in the subintervals *suitably* as explained below, so that the limit of the sum on the right-hand side of equation (A) can be evaluated easily. The *most convenient partition* is that which divides [*a*, *b*] into subintervals of *equal length*, say *h*. Such a partition is called as a *regular partition*.

If $a = x_0, x_1, x_2, x_3, \ldots, x_n = b$ are the points of division such that $a = x_0 < x_1 < x_2 < x_3 < \ldots < x_{n-1} < x_n = b$, then $x_1 - x_0 = (x_1 - a) = h, x_2 - x_1 = h, x_3 - x_2 = h, \ldots, x_n - x_{n-1} = h$.

Hence, the points of subdivision are, $a, a + h, a + 2h, a + 3h, \ldots, a + nh$.

There are *n* subintervals in [*a*, *b*] *each of length h*.

∴ The sum of the lengths of these subintervals must be *b* − *a*.

$$\therefore \quad nh = b - a \qquad \therefore \quad h = \frac{b-a}{n} = \|P\|$$

Now, as $n \to \infty$, $h \to 0$ (i.e., the length of each subinterval tends to zero, so that $\|P\| \to 0$).

(ii) We choose the points $\overline{x}_1, \overline{x}_2, \overline{x}_3, \ldots, \overline{x}_n$ as *the right-hand end point of each subinterval*, in computing a sum.[14]

Thus, $\overline{x}_1 = a + h = x_1, \overline{x}_2 = a + 2h = x_2, \overline{x}_3 = a + 3h = x_3, \ldots, \overline{x}_n = a + nh = x_n$.

---

[12] Unfortunately, the proof of this theorem is not simple. We, therefore, accept the theorem without proof. For the proof of this theorem, advanced texts on Calculus may be referred to.

[13] The meanings of *P*, $\|P\|$, Δx_i and $\overline{x}_i$ have already been explained earlier, in the text.

[14] By choosing each subinterval Δx_i of equal length, and the points $\overline{x}_1, \overline{x}_2, \overline{x}_3, \ldots, \overline{x}_n$ as the right-hand end point of each subinterval, helps in computing the sum $\sum_{i=1}^{n} f(\overline{x}_i)\Delta x_i$, easily.

Then, we have

$$\int_a^b f(x)dx = \lim_{||P|| \to 0} \sum_{r=1}^n f(\overline{x}_r)\Delta x_r$$

$$= \lim_{n \to \infty} \sum_{r=1}^n f(x_r) \cdot h$$

$$= \lim_{n \to \infty} \sum_{r=1}^n f(a + rh) \cdot h, \quad \text{where } h = \frac{b-a}{n}$$

$$= \lim_{n \to \infty} \sum_{r=1}^n h \cdot f(a + rh), \quad \text{where } h = \frac{b-a}{n}$$

$$\int_a^b f(x)dx = \lim_{n \to \infty} \left(\frac{b-a}{n}\right) \sum_{i=1}^n f\left(a + r\left(\frac{b-a}{n}\right)\right) \qquad \text{(B)}^{(15)}$$

By using the formula (B), we can compute the definite integral $\int_a^b f(x)dx$, as the limit of a sum.

For solving problems on *"integral as the limit of a sum"*, we shall require the following results, studied in earlier classes.

(1) $\sum_{r=1}^n (a \pm b_r) = \sum_{r=1}^n a_r \pm \sum_{r=1}^n b_r$

(2) $\sum_{r=1}^n k\, a_r = k \sum_{r=1}^n a_r$, where k is a constant, independent of r.

(3) $\sum_{r=1}^n k = nk$, where k is a constant.

(4) $\sum_{r=1}^n r = \frac{n(n+1)}{2}$

(5) $\sum_{r=1}^n r^2 = \frac{n(n+1)(2n+1)}{6}$

(6) $\sum_{r=1}^n r^3 = \frac{n^2(n+1)^2}{4}$

(7) If S_n denotes the sum of first n terms of a G.P. whose first term is "a" and common ratio is r, then

$$S_n = a\left(\frac{1-r^n}{1-r}\right), \text{ if } r < 1$$

and $\quad S_n = a\left(\frac{r^n - 1}{r - 1}\right), \text{ if } r > 1.$

At this stage, we state *the second fundamental theorem of Calculus, which links definite integral $\int_a^a f(x)dx$ to the antiderivative of $f(x)$.*

---

[15] If we choose the left-hand endpoint of each subinterval, then we will have the equation $\int_a^b f(x)dx = \lim_{n \to \infty} \left(\frac{b-a}{n}\right) \sum_{i=1}^n \left(a + (r-1)\left(\frac{b-a}{n}\right)\right)$, which is comparatively not so convenient. Of course, the result remains the same in both the cases.

It states that $\int_a^b f(x)\,\mathrm{d}x = \phi(b) - \phi(a)$, where $\int f(x)\,\mathrm{d}x = \phi(x)$ (which means ϕ is the antiderivative of f).

The proof is not given here. It is introduced in Chapter 6a, and its applications are discussed in Chapter 7a of Part II.

Remarks:

(i) It is convenient to introduce a special symbol for $\phi(b) - \phi(a)$. We write

$$\phi(b) - \phi(a) = [\phi(x)]_a^b.$$

Thus, we have $\int_a^b f(x)\,\mathrm{d}x = [\phi(x)]_a^b = \phi(b) - \phi(a)$.

(ii) The concepts of the slope of the tangent line (derivative) and the area of the curved region (definite integral) were known long back. Historically, the basic concepts of the definite integral were used by the ancient Greeks, principally Archimedes' (287–212 BC), more than 2000 years ago. That was many years before differential Calculus was discovered in the seventeenth century by Newton and Leibniz. The fact being that the concepts of *derivatives and definite integral* were known prior to the period of Newton and Leibniz, and that a number of mathematicians had contributed toward the development of the subject, the question is: *Why then do Newton and Leibniz figure so prominently in the history of Calculus?*

They do so because they understood and exploited the intimate relationship that exists between antiderivatives and definite integrals. It is this relationship that enables us to compute easily the exact values of many definite integrals without ever using Riemann sums. This connection is so important that it is called the second fundamental theorem of Calculus.[16]

Now, we proceed to evaluate some definite integrals by two methods: first as the limit of a sum and second by applying the *second fundamental theorem of Calculus*, which provides a very simple method to calculate definite integrals.

Illustrative Examples

Example (1): Express $\int_2^3 x\,\mathrm{d}x$ as *the limit of a sum* and hence evaluate.

Solution: Divide the interval $[2,3]$ into n *equal parts*. The length of each subinterval so obtained is $((3-2)/n) = (1/n)$, and the partition formed by the points is given by $P = \{2, (2 + (1/n)), (2 + (2/n)), \ldots, (2 + ((n-1)/n)), 3\}$.

Method (I): Let the sample points $\overline{x}_r = x_r = (2 + (r/n))$ $(r = 1, 2, 3, \ldots, n)$ (i.e., we choose *the right-hand endpoint* of each subinterval as the *sample point* $\overline{x}_r$ for computing the sum).

---

[16] In fact, based on the definition of area function $A(x)$ (to be introduced in the next chapter), there are two basic fundamental theorems to be discussed later. To understand the (second) fundamental theorem we have to go through the (first) fundamental theorem of Calculus.

Then, we have,

$$\int_a^b f(x)dx = \lim_{n \to \infty} \sum_{r=1}^{n} f(x_r) \cdot h, \text{ where } x_r = 2 + \frac{r}{n} \text{ and } h = \frac{b-a}{n} = \frac{1}{n}.$$

∴ Recall the Formula (B). We have

$$\int_a^b f(x)dx = \lim_{n \to \infty} \left(\frac{b-a}{n}\right) \sum_{i=1}^{n} f\left(a + r\left(\frac{b-a}{n}\right)\right)$$

$$\therefore \int_2^3 x\,dx = \lim_{n \to \infty} \frac{1}{n} \sum_{r=1}^{n} f\left(2 + \frac{r}{n}\right) = \lim_{n \to \infty} \frac{1}{n}\left[\sum_{r=1}^{n}\left(2 + \frac{r}{n}\right)\right] \quad \{\because f(x) = x$$

$$= \lim_{n \to \infty} \frac{1}{n}\left[2\sum_{r=1}^{n} 1 + \frac{1}{n}\sum_{r=1}^{n} r\right] = \lim_{n \to \infty} \frac{1}{n}\left[2.n + \frac{1}{n} \cdot \frac{n(n+1)}{2}\right]$$

$$= \lim_{n \to \infty} \left[2 + \frac{1}{2}\left(1 + \frac{1}{n}\right)\right] = 2 + \frac{1}{2} = \frac{5}{2} \quad \text{Ans.}$$

Method (II): Now, using the second fundamental theorem of integral Calculus:

$$\int_a^b f(x)dx = [\phi(x)]_a^b = \phi(b) - \phi(a)$$

where $\int f(x)dx = \phi(x)$, we get

$$\int_2^3 xdx = \left[\frac{x^2}{2}\right]_2^3 = \frac{3^2}{2} - \frac{2^2}{2}$$

$$= \frac{9}{2} - \frac{4}{2} = \frac{5}{2} \quad \text{Ans.}$$

Example (2) Express $\int_0^2 (3x + 5)dx$ as the limit of a sum and hence evaluate.

Solution: Divide the interval $[0,2]$ into n *equal parts*. The length of each subinterval $= \frac{2-0}{n} = \frac{2}{n}$. The partition so formed by the points, is given by

$$P = \left\{0, \left(0 + \frac{2}{n}\right), \left(0 + 2\left(\frac{2}{n}\right)\right), \left(0 + 3\left(\frac{2}{n}\right)\right), \dots, \left(0 + (n-1)\left(\frac{2}{n}\right)\right), \left(0 + n\left(\frac{2}{n}\right)\right)\right\}$$

$$= \left\{0, \left(1\left(\frac{2}{n}\right)\right), \left(2\left(\frac{2}{n}\right)\right), \dots, \left((n-1)\left(\frac{2}{n}\right)\right), 2\right\}$$

Let $\bar{x}_r = x_r = 0 + r \cdot \frac{2}{n} = \frac{2r}{n}$, $[r = 1, 2, 3, \dots, n]$ (i.e., we choose the *right-hand endpoint of each subinterval* as the sample point $\bar{x}_r$, for computing the sum).

Method (I): Now, we have by definition (i.e., the result "B"):

$$\int_a^b f(x)dx = \lim_{n \to \infty} \left(\frac{b-a}{n}\right) \sum_{r=1}^{n} f\left(a + r\left(\frac{b-a}{n}\right)\right)$$

where $a = 0$ and $b = 2$ so that $((b - a)/n) = 2/n$.

$$\therefore \int_0^2 (3x + 5)dx = \lim_{n \to \infty} \frac{2}{n} \sum_{r=1}^{n} f\left(0 + r \cdot \frac{2}{n}\right)$$

$$= \lim_{n \to \infty} \frac{2}{n} \sum_{r=1}^{n} f\left(\frac{2r}{n}\right)$$

$$= \lim_{n \to \infty} \frac{2}{n} \sum_{r=1}^{n} \left[3\left(\frac{2r}{n}\right) + 5\right], \quad \{\because \; f(x) = 3x + 5$$

$$\therefore \int_0^2 (3x + 5)dx = \lim_{n \to \infty} \frac{2}{n} \left[\frac{6}{n} \sum_{r=1}^{n} r + 5 \sum_{r=1}^{n} 1\right]$$

$$= \lim_{n \to \infty} \frac{2}{n} \left[\frac{6}{n} \cdot \frac{n(n + 1)}{2} + 5n\right]$$

$$= \lim_{n \to \infty} \frac{2}{n} \left[\frac{3}{n} \cdot n^2\left(1 + \frac{1}{n}\right) + 5n\right] \quad \text{Note that } (n + 1) = n\left(1 + \frac{1}{n}\right)$$

$$= \lim_{n \to \infty} \left[6\left(1 + \frac{1}{n}\right) + 10\right] = 6 + 10 = 16 \quad \text{Ans.}$$

Method (II): Using the fundamental theorem of integral Calculus:

$$\int_0^2 (3x + 5)dx = \left[3\frac{x^2}{2} + 5x\right]_0^2, \quad \text{where } \left(3\frac{x^2}{2} + 5x\right) \text{ is antiderivative of } (3x + 5)$$

$$= \left(3 \cdot \frac{2^2}{2} + 5(2)\right) - (3.0 + 5.0)$$

$$= 6 + 10 - 0 = 16 \quad \text{Ans.}$$

Example (3): Find $\int_0^2 (x^2 + 1)dx$ as the limit of a sum.

Solution: Divide the interval $[0,2]$ into n *equal parts*. The length of each subinterval $= ((2 - 0)/n) = 2/n$. The partition so formed by the points, is given by

$$P = \left\{0, \left(0 + \frac{2}{n}\right), \left(0 + 2\left(\frac{2}{n}\right)\right), \left(0 + 3\left(\frac{2}{n}\right)\right), \dots, \left(0 + (n - 1)\left(\frac{2}{n}\right)\right), \left(0 + n\left(\frac{2}{n}\right)\right)\right\}$$

$$= \left\{0, \left(1\left(\frac{2}{n}\right)\right), \left(2\left(\frac{2}{n}\right)\right), \dots, \left((n - 1)\left(\frac{2}{n}\right)\right), 2\right\}$$

Let $\overline{x}_r = x_r = 0 + r\left(\frac{2}{n}\right) = \frac{2r}{n}$, (i.e., we choose the *right-hand endpoint of each subinterval* as the sample point $\overline{x}_r$, for computing the sum).

Method (I): We have by definition (i.e., the result (B)):

$$\int_a^b f(x)dx = \lim_{n \to \infty} \left(\frac{b-a}{n}\right) \sum_{r=1}^n f\left(a + r\left(\frac{b-a}{n}\right)\right)$$

Here $a=0$, $b=2$, $f(x)=x^2 + 1$, $h = \dfrac{b-a}{n} = \dfrac{2-0}{n} = \dfrac{2}{n}$

$$\therefore \int_0^2 (x^2 + 1)dx = \lim_{n \to \infty} \left(\frac{2}{n}\right) \sum_{r=1}^n f\left(0 + r\left(\frac{2}{n}\right)\right)$$

$$= \lim_{n \to \infty} \frac{2}{n} \left[\sum_{r=1}^n f\left(\frac{2r}{n}\right)\right]$$

$$= \lim_{n \to \infty} \frac{2}{n} \sum_{r=1}^n \left[\left(\frac{2r}{n}\right)^2 + 1\right] \quad \{\because \quad f(x) = x^2 + 1$$

$$= \lim_{n \to \infty} \frac{2}{n} \sum_{r=1}^n \left[\frac{4r^2}{n^2} + 1\right] = \lim_{n \to \infty} \frac{2}{n} \left[\frac{4}{n^2} \sum_{r=1}^n r^2 + \sum_{r=1}^n 1\right]$$

$$= \lim_{n \to \infty} \left[\frac{8}{n^3} \cdot \frac{n(n+1)(2n+1)}{6} + \frac{2}{n} \cdot n\right]$$

$$= \lim_{n \to \infty} \left[\frac{8}{6n^3} \cdot n^3\left(1 + \frac{1}{n}\right)\left(2 + \frac{1}{n}\right) + 2\right] = \left[\frac{4}{3}(1 + 0)(2 + 0) + 2\right] = \frac{14}{3} \quad \text{Ans.}$$

Method (II): Using the second fundamental theorem of integral Calculus.

$$\int_0^2 (x^2 + 1)dx = \left[\frac{x^3}{3} + x\right]_0^2, \text{ where } \left(\frac{x^3}{3} + x\right) \text{ is the antiderivative of } (x^2 + 1)$$

$$= \left[\left(\frac{(2)^3}{3} + 2\right) - 0\right]$$

$$= \frac{8}{3} + 2$$

$$= \frac{14}{3} \quad \text{Ans.}$$

Example (4): Evaluate $\int_0^2 e^x dx$, using the definition of a definite integral as the limit of a sum.

Solution: Divide the interval $[0,2]$ into n *equal parts* so that we get each subinterval of the length $((2 - 0)/n) = 2/n$. Then, the partition formed by the points is given by $P = \{0, (2/n),$ $(4/n), (6/n), \dots, ((2(r-1))/n), 2\}$.

Method (I): Let $\bar{x}_r = x_r = 0 + r\left(\frac{2}{n}\right) = \frac{2r}{n}$

We have by definition (i.e., the result "B")

$$\int_a^b f(x)dx = \lim_{n \to \infty} \left(\frac{b-a}{n}\right) \sum_{r=1}^n f\left(a + r\left(\frac{b-a}{n}\right)\right).$$

Here $a = 0$, $b = 2$, $f(x) = e^x$, $((b-a)/n) = 2/n$

$$\therefore \int_0^2 e^x dx = \lim_{n \to \infty} \frac{2}{n} \sum_{r=1}^n f\left(0 + r\left(\frac{2}{n}\right)\right)$$

$$= \lim_{n \to \infty} \frac{2}{n} \sum_{r=1}^n f\left(\frac{2r}{n}\right)$$

$$= \lim_{n \to \infty} \frac{2}{n} \sum_{r=1}^n e^{2r/n}, \quad [\because \ f(x) = e^x]$$

$$= \lim_{n \to \infty} \frac{2}{n} \left[e^{2/n} + e^{4/n} + e^{6/n} + \cdots + e^{2n/n}\right]$$

The sum in the square bracket is a geometric series with the first term $= e^{2/n}$ *and the common ratio* $(r') = \frac{e^{4/n}}{e^{2/n}} = e^{(4/n - 2/n)} = e^{2/n}$.

$$\therefore \text{This sum } (S_n) = e^{2/n} \left[\frac{1 - (r')^n}{1 - r'}\right]$$

$$= e^{2/n} \left[\frac{1 - (e^{2/n})^n}{1 - e^{2/n}}\right]$$

$$\therefore S_n = \frac{e^{2/n}(1 - e^2)}{1 - e^{2/n}}$$

$$\therefore \int_0^2 e^x dx = \lim_{n \to \infty} \frac{2}{n} \left[\frac{(1 - e^2) \cdot e^{2/n}}{1 - e^{2/n}}\right] = L \quad \text{(say)}$$

Put $(2/n) = t$ on right-hand side and note that as $n \to \infty$, $t \to 0$

We get

$$L = \lim_{t \to 0} t\left[\frac{(1 - e^2)e^t}{1 - e^t}\right] = \lim_{t \to 0} \frac{(1 - e^2)e^t}{(1 - e^t)/t}, \quad \text{[Imp step]}$$

$$= \lim_{t \to 0} \left[\frac{-(e^2 - 1)e^t}{-(e^t - 1)/t}\right]$$

$$= \frac{(e^2 - 1)e^0}{\log_e e} = (e^2 - 1) \cdot 1 \qquad \left[\begin{array}{l} \because \ \lim_{x \to 0} \dfrac{a^x - 1}{x} = \log_e a \\[2mm] \therefore \ \lim_{t \to 0} \dfrac{e^t - 1}{t} = \log_e e = 1 \end{array} \right]$$

$$\therefore \quad \int_0^2 e^x dx = e^2 - 1 \quad \text{Ans.}$$

Method (II): Using the fundamental theorem of integral Calculus:

$$\int_0^2 e^x dx = [e^x]_0^2, \text{ where } e^x \text{ is the antiderivative of } e^x$$

$$= e^2 - e^0$$

$$= (e^2 - 1) \quad \text{Ans.}$$

The four examples are meant to illustrate the theory behind the concept of *definite integrals, as the limit of a sum.* Also we have seen that by applying the second fundamental theorem of Calculus, we can compute, very easily, the exact values of these definite integrals without using Riemann sums. Hence, we now dispense with the terminology of antiderivatives and antidifferentiation and begin to call the expression $\int f(x) dx$ *an indefinite integral*—the term derived from *the definite integral.* Accordingly, the process of evaluating both *an indefinite integral* or the *definite integral* is called *integration.* This is what we had pointed out in Chapter 1 of this volume.

In terms of the symbol for indefinite integrals, we may write the conclusion of the second fundamental theorem as: $\int_a^b f(x) dx = \left[\int f(x) dx \right]_a^b$.

Note that, for applying the second fundamental theorem, *an important requirement* is *to find the indefinite integral* $\int f(x) dx$, by using any suitable technique that we have learnt in previous chapters.

Note: The distinction between *an indefinite integral and the definite integral should be emphasized. The indefinite integral* $\int f(x) dx$ *represents (jointly) all functions whose derivative is f(x).* However, *the definite integral* $\int_a^b f(x) dx$ is a number whose value depends on the *function f and the numbers a and b, and is defined as the limit of a Riemann sum.*

Remark: Note that the definition of the definite integral makes no reference to differentiation.

Important Note: *Integral Calculus* (like the differential Calculus) has important applications in situations where the quantities involved vary. We know that the area of the rectangular region changes if one or both of its dimensions are changed. If we consider the shaded region of Figure 5.3, wherein the height [of the curve $y = f(x)$] varies as we travel across the region from left to right, then the area of the region R changes continuously. This is the type of situation where integral Calculus comes into play. More complicated situations are considered in Chapter 8a of this book.

6a The Fundamental Theorems of Calculus

6a.1 INTRODUCTION

Until now, the limiting processes of the *derivative* and *definite integral* have been considered as *distinct concepts*. We shall now bring these fundamental ideas together and establish the relationship that exists between them. As a result, definite integrals can be evaluated more efficiently.

We have defined the definite integral $\int_a^b f(x)dx$, as the *limit of a sum* and have had some practice in estimating the integral. Calculating definite integrals this way is *always tedious, usually difficult, and sometimes impossible*.[1]

Since evaluation of the *definite integral* $\int_a^b f(x)dx$ has a great variety of important applications, it is highly desirable to have an easy way to compute $\int_a^b f(x)dx$.

The purpose of this section is to develop a *general method* for evaluating $\int_a^b f(x)dx$ that does not necessitate computing various sums. The method will allow us to evaluate many (but not all) of the definite integrals that arise in applications. It turns out that the exact value of $\int_a^b f(x)dx$ can be easily found if we can compute $\int f(x)dx$ [i.e., if we can find *the antiderivative* of $f(x)$].

6a.2 DEFINITE INTEGRALS

In the previous chapter, we evaluated certain definite integrals using two methods: first- *as the limit of a sum* (which is based on the definition of definite integral) and *second- by applying the second fundamental theorem of Calculus*, for which it is only necessary (as we will see shortly) that one should be able to compute $\int f(x)dx$ to evaluate $\int_a^b f(x)dx$.

Having experienced the convenience in estimating exact values of definite integrals by the second method, one should appreciate the elegance and beauty of such easy computations.[2]

The definite integral 6a-The fundamental theorems of the Calculus and their applications (Differentiation and integration as inverse processes and the MVT for integrals)

[1] In fact, we have been able to evaluate a few definite integrals directly from the definition (as the limit of a sum) only because we have nice formulas for $1 + 2 + 3 + \cdots + n$, $1^2 + 2^2 + 3^2 + \cdots + n^2$, and so on.

[2] This method is of great practical importance, since it enables us to calculate not only areas, but also volumes, lengths of curves, centers of mass, moments of inertia, and so on, which are capable of being expressed in the form $\sum_{x=a}^{x=b} f(x)\Delta x$.

Introduction to Integral Calculus: Systematic Studies with Engineering Applications for Beginners, First Edition.
Ulrich L. Rohde, G. C. Jain, Ajay K. Poddar, and A. K. Ghosh.
© 2012 John Wiley & Sons, Inc. Published 2012 by John Wiley & Sons, Inc.

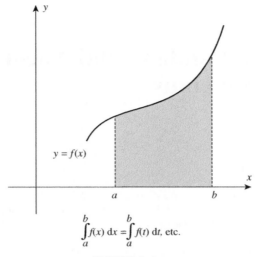

$$\int\limits_a^b f(x)\,dx = \int\limits_a^b f(t)\,dt, \text{ etc.}$$

FIGURE 6a.1

Now, we shall show *why antidifferentiation enables us to find the area under the graph of the function f(x).* The *trick* is to consider *the integral function* $\int_a^x f(x)dx$ (to be discussed shortly). Let $\int_a^x f(x)dx = A(x)$, then we shall show that $A'(x) = f(x)$, and this will remove the mystery. If the *derivative of the integral function* $\int_a^x f(x)dx$ is $f(x)$, then surely to find $A(x)$, we should find *an antiderivative* of $f(x)$ [i.e., we must evaluate $\int f(x)dx$].

To understand the approach of Newton and Leibniz in developing the two theorems, we use the Integrability Theorem, which states that *if f is continuous on [a, b], then* $\int_a^b f(x)dx$ *exists*. We begin our development of these theorems by discussing *definite integrals having a variable upper limit*.

Let the function f be *continuous on the closed interval [a, b]*. Then, the *value of the definite integral $\int_a^b f(x)dx$ depends only* on f and the numbers "*a*" and "*b*", and *not* on the symbol x, used here as the variable of integration. In other words, the definite integrals $\int_a^b f(x)dx$, $\int_a^b f(t)dt$, and $\int_a^b f(u)du$, and so on, represent the same (closed) area from a to b (Figure 6a.1). For the present, we will assume that $a < b$.

Let us now *use the symbol x to represent a number in the closed* interval $[a, b]$. Then, because f is *continuous* on $[a, b]$, *it is continuous* on $[a, x]$ and $\int_a^x f(x)dx$ exists. It represents the area enclosed by the graph of f and the x-axis, from a to x (Figure 6a.2).

Furthermore, *this definite integral is a unique number whose value depends on x*, that is the upper limit of the integral. It is a new function of x (in the form of a definite integral, with a variable upper limit). We call it the area function and denote it by $A(x)$.

In order to use x *as a variable in our discussion*, we write it as an upper limit (of a definite integral) and replace the expression $f(x)dx$ by the expression $f(t)dt$. Thus, we get the definite integral $\int_a^x f(t)dt$ [in place of $\int_a^x f(x)dx$] that clearly indicates the upper limit x, avoiding the confusion with the variable of integration, which is now t.

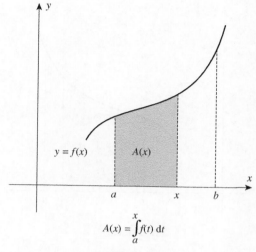

$$A(x) = \int\limits_{a}^{x} f(t)\, dt$$

FIGURE 6a.2

6a.3   THE AREA OF FUNCTION $A(x)$

We have defined $\int_a^b f(x)\mathrm{d}x$ as the area of the region bounded by the curve $y = f(x)$, the ordinates $x = a$ and $x = b$, and the x-axis (Figure 6a.1). Since x is a point in $[a, b]$, $\int_a^x f(x)\mathrm{d}x$ *represents the area of the shaded region in* Figure 6a.2. Here, it is assumed that $f(x) \geq 0$ for $x \in [a, b]$.

Thus, $A(x)$ defines *a function of x which is the variable upper limit of the integral function* $\int_a^x f(x)\mathrm{d}x$, whose domain is all numbers in $[a, b]$ and whose function value *at any number x* in $[a, b]$ is given by

$$A(x) = \int\limits_{a}^{x} f(x)\mathrm{d}x \tag{1}$$

Note that, f is *continuous* on the interval $[a, b]$ and since $f(x) \geq 0$, *its graph does not fall below the x-axis*.

6a.3.1   First Fundamental Theorem of Calculus

From the definition of the area function $A(x)$, we state its two properties immediately:

(i) $A(a) = \int_a^a f(x)\mathrm{d}x = 0$, *since there is no area from a to a.*

(ii) $A(b) = \int_a^b f(x)\mathrm{d}x$, *represents the area from a to b.*

If x is increased by h units, then the area of the shaded region $= A(x + h)$, as shown in Figure 6a.3. Hence, the *difference of areas* in Figures 6a.3 and 6a.2 will be $A(x + h) - A(x)$, as shown by the area of the shaded region in Figure 6a.4.

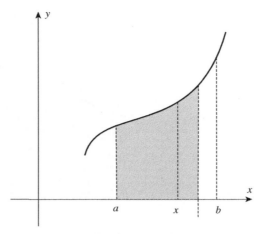

FIGURE 6a.3

The area of the shaded region in Figure 6a.4 is the same as *the area of a rectangle* in Figure 6a.5, whose *base is h and height is some value $\bar{y}$ between f(x) and f(x + h).* Thus, the area of this rectangle is, *on the one hand,* $A(x + h) - A(x)$, and *on the other hand* it is $h \cdot \bar{y}$.

Therefore, we have $A(x + h) - A(x) = h \cdot \bar{y}$ or $(A(x + h) - A(x))/(h) = \bar{y}$.

As $h \to 0$, then $\bar{y}$ *approaches the number f(x)* [as clear from Figure 6a.5], and so,

$$\lim_{h \to 0} \frac{A(x + h) - A(x)}{h} = f(x) \tag{2}$$

But *the left-hand side of (2) is merely the derivative of A(x).* Thus, equation (2) becomes $A'(x) = f(x)$.

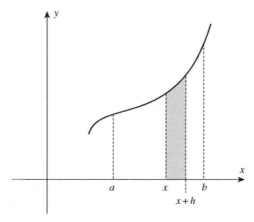

FIGURE 6a.4

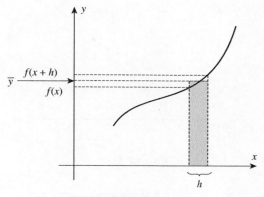

FIGURE 6a.5

Note that equation (2) can also be expressed as

$$A'(x) = \frac{d}{dx} \int_a^x f(x)dx = f(x) \tag{I}$$

This result is a crucial equation and it is so important that it is called the first fundamental theorem of Calculus.

We conclude that the area function $A(x)$ has the additional property of having its derivative $A'(x)$ as $f(x)$. In other words, $A(x)$ is an antiderivative of $f(x)$.

6a.3.2   The Background for the Second Fundamental Theorem

Now, using the first fundamental theorem of Calculus, we try to understand the second fundamental theorem that is useful in evaluating definite integrals. Suppose $\phi(x)$ is *any antiderivative of $f(x)$*, then we have $\phi'(x) = f(x)$.

Since both $A(x)$ and $\phi(x)$ are *antiderivatives of the same function*, we conclude that they must differ by a constant c.

$$\therefore A(x) - \phi(x) = c$$
$$\text{or} \quad A(x) = \phi(x) + c \tag{3}$$

Let us apply the property (i) of $A(x)$ to equation (3).[3]

Since $A(a) = 0$, evaluating both sides of equation (3), when $x = a$, we get

$$0 = \phi(a) + C$$
$$\therefore \quad C = -\phi(a)$$

---

[3] This equation supports our observation that we made, as property (i) of the area function $A(x)$, defined on $[a, b]$, that $A(a) = \int_a^a f(x)dx = 0$

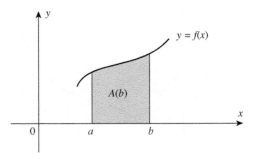

FIGURE 6a.6

Thus, equation (3) becomes

$$A(x) = \phi(x) - \phi(a) \qquad (4)$$

Note that, equation (4) defines *the area function $A(x)$, in terms of an antiderivative $\phi(x)$, of the given function $f(x)$.*

If $x = b$, then from equation (4), we get

$$A(b) = \phi(b) - \phi(a) \qquad (5)^{[4]}$$

But recall that $A(b)$ is the area from a to b (Figure 6a.6).

Since the area of this region can also be obtained *as the limit of a sum*, we can also refer to it by $\int_a^b f(x)\mathrm{d}x$. Hence,

$$A(b) = \int_a^b f(x)\mathrm{d}x \qquad (6)$$

From equations (5) and (6), we get

$$\int_a^b f(x)\mathrm{d}x = \phi(b) - \phi(a) \qquad (7)$$

Equation (7) expresses the relationship between a *definite integral $\int_a^b f(x)\mathrm{d}x$* and the *difference $\phi(b) - \phi(a)$*, where ϕ is an *antiderivative* of $f(x)$. It implies that, to find $\int_a^b f(x)\mathrm{d}x$ it is sufficient to find an antiderivative of $f(x)$ [say $\phi(x)$], and subtract its value at the lower limit "a" from its value at the upper limit "b".

Our result can be stated *more generally* as follows.

---

[4] Recall that $A(b)$ is the area under f from a to b [property (ii) of $A(x)$], that is $A(b) = \int_a^b f(x)\mathrm{d}x = \phi(b) - \phi(a)$, where $\phi(x) = \int f(x)\mathrm{d}x$

6a.3.3   Second Fundamental Theorem of Integral Calculus

If f is *continuous on the interval* $[a, b]$ and $\phi(x)$ is any antiderivative of $f(x)$ there, then

$$\int_a^b f(x)\mathrm{d}x = \phi(b) - \phi(a) \tag{II}^{(5)}$$

Remark: It is crucial that we understand the *distinction* between a *definite* integral and an *indefinite* integral. The *definite integral* $\int_a^b f(x)\mathrm{d}x$ *is a number defined to be the limit of a sum.* The fundamental theorem says that an antiderivative of $f(x)$ [i.e., $\int f(x)\mathrm{d}x$], which is *related to the differentiation process*, and can be used to determine the limit $\int_a^b f(x)\mathrm{d}x$.

Note: The above discussion is very useful for understanding the meaning of the second fundamental theorem of integral Calculus. However, it is not the proof of the theorem. We shall now prove this theorem, using the definition of a definite integral, as the limit of a sum.

6a.4   STATEMENT AND PROOF OF THE SECOND FUNDAMENTAL THEOREM OF CALCULUS

Let f be continuous on $[a, b]$ and let ϕ be *any antiderivative* of f there. Then,

$$\int_a^b f(x)\mathrm{d}x = \phi(b) - \phi(a)$$

Proof: Let P be an *arbitrary partition* of $[a, b]$ given by

$$P = \{a = x_0 < x_1 < x_2 < \cdots < x_{n-1} < x_n = b\}$$

[Note that $\phi(a) = \phi(x_0)$ and $\phi(b) = \phi(x_n)$.]
    Then, by using the standard *"subtract and add" trick*, we get

$$\phi(b) - \phi(a) = \phi(x_n) - \phi(x_{n-1}) + \phi(x_{n-1}) + \cdots + \phi(x_1) - \phi(x_0)$$

$$= \sum_{i=1}^{n} [\phi(x_i) - \phi(x_{i-1})]$$

By the *mean value theorem for first derivatives applied to* ϕ, *on the interval* $[x_{i-1}, x_i]$,[6]
    $\phi(x_i) - \phi(x_{i-1}) = \phi'(\bar{x}_i)(x_i - x_{i-1}) = \phi'(\bar{x}_i)\Delta x_i$ for *some choice* of $\bar{x}_i$ in the open interval (x_{i-1}, x_i).

---

[5] Using this theorem, it is easy to prove that $\int_a^b f(x)\mathrm{d}x = -\int_b^a f(x)\mathrm{d}x$. If $\int_a^b f(x)\mathrm{d}x$ is looked upon as an area under the curve $y = f(x)$ from a to b $\int_b^a f(x)\mathrm{d}x$, then must be considered as the area with the same magnitude, but opposite in sign.
[6] **Note:** We give below, for convenience, the statement of the *MVT for first derivatives*: If f is continuous on a closed interval $[a, b]$ and differentiable on its interior (a, b), then there is at least one point c in (a, b) where $\frac{f(b)-f(a)}{b-a} = f'(c)$ or $f(b) - f(a) = f'(c) \cdot (b - a)$.

Thus,

$$\phi(b) - \phi(a) = \sum_{i=1}^{n} f(\bar{x}_i)\Delta x_i$$

On the left-hand side we have a *constant*; and on the right-hand side we have a *Riemann sum for the function f on* [a, b].

When we take limits of both sides as $\|P\| \to 0$, we obtain

$$\phi(b) - \phi(a) = \lim_{\|P\| \to 0} \sum_{i=1}^{n} f(\bar{x}_i)\Delta x_i$$

$$= \int_{a}^{b} f(x)dx \qquad \text{By definition of the definite integral}$$

$$(\text{Proved})$$

Note (1): *The first fundamental theorem of the Calculus* says that, if a function f is *continuous* on [a, b] and x is an arbitrary point in [a, b], then the definite integral $\int_{a}^{x} f(t)dt$ is a *function of the variable upper limit* x, and its derivation is given by

$$\frac{d}{dx}\left[\int_{a}^{x} f(x)dx\right] = f(x) \qquad \text{(III)}$$

In other words, *the derivative of a definite integral* $\int_{a}^{b} f(t)dt$, with respect to its variable upper limit x, is *the integrand evaluated at the upper limit*. (Note our use of t rather then x as the dummy variable to avoid confusion with the upper limit.)

One *theoretical consequence* of this theorem is that every *continuous function f* has *an antiderivative* ϕ, given by

$$\phi(x) = \int_{a}^{x} f(t)dt$$

In other words, for any continuous function f *we can always write its antiderivative as* $\int_{a}^{x} f(t)dt$. However, this fact is not helpful in getting a nice formula for any particular antiderivative.[7]

6a.5   DIFFERENTIATING A DEFINITE INTEGRAL WITH RESPECT TO A VARIABLE UPPER LIMIT

Now as an application of result (III), we give below a variety of problems involving differentiation of a definite integral function, with respect to a variable upper limit. The result (III) implies that to compute $(d/dx)\int_{a}^{x} f(t)dt$, we do not require to compute $\int f(x)dx$. (*Why?*)

---

[7] Shortly in Chapter 6b, we will show that the special integral function $\int_{1}^{x}\frac{1}{t}dt$, $(x > 0)$, defines the function In x (i.e., $\log_e x$), and this further indicates the power of Calculus.

Example (1): Find $\frac{d}{dx}\left[\int_1^x t^2\,dt\right]$

Solution: We can find this derivative in *two ways*.

(a) *Hard Way:* that is, by evaluating the integral and then taking the derivative.

$$\int_1^x t^2\,dt = \left[\frac{t^3}{3}\right]_1^x = \frac{x^3}{3} - \frac{1}{3}$$

$$\therefore \quad \frac{d}{dx}\left[\int_1^x t^2\,dt\right] = \frac{d}{dx}\left[\frac{x^3}{3} - \frac{1}{3}\right] = x^2$$

(b) *Easy Way:* By the result (I)

$$\frac{d}{dx}\left[\int_1^x t^2\,dt\right] = x^2$$

Example (2): Find $\frac{d}{dx}\left[\int_2^x \frac{t^{3/2}}{\sqrt{t^2+17}}\,dt\right]$

Solution: Note that in this example, it is not possible to find the indefinite integral of $(t^{3/2}/\sqrt{t^2+17})$ (why?). Therefore, we cannot solve this problem by first evaluating the integral. However, by using the result (I), it is a trivial problem.

$$\frac{d}{dx}\left[\int_2^x \frac{t^{3/2}}{\sqrt{t^2+17}}\,dt\right] = \frac{x^{3/2}}{\sqrt{x^2+17}} \qquad \text{Ans.}$$

Example (3): Find $\frac{d}{dx}\left[\int_x^4 \tan^2 t \cos t\,dt\right]$

Solution: *Observe that in this problem x is the lower limit, rather than the upper limit.* We handle this difficulty as follows:

$$\frac{d}{dx}\left[\int_x^4 \tan^2 t \cos t\,dt\right] = \frac{d}{dx}\left[-\int_4^x \tan^2 t \cos t\,dt\right]$$

$$= -\frac{d}{dx}\left[\int_4^x \tan^2 t \cos t\,dt\right]$$

$$= -\tan^2 x \cdot \cos x \qquad \text{Ans.}$$

Example (4): Find $\frac{d}{dx}\left[\int_0^{x^2} (3t - 1)\,dt\right]$

Solution: Now, we have a new complication; the upper limit is x^2 *rather than* x. In order to apply the result (I), we need x there. This problem is handled by the *Chain Rule*. We may think of the *expression in brackets as*

$$\int_0^u (3t - 1)dt, \quad \text{where } u = x^2$$

By Chain Rule, the derivative of this composite function, with respect to x is

$$\frac{d}{du} \left[\int_0^u (3t - 1)dt \right] \cdot \frac{d}{dx}(u) = (3u - 1)2x$$
$$= (3x^2 - 1)(2x) = 6x^3 - 2x \qquad \text{Ans.}$$

Example (5): Find $\frac{d}{dx} \left[\int_{2x}^5 \sqrt{t^2 + 2t} \, dt \right]$

Solution: Here, we first interchange limits and then use the result (I) in conjunction with the *Chain Rule*.

$$\frac{d}{dx} \left[\int_{2x}^5 \sqrt{t^2 + 2} \, dt \right] = \frac{d}{dx} \left[- \int_5^{2x} \sqrt{t^2 + 2} \, dt \right]$$
$$= -\sqrt{(2x)^2 + 2} \cdot \frac{d}{dx}(2x)$$
$$= -\sqrt{4x^2 + 2} \cdot (2) = -2\sqrt{4x^2 + 2} \qquad \text{Ans.}$$

6a.5.1   Differentiation and Integration as Inverse Processes

The first fundamental theorem of Calculus tells that

$$\frac{d}{dx} \left[\int_a^x f(t)dt \right] = f(x) \qquad \qquad \text{(IV)}$$

where $\int f(t)dt = \phi(t)$.[8]

   This formula tells us that if we start with a continuous function f, integrate it to obtain $\int_a^x f(t)dt$, and *then* differentiate, the result is the *original function f*. Thus, *the differentiation has nullified the integration*. On the other hand, if we start with a function F (having a *continuous derivative*), first differentiate F to get F', and then integrate $F'(x)$ (from a to x), we obtain $\int_a^x F'(t)dt$.

---

[8] Note that $\frac{d}{dx} \left[\int_a^x f(t)dt \right] = \frac{d}{dx}[\phi(x) - \phi(a)] = \phi'(x) = f(x)$

But, by the second fundamental theorem of Calculus,

$$\int_a^x F'(t)dt = F(x) - F(a) \tag{V}$$

So we obtain the original function F *altered by at most a constant* $[F(a)]$.

This time the integration has essentially nullified the differentiation.

Thus, the two basic processes of Calculus, *differentiation and integration, are inverses of each other.* Furthermore, whenever *we know the derivative F' of a function F*, (II) *gives an integration formula.*

For example, we know already that

$$\frac{d}{dx} \sin x = \cos x$$

Therefore, (V) tells us that

$$\int_{\pi/4}^x \cos t\, dt = \sin x - \sin\frac{\pi}{4}$$

$$= \sin x - \frac{1}{\sqrt{2}} = \sin x - \frac{\sqrt{2}}{2}$$

Similarly, we have

$$\frac{d}{dx} \tan x = \sec^2 x$$

$$\therefore \quad \int_{\pi/4}^x \sec^2 t\, dt = \tan x - \tan\frac{\pi}{4} = \tan x - 1$$

In physics, the *velocity* of a particle moving along a *straight line* is *the derivative* of the *position function*. If we use

t for the *independent variable* representing time,

f for the *position function*,

v for *velocity*, and

s for *variable of integration*

Then we can write

$$f(t) - f(t_0) = \int_{t_0}^t v(s)ds \tag{L}$$

In (L) the number t_0 is *arbitrary*, and it plays the same role as "a" in (V) above. In applications, t_0 is usually a special instant of time. *When t_0 is the moment at which motion begins*, it is called the *initial time*.

The *acceleration* "*a*" of a particle is the *derivative of the velocity*. Hence, we obtain

$$v(t) - v(t_0) = \int_{t_0}^{t} a(s)\mathrm{d}s \tag{M}$$

6a.5.2   The Mean Value Theorem for Definite Integrals (or Simply Integrals)

We have already studied the *mean value theorem (MVT) for derivatives* in Chapter 20 of Part I, and observed that it plays an important role in Calculus. There is a theorem by the same name for *integrals*. Although, the MVT for definite integrals is not as attractive as the MVT for derivatives (due to its fewer applications), it is still worth knowing. Geometrically, it says that (in Figure 6a.7), *the area under the curve is equal to the area of the shaded rectangle*, where "*c*" is some number, in [*a*, *b*].

6a.5.2.1   Theorem: (Mean Value Theorem for Integrals)   If *f* is *continuous* on [*a*, *b*], there is a number between *a* and *b*, such that

$$\int_{a}^{b} f(t)\mathrm{d}t = f(c)(b - a)$$

Proof: Let

$$F(x) = \int_{a}^{x} f(t)\mathrm{d}t \quad a \le x \le b \tag{A}$$

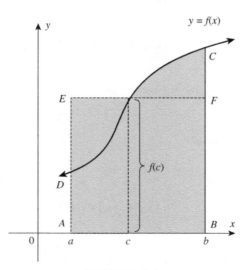

FIGURE 6a.7

Then, we can say that F is *differentiable at each* $x \in [a, b]$, and that

$$F'(x) = f(x) \tag{B}$$

(From the first fundamental theorem of Calculus.)

By the *mean value theorem for derivatives*, applied to F, there is a point c in (a, b) such that

$$F(b) - F(a) = F'(c)(b - a) \tag{C}$$

that is,

$$\int_a^b f(t)\mathrm{d}t - 0 = F'(c)(b - a) \tag{D}$$

$$\left[\because \quad F(b) = \int_a^b f(t)\mathrm{d}t \quad \text{and} \quad F(a) = \int_a^a f(t)\mathrm{d}t = 0 \right]$$

But by (A), we have $F'(c) = f(c)$.

Using this relation in (D), we get

$$\int_a^b f(t)\mathrm{d}t - 0 = f(c)(b - a) \qquad \text{(Proved)} \tag{E}$$

Note that, if we solve (E) for $f(c)$, we get

$$f(c) = \frac{\int_a^b f(t)\mathrm{d}t}{b - a}$$

The number $(\int_a^b f(t)\mathrm{d}t)/(b - a)$ is called the *mean value* (or average value) of f on $[a, b]$. *To see why it has this name, consider a regular partition*

$$P : \quad x_0 < x_1 < x_2 < \cdots < x_n = b$$

with $\Delta x = (b - a)/n$.

The average of the n values $f(x_1)f(x_2)\ldots f(x_n)$ is

$$\frac{f(x_1) + f(x_2) + \cdots + f(x_n)}{n} = \sum_{i=1}^n f(x_i)\frac{1}{n}$$

$$= \frac{1}{(b - a)} \sum_{i=1}^n f(x_i)\frac{(b - a)}{n}$$

$$= \frac{1}{(b - a)} \sum_{i=1}^n f(x_i) \Delta x$$

The sum in the last expression is a *Riemann Sum* for f on $[a, b]$, and therefore approaches $\int_a^b f(x)dx$ as $n \to \infty$.

Thus, $\left(\int_a^b f(x)dx\right)/(b - a)$ appears as the *natural extension* of the *familiar notion of average value*. (It is also called the *arithmetic mean value of a function*.) We now define *the arithmetic mean value of a function*.

6a.5.2.2   The Arithmetic Mean Value of a Function

Definition: The arithmetic mean value (or simply, the mean) y_m of a *continuous function* $y = f(x)$ on the interval $[a, b]$ is the ratio of *the definite integral* of this function (from a to b) to the length of the interval:

$$y_m = \frac{\int_a^b f(x)dx}{b - a}$$

(Justification to this definition is already given in the MVT for integral in Section 6a.5.2.1.)

Now, we shall prove the following theorem using the *mean value theorem for integrals*.

6a.5.2.3   Theorem: An Integral with Variable Upper Limit is an Antiderivative of its Integrand   Note that this is the statement of *the first fundamental theorem of Calculus*.[9]

Proof: Given an integral function $I(x) = \int_a^x f(x)dx$, with variable upper limit x.

$$\text{To prove} \quad \left[\int_a^x f(x)dx\right]' = f(x) \tag{VI}$$

Let us give an increment Δx to x. Then, the new value of the function $I(x)$ is given by

$$I(x + \Delta x) = \int_a^{x+\Delta x} f(x)dx$$

Let,

$$\Delta I = I(x + \Delta x) - I(x)$$

$$= \int_a^{x+\Delta x} f(x)dx - \int_a^x f(x)dx$$

Now, expressing the first integral on the right-hand side by breaking the integral, we get

$$\Delta I = \int_a^x f(x)dx + \int_x^{x+\Delta x} f(x)dx - \int_a^x f(x)dx$$

$$= \int_x^{x+\Delta x} f(x)dx$$

---

[9] We have already proved the theorem, using the definition of area function $A(x) = \int_a^x f(t)dt$. Now, we shall prove it, simply by using the definition of the definite integral and the mean value theorem for integrals.

According to *the mean value theorem for integrals*, the above integral is presentable as

$$\Delta I = f(\xi)(\overline{x + \Delta x} - x) = f(\xi)\Delta x \tag{VII}$$

where ξ is a point lying between x and $(x + \Delta x)$

By the definition of the derivative, by using (VII) we have

$$\lim_{\Delta x \to 0} \frac{\Delta I}{\Delta x} = \lim_{\Delta x \to 0} \frac{f(\xi)\Delta x}{\Delta x}$$

$$= \lim_{\Delta x \to 0} f(\xi)$$

But if $\Delta x \to 0$, the $x + \Delta x$ tends to x, and therefore $\xi \to x$, which implies, since $f(x)$ is a continuous function, that

$$\lim_{\Delta x \to 0} f(\xi) = \lim_{\xi \to x} f(\xi) = f(x) \tag{VIII}$$

since f is a continuous function. *This is what we wanted to prove.*

6a.5.3   Geometrical Interpretation of MVT

From the theorem $(d/dx) \int_a^x f(t)dt = f(x)$, it follows that

$$d \int_a^x f(t)dt = f(x)dx$$

Now, we can give the geometrical interpretations of the above statements of the theorem.

[The geometrical interpretation of the theorem $(d/dx) \int_a^b f(t)dt = f(x)$ was discussed in Section 6a.3.1.] *Here we explain it once more.*

$$\text{Let } I(x) = \int_a^x f(t)dt$$

The function $I(x)$ expresses the *variable area* of the *curvilinear trapezoid* with variable base $[a, x]$ bounded by the curve $y = f(x)$.[10]

The assertion of the theorem implies that the derivative of the area of the trapezoid with respect to the abscissa x *is equal to the ordinate of the line segment AB* $[=f(x)]$ *bounding the trapezoid*, which represents the height of the graph at x. In other words, *the differential of the area of the trapezoid is equal to the area of the rectangle ABDE with sides equal, respectively, to the base of the trapezoid and to the ordinate of the line $y = f(x)$ at the point x* (Figure 6a.8).

Note (2): The practical importance of the *integral Calculus* as well as of *differential Calculus, lies in its ability to handle situations in which quantities are varying continuously.*

---

[10] Note that as x varies in $[a, x]$, the area $I(x)$ varies.

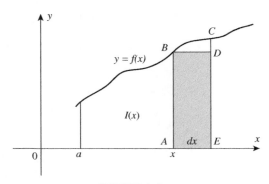

FIGURE 6a.8

This can easily be visualized by considering the graph of a *continuous function* $y = f(x)$, with $f(x) > 0$, on a closed interval $[a, b]$. The graph of any such function obviously lies above the x-axis. If it is different from a straight line, then the slope of the tangent line to the curve (which is the derivative of f) varies continuously with $x \in [a, b]$, and for any value of x (say x_1), it can be easily computed using differential Calculus (Figure 6a.9).

Also, *the area of the curved region below the graph of $y = f(x)$ and above the x-axis* (i.e., the definite integral of f) from a to x ($a \leq x \leq b$), varies continuously with $x \in [a, b]$ and is given by $A(x) = \int_a^x f(x) dx$, and for any value x (say x_1), it can be easily computed using integral Calculus (Figure 6a.10).

Remark: Because of the *connection between definite integrals* and *antiderivatives*, it is logical to use the integral sign "$\int$" in the notation $\int f(x) dx$ for an antiderivative. Now it must also be clear that we may dispense with *the terminology of derivatives and antidifferentiation*, and begin to call the expression $\int f(x) dx$ as *an indefinite integral* (which is the term derived from the concept of the definite integral). Once this idea is clear, the process of evaluating an

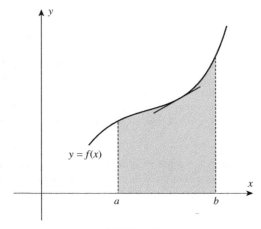

FIGURE 6a.9

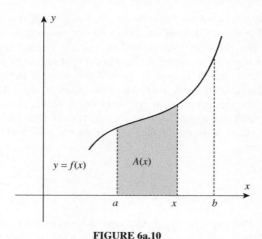

FIGURE 6a.10

indefinite integral $\int f(x)\mathrm{d}x$ and the definite integral $\int_a^b f(x)\mathrm{d}x$ are identical from computation point of view. Of course, the two concepts are entirely different. It is for this reason that *the process of evaluating an indefinite integral* $\left[\int f(x)\mathrm{d}x\right]$ or the *definite integral* $\left[\int_a^b f(x)\mathrm{d}x\right]$ is called *integration*.

The distinction between an indefinite integral and a definite integral should be emphasized.

(i) The *indefinite integral* $\int f(x)\mathrm{d}x$ represents *all functions whose derivative is f(x)*. However, the definite integral $[\int_a^b f(x)\mathrm{d}x]$ *is a number whose value depends on the function f and the numbers a and b*, and it is defined as the limit of a *Riemann Sum*.

Remark: We emphasize that the definition of the definite integral makes no reference to differentiation.

Thus, we can differentiate both an indefinite integral and the definite integral in the form $\int_a^x f(t)\mathrm{d}t$. Of course, $(\mathrm{d}/\mathrm{d}x) \int_a^b f(t)\mathrm{d}t = 0$, always [*Why?*].

(ii) The indefinite integral involves *an arbitrary constant c*.

For instance $\int x^2 \, \mathrm{d}x = \frac{x^3}{3} + c$

The arbitrary constant c, as we know, is called *constant of integration*.

In applying *the second fundamental theorem of Calculus* to evaluate a definite integral, we do not need to include the arbitrary constant c in the expression $g(x)$, *because the theorem permits us to select any antiderivative, including the one for which* $c = 0$.

Note (3): The *second fundamental theorem of Calculus* provides a key method to find *the definite integral* with the aid of *antiderivatives*. It links the process of *integration* with that of *differentiation*. For this reason, some authors call this theorem the *fundamental theorem of integral Calculus*. Some authors also call it the *Newton–Leibniz Theorem*, because they were the first to establish a relationship between *integration* and *differentiation*, thus making possible the rule for evaluating definite integrals, *avoiding summation.*

It is only after this theorem was established, that the definite integral acquired its present significance. It greatly expanded the field of the applications of the definite integral, because mathematics obtained a general method for solving various problems of a particular type, and so could considerably extend the range of applications of the definite integrals to technology, mechanics, astronomy, and so on. This is better appreciated when the theorem is applied to compute not only *area*s, but also quantities like *volumes, length of curves, centers of mass, moments of inertia,* and so on, which are capable of being expressed in the form

$$\sum_{x=a}^{x=b} f(x)\,\Delta x \ ^{(11)}$$

Archimedes (287–212 BC) must be regarded as one of the greatest mathematicians of recorded history, for this work alone, which was nearly 2000 years ahead of his time. [Refer to Chapter 5, Section 5.1.1.]

When Newton (1642–1727) and Leibniz (1616–1716) appeared on the scene, it was the natural time for Calculus to be developed, as evidenced by their simultaneous, but independent achievements in the field.

Newton and Leibniz had the analytic geometry of Fermat (1601–1665) and Descartes (1596–1650) on which to build Calculus. This was not available to Archimedes.

---

[11] Differentiation and integration arose from apparently unrelated problems of geometry. The problem of the tangent line led us to derivatives and the problem of area to integration. It was only after mathematicians had worked for centuries with derivatives and integrals separately, that Isaac Barrow (1630–1677), who was Newton's teacher, discovered and proved the Fundamental Theorem of Calculus. His proof was completely geometric, and his terminology far different from ours. Beginning with the work of Newton and Leibniz, the theorem grew in importance, eventually becoming the cornerstone for the study of integration.*Calculus with Analytic Geometry* (Alternate Edition) by Robert Ellis and Denny Gulick (p. 263), HBJ Publishes, USA, 1988.

6b The Integral Function $\int_1^x \frac{1}{t}\, dt,\, (x > 0)$ Identified as $\ln x$ or $\log_e x$

6b.1 INTRODUCTION

The definition of the logarithmic function that we encountered in algebra was *based on exponents*, and the properties of logarithms were then proved from corresponding properties of exponents. It is useful to review these properties and revise how we learned them in our algebra course.

One property of exponents is

$$a^x \cdot a^y = a^{x+y} \tag{1}$$

Let us discuss the following cases:

(i) If the exponents x and y are *positive integers* and if *a is any real number*, then (1) *follows from the definition of positive integer exponent and mathematical induction.*

(ii) If the exponents are allowed to be *any integer*, either *positive, negative*, or *zero*, and $a \neq 0$, then (1) *will hold if zero exponent and negative integer exponent are defined* by

$$a^0 = 1$$

and

$$a^{-n} = \frac{1}{a^n}, \quad n > 0 \tag{2}$$

(iii) If the exponents are rational numbers and $a \geq 0$, then (1) holds when $a^{m/n}$ is defined by

$$a^{m/n} = \sqrt[n]{a^m} \tag{3}$$

(iv) *It is not quite so simple to define a^x when x is an irrational number.*

For example, what is meant by $a^{\sqrt{2}}$? Stated simply, we use an approximation method. First, $a^{\sqrt{2}}$ is approximately $a^{1.4} = a^{7/5} = \sqrt[5]{a^7}$, which is defined.

6b-The logarithm defined using Calculus. The integral function $\int_1^x \frac{1}{t}\, dt,\, (x > 0)$ identified as natural logarithm $\ln x$ or $\log_e x$ and the definition of natural exponential function $\exp(x)$ or e^x as inverse of $\ln x$.

Introduction to Integral Calculus: Systematic Studies with Engineering Applications for Beginners, First Edition. Ulrich L. Rohde, G. C. Jain, Ajay K. Poddar, and A. K. Ghosh.
© 2012 John Wiley & Sons, Inc. Published 2012 by John Wiley & Sons, Inc.

Better approximations are $a^{1.41} = \sqrt[100]{a^{141}}$ and $a^{1.414}$. In this way, the meaning of $a^{\sqrt{2}}$ becomes clear.

Based on the assumption that a^x exists if a is any positive number and x is any real number, we agreed to write the equation

$$a^x = N \tag{4}$$

where *a is any positive number except 1 and N is any positive number*. The definition of the logarithmic function was then based on the above equation, which can be solved for x, and *x is uniquely determined by*

$$x = \log_a N \tag{5}$$

From this *definition (of the logarithmic function) and properties of exponents*, the following properties of logarithms were proved

$$\log_a 1 = 0 \tag{I}$$

$$\log_a m \cdot n = \log_a m + \log_a n \tag{II}$$

$$\log_a \frac{m}{n} = \log_a m - \log_a n \tag{III}$$

$$\log_a m^n = n \log_a m \tag{IV}$$

$$\log_a a = 1 \tag{V}$$

The power of Calculus, both *that of derivatives and integrals has been amply demonstrated. Shortly, we shall be defining the logarithmic function as an integral*, indicating one more application of Calculus.

We begin by observing a *peculiar gap* in our knowledge of derivatives.

$$\frac{d}{dx}\left(\frac{x^3}{3}\right) = x^2, \quad \frac{d}{dx}\left(\frac{x^2}{2}\right) = x^1, \quad \frac{d}{dx}(x) = 1 = x^0, \quad \frac{d}{dx}(???) = x^{-1},$$

$$\frac{d}{dx}(-x^{-1}) = x^{-2}, \quad \frac{d}{dx}\left(-\frac{x^{-2}}{2}\right) = x^{-3}$$

Here we ask the question: *Is there no function whose derivative is $(1/x)$? Alternatively, is there no function that equals $\int(1/x)dx$? We will most certainly reach this conclusion if we restrict our attention to the functions studied so far.* However, we are about to launch the process of defining a new function to fill the gap noticed above.

For the time being, we just accept the fact that we are going to define a new function and study its properties.

Recall the formula $\int x^n \, dx = ((x^{n+1})/(n+1)) + c$, $n \neq -1$. *This formula does not hold when $n = -1$.* To evaluate $\int x^n dx$ for $n = -1$, we need a function whose derivative is $1/x$.

In other words, to evaluate $\int(1/x)dx$, we must obtain a function $\phi(x)$ such that $(d/dx)\phi(x) = (1/x)$. Then, obviously, we can say that $\int(1/x)dx = \phi(x)$. However, we are neither aware of any such function nor are we able to guess it. Of course, in Chapter 13a of Part I,

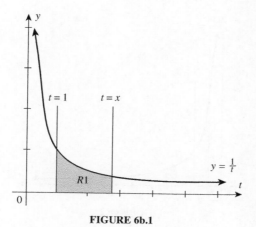

FIGURE 6b.1

we have indicated that $(d/dx)\log_e x = (1/x)$. Now the question is: *Can we obtain this function by any other method and establish that "$\log_e x$" is the only function whose derivative is $1/x$?* In this chapter, we shall prove this.

The first fundamental theorem of Calculus (discussed in Chapter 6a) gives us one such method. It is (the definite integral)

$$\int_a^x \frac{1}{t}\,dt$$

where "a" *can be any real number* having the *same sign* as $x^{(1)}$

To interpret such a function, we consider the special case of this function denoted by

$$\ln x \int_1^x \frac{1}{t}\,dt \tag{i}$$

Let R_1 be the region bounded by the curve $y = 1/t$, by the t-axis, *on the left by the line $t = 1$,* and *on the right by the line $t = x$,* where $x > 1$. This region R_1 is shown in Figure 6b.1.

The measure of the area of R_1 is a function of x; call it $A(x)$ and *define it as a definite integral* as given by

$$A(x) \int_1^x \frac{1}{t}\,dt$$

Now, *consider this integral* if $0 < x < 1$. *It can be easily shown* (using the second fundamental theorem of Calculus) that

$$\int_1^x \frac{1}{t}\,dt = -\int_x^1 \frac{1}{t}\,dt$$

---

[1] Now, what remains is to assign a name to this function.

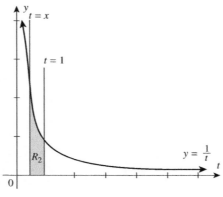

FIGURE 6b.2

From the above, it follows that the integral $\int_1^x (1/t)dt$ represents the measure of the area of region R_2, *bounded by the curve* $y = 1/t$, and the t-axis, on the left by line $t = x$, and on the right by the line $t = 1$.[2]

Thus, the integral $\int_1^x (1/t)dt$ is the negative of the measure of the area of the region R_2 shown in Figure 6b.2. If $x = 1$, the integral $\int_1^x (1/t)dt$ becomes $\int_1^1 (1/t)dt$, which equals zero. (By definition, see Chapter 5, Section 5.5.4.)

In this case, the left and right boundaries of the region are the same and so the measure of the area is 0. (This fact can also be proved using the second fundamental theorem of Calculus.)

Thus, the *integral* $\int_1^x (1/t)dt$, for $x > 0$ *can be interpreted in terms of the measure of the area of a region*. The value of this integral depends on (the upper limit x) and is used to define the *natural logarithmic function*, denoted by $\ln x$. We write,

$$\ln x = \int_1^x \frac{1}{t}dt, \quad x > 0$$

(Though, we have given the name natural logarithmic function to $\ln x$, we have to justify this name. For this purpose, we must check whether $\ln x$ satisfies all the properties of logarithms, as listed in Section 6b.1.1.)

6b.2 DEFINITION OF NATURAL LOGARITHMIC FUNCTION

The *natural logarithmic function* is defined by

$$\ln x = \int_1^x \frac{1}{t}dt, \quad x > 0$$

The *domain of the natural logarithmic function is the set of all positive numbers*. We read "$\ln x$" as "*the natural logarithm of x*".

---

[2] Note that, here x varies in the interval $(0, x)$, from 1 to 0. The area to the left of the ordinate $t = 1$ is taken as negative (so that, it is represented by a negative number) and that to the right of line $t = 1$, is taken as positive.

6b.3   THE CALCULUS OF $\ln x$

From the *first fundamental theorem of Calculus*, we have

$$\frac{d}{dx}(\ln x) = \frac{d}{dx}\left[\int_1^x \frac{1}{t}dt\right] = \frac{1}{x}$$

Thus, $\ln x$ is a differentiable function with

$$\frac{d}{dx}(\ln x) = \frac{1}{x} \tag{ii}$$

Since $(1/x) > 0$ for $x > 0$, *we also conclude that* $\ln x$ *is strictly increasing.*

We know that *corresponding to each differentiation formula, there is an integration formula.* Thus from (ii), we can write

$$\int \frac{1}{x}dx = \ln x + C \tag{iii}$$

which is the indefinite integral form of (ii).

[It must be clearly understood that, whereas the function $\ln x$ is defined by the definite integral $\int_1^x (1/t)dt$, $x > 0$, the indefinite integral form of $\ln x$ is given by equation (iii).]

From the result (ii) and the chain rule, we have the following theorem

Theorem (A):   If u is a differentiable function of x and $u(x) > 0$, then

$$\frac{d}{dx}(\ln u) = \frac{1}{u}\frac{d}{dx}(u)$$

Example (1):   Find $f'(x)$ if $f(x) = \ln(5x^2 - 2x + 7)$

Solution: From Theorem (A),

$$f'(x) = \frac{1}{5x^2 - 2x + 7}(10x - 2)$$

$$= \frac{10x - 2}{5x^2 - 2x + 7} \qquad \text{Ans.}$$

Example (2):   Find $f'(x)$ if $f(x) = \ln(\sin x)$, $\sin x > 0$

Solution: Using Theorem (A), we get

$$\frac{d}{dx}[\ln(\sin x)] = \frac{1}{\sin x} \cdot \frac{d}{dx}(\sin x)$$

$$= \frac{\cos x}{\sin x} = \cot x, \quad \sin x > 0 \qquad \text{Ans.}$$

Example (3):   Show that $\dfrac{d}{dx}\ln|x| = \dfrac{1}{x}$, $\quad x \neq 0$

Solution: Here we have to consider two cases.

(a) If $x > 0$, $|x| = x$,   and so

$$\frac{d}{dx}(\ln|x|) = \frac{d}{dx}(\ln x) = \frac{1}{x}$$

(b) If $x < 0$, $|x| = -x$,   and so

$$\frac{d}{dx}(\ln|x|) = \frac{d}{dx}(\ln(-x)) = \frac{1}{-x} \cdot \frac{d}{dx}(-x) \quad \text{[Note that for } x < 0, \ -x > 0]$$

$$= \left(\frac{1}{-x}\right)(-1) = \frac{1}{x} \qquad \text{Ans.}$$

Example (3) tells that

$$\int \frac{1}{x}dx = \ln|x| + c, \quad x \neq 0$$

This result fills the long-standing gap in the power rule for integration. If r is any rational number then, we can now write

$$\int u^r \, du = \begin{cases} \dfrac{u^{r+1}}{r+1} + c, & \text{if } r \neq -1 \\[2mm] \ln|u| + c, & \text{if } r = -1 \end{cases}$$

Example (4): Find $\dfrac{d}{dx}\left[\ln(x^2 - x - 2)\right]$

Solution: *This problem makes sense,* provided $x^2 - x - 2 > 0$.

Now $x^2 - x - 2 = (x + 1)(x - 2)$, which is positive provided *both the factors* are either negative or positive. This condition is satisfied provided $x < -1$ or $x > 2$. Thus, the domain of $\ln(x^2 - x - 2)$ is $(-\infty, -1) \ U(2, \infty)$. *On this domain,*

$$\frac{d}{dx}\left[\ln(x^2 - x - 2)\right] = \frac{2x - 1}{x^2 - x - 2} \qquad \text{Ans.}$$

Now we show that the natural logarithmic function obeys the properties of logarithms that we learnt in algebra, and listed in Section 6b.1. [This will also justify the name given to the function $\int_1^x (1/t)dt$, $(x > 0)$.]

Now, it is proposed to prove the following theorem, which gives the properties of $\ln x$ (except the property $\log_a a = 1$).

Theorem (B): If a and b are *positive* numbers and r is any *rational* number, then

(I) $\ln 1 = 0$

(II) $\ln(a \cdot b) = \ln a + \ln b$

(III) $\ln\left(\frac{a}{b}\right) = \ln a - \ln b$

(IV) $\ln(a^r) = r(\ln a)$

These relationships make $\ln x$ a very important function. These properties of $\ln x$ support our use of the name "logarithm" for the function $\ln x$.[3]

Let us prove these properties.

---

[3] Of course, we have not yet talked about the base of the natural logarithm.

Proof: **(I)** To prove: $\ln 1 = 0$

From the definition of $\ln x$, we have $\ln x = \int_1^x \frac{1}{t} dt$

$$\therefore \quad \ln 1 = \int_1^1 \frac{1}{t} dt = 0$$

We now prove that the natural logarithm of the product of two positive numbers is the sum of their natural logarithms.

(II) If a and b are two *positive numbers* then

$$\ln(a \cdot b) = \ln a + \ln b$$

Proof: Consider the function $f(x) = \ln(ax)$

For $x > 0$ (and $a > 0$, given), we have

$$\frac{d}{dx}(\ln ax) = \frac{1}{ax} \cdot \frac{d}{dx}(ax), \qquad \text{[By Theorem (A)]}$$

$$= \frac{1}{ax} \cdot a = \frac{1}{x} \quad \text{and} \quad \frac{d}{dx}(\ln x) = \frac{1}{x}$$

The derivatives of $\ln(ax)$ and $\ln x$ are therefore equal. Thus, their values differ only by a constant, that is, $\ln ax = \ln x + C$

To evaluate C, we put $x = 1$, which gives $\ln a = C$ (since $\ln 1 = 0$).

Thus, $\ln ax = \ln x + \ln a$

Now, put $x = b$, we get $\ln(a \cdot b) = \ln a + \ln b$.   (Proved)

(III) If a and b are two positive numbers then

$$\ln\left(\frac{a}{b}\right) = \ln a - \ln b$$

Proof: We have, shown above that

$$\ln(a \cdot b) = \ln a + \ln b$$

Replacing a by $1/b$, we get on the right-hand side

$$\ln\frac{1}{b} + \ln b = \ln\left(\frac{1}{b} \cdot b\right), \qquad [\because \quad \ln a + \ln b = \ln(a \cdot b)]$$

$$= \ln 1 = 0 \quad \text{Thus we get} \quad \ln\frac{1}{b} = -\ln b$$

Now consider, $\ln\left(\frac{a}{b}\right) = \ln\left(a \cdot \frac{1}{b}\right) = \ln a + \ln\frac{1}{b}$

$$= \ln a - \ln b \qquad \text{(Proved)}$$

(IV) If a and b are two positive numbers and r is any *rational number*, then

$$\ln(a^r) = r(\ln a)$$

Proof: For $x > 0$, we have

$$\frac{d}{dx}(\ln x^r) = \frac{1}{x^r} \cdot \frac{d}{dx}(x^r) \quad \text{[By Theorem (A)]}$$

$$= \frac{1}{x^r} \cdot r \cdot x^{r-1} \tag{iv}$$

$$= \frac{r}{x}$$

and

$$\frac{d}{dx}(r \ln x) = r \cdot \frac{1}{x}$$
$$= \frac{r}{x} \qquad \text{(v)}$$

Thus, *derivatives of $\ln(a^r)$ and $r(\ln x)$ are equal.* It follows that, $\ln(a^r)$ and $r(\ln x)$ must differ only by a constant C.

$$\therefore \quad \ln x^r = r \ln x + C \qquad \text{(vi)}$$

To find C, we put $x = 1$ and get

$$\ln 1^r = r \ln 1 + C$$

But $\ln 1 = 0$; hence $C = 0$.

Replacing C by 0, we get from (vi) above

$$\ln x^r = r \ln x \qquad \text{(Proved)}$$

The properties of the natural logarithmic function can be used to simplify the work involved in differentiating complicated expressions involving products, quotients, and powers as can be seen from the following solved examples.

Example (5): Let us differentiate $\ln\left(\sqrt{(x^2 + 1)(2x + 3)}\right)$

Solution: $\dfrac{d}{dx}\left[\ln\left(\sqrt{(x^2 + 1)(2x + 3)}\right)\right]$

$$= \frac{d}{dx}\left[\frac{1}{2}\left\{\ln(x^2 + 1) + \ln(2x + 3)\right\}\right]$$

$$= \frac{1}{2}\left[\left(\frac{1}{x^2 + 1} \cdot 2x\right) + \frac{1}{2x + 3} \cdot 2\right]$$

$$= \frac{x}{x^2 + 1} + \frac{1}{2x + 3}, \quad x > -\frac{3}{2} \qquad \text{Ans.}$$

[Note that the denominator in the second term will not be zero if $x > (-(3/2))$.]

Example (6): Find $\dfrac{dy}{dx}$ if $y = \ln\sqrt[3]{(x - 1)/x^2}$, $\quad x > 1$

Solution: Our problem becomes easier if we first use the properties of natural logarithm to simplify y.

$$y = \ln\left(\frac{x - 1}{x^2}\right)^{1/3} = \frac{1}{3}\ln\left(\frac{x - 1}{x^2}\right)$$

$$= \frac{1}{3}\left[\ln(x - 1) - \ln x^2\right]$$

$$= \frac{1}{3}\left[\ln(x - 1) - 2\ln x\right]$$

$$\therefore \quad \frac{dy}{dx} = \frac{1}{3}\left[\frac{1}{x - 1} - \frac{2}{x}\right] = \frac{1}{3}\left[\frac{2 - x}{x^2 - x}\right] \qquad \text{Ans.}$$

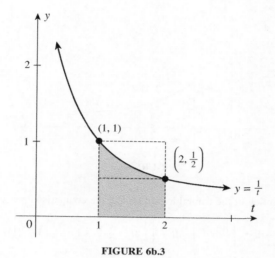

FIGURE 6b.3

Note (1): The process illustrated in Examples (5) and (6) is called logarithmic differentiation. In this process, the expressions involving quotients or powers are reduced to simple sums and products of functions. The procedure involves taking the natural logarithm of each side of the given function and then using the properties of logarithms.[4]

The following procedure is useful in understanding the inequality $0.5 < \ln 2 < 1$.

Solution: Here, we can write

$$\ln 2 = \int\limits_{1}^{2} \frac{1}{t}\,dt.$$

The above definite integral can be interpreted as the measure of the area of the shaded region appearing in Figure 6b.3.

From this figure, we observe that ln 2 is *between the measures of the areas of the rectangles*, each having base of length 1 unit and the altitudes of lengths 1/2 and 1 unit, which tells that, $0.5 < \ln 2 < 1$.[5]

Note (2): The number 0.5 is a lower bound of ln 2 and 1 is an upper bound. In a similar manner, *we can obtain a lower and upper bound for the natural logarithm of any positive real number.* (Later on, it will be convenient to compute the natural logarithm of any positive real number to any number of decimal places (to achieve the desired accuracy), by expanding the function ln x into an infinite series, as discussed in Chapter 22 of Part I.)

Remark: We have not yet established the property of logarithms, which states that $\log_a a = 1$, because so far we do not have a base for natural logarithm. To obtain the base for the natural logarithm we now show that $\ln 4 > 1$. [Once any base "a" is obtained for the (natural) logarithmic function "ln x" we must check that $a > 0$, $a \neq 1$.]

---

[4] This process was developed in 1697 by Johann Bernoulli (1667–1748). We have discussed this method and solved a good number of problems in Chapter 15a of Part I.

[5] This inequality can also be obtained analytically, using the mean value theorem for integrals introduced in Chapter 6a.

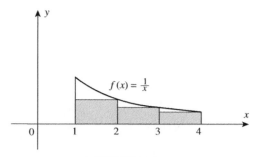

FIGURE 6b.4

Solution: By definition of the natural logarithm and the comparison property, we have

$$\ln 4 = \int_1^4 \frac{1}{t}dt = \int_1^2 \frac{1}{t}dt + \int_2^3 \frac{1}{t}dt + \int_3^4 \frac{1}{t}dt = \frac{1}{2}(2-1) + \frac{1}{2}(3-2) + \frac{1}{2}(4-3)$$

$$\geq \frac{1}{2}(2-1) + \frac{1}{3}(3-2) + \frac{1}{4}(4-3)$$

$$\geq \frac{1}{2} + \frac{1}{3} + \frac{1}{4}(= 13/12)$$

$$\geq 1.$$

Thus, we have shown that $\ln 4 > 1$. We have seen that $\ln 1 = 0$, $\ln 2 < 1$ and $\ln 4 > 1$ (Figure 6b.4). Intuitively, we can guess that there exists a number between 1 and 4 (let us call it "e"), such that $\ln e = 1$. It remains to show that such a number is in *the domain of* $\ln x$, *and it is unique.*[6]

    Now note that, $\ln x$ *is differentiable and hence continuous. Therefore, from the intermediate value theorem, it follows that there exists a number between 1 and 4* (denoted by "e"), *such that*

$$\ln e = 1$$

Also, since the function $\ln x$ *is strictly increasing, e is unique. It is the value of x for which the area of the shaded region in* Figure 6b.5 *is 1.*

    Further, the equality $\ln e = 1$, suggests that the base of natural logarithmic function "$\ln x$" must be the number "e".[7]

    In Chapter 13a of Part I, we have discussed at length about the number e, its origin, its value, and its properties. *There, we have also seen that* $\log_e x$ *is a new function such that*

$$\frac{d}{dx}\log_e x = \frac{1}{x}$$

This is consistent with what we have seen in the case of "$\ln x$". Thus, we identify the function $\ln x$ with $\log_e x$. The number "e", as we know, is an *irrational number* that has the *non-terminating* decimal expansion

$$e = 2.71828182845904523536$$

The symbol e was first adopted for this number by the great Swiss mathematician Leohard Euler. It has come to occupy a special place both, in mathematics and in its applications.

---

[6] Once this is shown, it will be justified to identify $\ln x$ as the logarithmic function to the base "e".

[7] From the properties of exponents and definition of logarithms, we know that for $a > 0$, $a^1 = a$, which also means that logarithm of "a" to the same base "a" is 1 [i.e., $\log_a a = 1$].

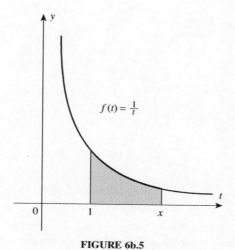

$f(t) = \frac{1}{t}$

FIGURE 6b.5

Now let us consider the following examples.

Example (7): Evaluate $\int_2^6 \frac{1}{x}dx$ in terms of logarithms

Solution: $\int_2^6 \frac{1}{x}dx = [\ln x]_2^6 = \ln 6 - \ln 2 = \ln\frac{6}{2} = \ln 3$.

Example (8): Express $\int_{-8}^{-7} \frac{1}{x}dx$ in terms of logarithms

Solution: $\displaystyle\int_{-8}^{-7} \frac{1}{x}dx = [\ln x]_{-8}^{-7}$

$$= \ln|-7| - \ln|-8|$$

$$= \ln\frac{7}{8} \quad \text{Ans.}$$

Example (9): Find the exact value of $\int_0^2 \frac{x^2+2}{x+1}dx$

Solution: Because $\frac{x^2+2}{x+1}$ is an improper fraction, we divide the numerator by the denominator and obtain

$$\frac{x^2+2}{x+1} = x - 1 + \frac{3}{x+1}$$

Therefore,

$$\int_0^2 \frac{x^2+2}{x+1}dx = \int_0^2 x - 1 + \frac{3}{x+1}dx$$

$$= \frac{1}{2}x^2 - x + 3\ln|x+1|\Big]_0^2$$

$$= 2 - 2 - 3\ln 3 - 3\ln 1$$

$$= 3\ln 3 - 3.0$$

$$= 3\ln 3 = \ln 3^3 = \ln 27 \quad \text{Ans.}$$

Example (10): Evaluate $\int \frac{\ln x}{x} dx$

Solution: Let $\ln x = t$.

We get

$$\frac{1}{x} dx = dt$$

Therefore,

$$\int \frac{\ln x}{x} dx = \int t \, dt = \frac{t^2}{2} + c^{(8)}$$

$$= \frac{1}{2}(\ln x)^2 + c \quad \text{Ans.}$$

Remark: Here we have evaluated the integral $\int (\ln x/x)dx$ by the *method of substitution discussed in Chapter* 3a. To obtain the formulas of indefinite integral of trigonometric functions $\tan x$, $\cot x$, $\sec x$, and $\csc x$, *we had assumed the result* $\int (1/x)dx = \log e|x| + c$, and applied it to $\int (f'(x)/f(x))dx$ to obtain the result $\log_e |f(x)| + c$. *If this had not been done, we would have had to wait till this point for obtaining the above-mentioned formulas, because they involve the natural logarithmic function.*

Also, we have seen in Chapter 13a of Part I, *that the common logarithm is a multiple of the natural logarithm and vice versa, given by the following relations*:

$$\log_{10} x = \log_{10} e \cdot \log_e x = 0.4343 \cdot \log_e x$$

and

$$\log_e x = \log_e 10 \cdot \log_{10} x = 2.3026 \cdot \log_{10} x$$

Thus, we can always convert back and forth between the natural and the common logarithm of the same number.[9]

6b.4 THE GRAPH OF THE NATURAL LOGARITHMIC FUNCTION $\ln x$

Now, we analyze the graph of the natural logarithm function $\ln x$ and try to sketch it. First note that, $\ln 1 = 0$ $\left(\because \int_1^1 (dt/t) = 0 \right)$. Next, $\ln|x|$ is a *differentiable function* with

$$\frac{d}{dx} \ln|x| = \frac{1}{x}$$

since $(1/x) > 0$ for $x > 0$, $\ln x$ is *strictly increasing*, also it is clear from $(1/x) > 0$ that $(d^2/dx^2)\ln x = -\frac{1}{x^2} < 0$.

---

[8] Remember that, whereas evaluation of $\int (\ln x/x)dx$ has been very simple, the integral $\int \ln x \, dx$ cannot be evaluated by the method of substitution. Recall that, $\int \ln x \, dx$ was evaluated by the method of integration by parts [Chapter 4a].
[9] To remember the above relationship, it is useful to keep in mind the algebraic identity $x/10 = (x/e) \cdot (e/10)$. It helps in writing $\log_{10} x = \log_{10} e \cdot \log_e x$ and so on.

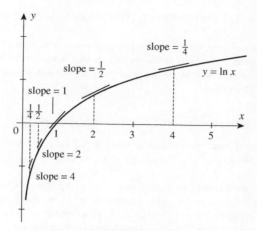

FIGURE 6b.6

Therefore (as we have noted earlier in Chapter 19a), the graph of $\ln x$ is concave downward on $(0, \infty)$. It can also be proved that

$$\lim_{x \to +\infty} \ln x = +\infty \quad \text{and} \quad \lim_{x \to 0^+} \ln x = -\infty \text{[10]}$$

With this information, and by plotting a few points with segments of tangent lines at the points, we can sketch the graph of the natural logarithmic function by hand, as shown in Figure 6b.6, where we have plotted the points having abscissas $\frac{1}{4}, \frac{1}{2}$, 1, 2, 4.

The slope of the tangent line is found from the formula

$$\frac{d}{dx} \ln |x| = \frac{1}{x}$$

From the property $\lim_{x \to 0^+} \ln x = -\infty$, we conclude that the graph of $\ln x$ is asymptotic to the negative part of the y-axis through the fourth quadrant.

In summary, the natural logarithmic function "$\ln x$", satisfies the following properties, as can be seen from its graph, and discussed earlier.

(a) The *domain* is the set of all positive numbers.

(b) The *range* is the set of all real numbers (since $\ln x \to +\infty$, as $x \to +\infty$).

(c) The function is *increasing* on its entire domain [since$(d/dx)(\ln x) = (1/x) > 0$ for all $x > 0$].

(d) The function is *continuous* at all numbers in its domain (since $\ln x$ is differentiable for all $x > 0$).

(e) The graph of the function is *concave downward* at all points (since $\ln x < 0$, for all $x > 0$. See Chapter 19a for second derivative test).

(f) The graph of the function is asymptotic to the negative part of the y-axis through the fourth quadrant (since $\lim_{x \to 0^+} \ln x = -\infty$).

[10] *The Calculus 7 of a Single Variable* by Louis Leithold (pp. 445–447), Harper Collins College Publishers.

6b.5   THE NATURAL EXPONENTIAL FUNCTION [exp(x) OR e^x]

We know that the natural logarithmic function is increasing on its entire domain, therefore by the definition of an inverse function, it has an inverse that is also an increasing function.

The inverse of $\ln(x)$ is called the *natural exponential function*, denoted by $\exp(x)$ (or e^x), which we now formally define.

6b.10.1   Definition of the Exponential Function exp(x) or e^x

The natural exponential function is the inverse of the natural logarithmic function.

It is therefore defined by $e^x = y$, if and only if $x = \ln y$.

Note: We agree that the expressions $\exp(x)$ and e^x stand for the same functions, as also the expressions $\ln x$ and $\log x$ mean the same function at "x". It is also clear that for $x = 1$, the exponential function e^x has the value "e" [or we say that the number "e" is the value of the (natural) exponential function at 1]. We can also write that $e = \exp(1)$. Note that, the exponential function e^x *is defined for all values of x and that its range is the set of all positive numbers.*

We have studied the properties of the natural exponential function in Chapter 13a of Part I. All the properties of exponential function studied there can be established using the definition of the logarithmic function, discussed in this chapter. It will be seen that these properties are consistent with the properties of exponents learnt in algebra.

7a Methods for Evaluating Definite Integrals

7a.1 INTRODUCTION

In Chapter 5, we have introduced the following concepts/ideas:

- the concept of *area*,
- the meaning of the *definite integral* as an area,
- the idea of the *definite integral as the limit of a sum*,
- the concept of *Riemann sums* and the *analytical definition of definite integral as the limit of Riemann sums*.
- the symbol $\int_a^b f(x)\mathrm{d}x$ for the *definite integral* of a (continuous) function $f(x)$ defined on a closed interval $[a, b]$,
- the statement of the integrability theorem, and
- *the statement of the second fundamental theorem of Calculus*, which links the definite integral $\int_a^b f(x)\mathrm{d}x$ with the antiderivative $\int f(x)\mathrm{d}x$.

Historically, methods of computing areas of certain regions were developed by the ancient Greeks. Such an area was called *the integral*. The symbol "$\int$" [with positive numbers a and b $(a < b)$] was used to indicate the measure of the area in question. Thus, the term "*integral*" and the symbol "$\int$" *were in use prior to the discovery of differential Calculus*.

The concept of derivatives was discovered in the seventeenth century and the methods for finding the derivatives of various functions, were developed then. Simultaneously, mathematicians developed the concept of the antiderivative of a function and the methods for finding antiderivatives. *The methods for computing derivatives of functions, together with those for computing antiderivatives constituted the subject of Calculus.*

Of course, a number of mathematicians have contributed through the centuries, towards the development of Calculus. However, Newton and Leibniz understood and exploited the intimate relationship that exists between antiderivatives and definite integrals. This relationship, which is known as *the second fundamental theorem of Calculus*, was introduced in Chapter 5. Also, some examples were solved to show how the theorem provides a very simple method for computing definite integrals.

7a-Methods for evaluating definite integrals using antiderivatives (Application of the second fundamental theorem of Calculus)

Introduction to Integral Calculus: Systematic Studies with Engineering Applications for Beginners, First Edition. Ulrich L. Rohde, G. C. Jain, Ajay K. Poddar, and A. K. Ghosh.
© 2012 John Wiley & Sons, Inc. Published 2012 by John Wiley & Sons, Inc.

The second fundamental theorem of Calculus says that, to evaluate the definite integral $\int_a^b f(x)dx$, we should compute the antiderivative of the function $f(x)$ (which is the integrand in the definite integral). The dependence of the definite integral on the evaluation of the antiderivative suggested that we identify the term antiderivative by the name "*indefinite integral*". Accordingly, *the old term "integral"* (which was in use prior to the discovery of derivative) was renamed "*the definite integral*".

The above discussion is a repetition of the contents discussed in previous chapters. However, such a repetition is expected to help us easily understand the methods and the material to be discussed in this chapter.

The underlying approach (for finding areas, as discussed in Chapter 5) is largely based on geometrical considerations and partly on our intuition. The procedure involved is lengthy and to some people it may appear dull. However, it is interesting to see how ideas are introduced for computing areas bounded by known curves.

The idea of the area function $A(x)$ introduced in Chapter 6a, helps in understanding *the first fundamental theorem of Calculus*, and in establishing the second fundamental theorem of Calculus. Both these theorems are then applied to show that *differentiation and integration are inverse processes*. Also, these theorems are very useful in proving many other important results in Calculus. We again come back to our discussion on the second fundamental theorem of Calculus.

If $f(x)$ is a *continuous function and* $f(x) > 0$ on $[a, b]$, such that the graph of the function is a *continuous curve above the x-axis*, then the area of the region bounded by the curve $y = f(x)$, the ordinates $x = a$ and $x = b$, and the portion of x-axis from the point a to the point b is denoted by $A = \int_a^b f(x)dx$. This is the *simplest plane region* and the symbol $\int_a^b f(x)dx$ measures the *shaded area*.

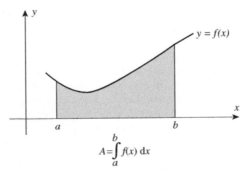

$$A = \int_a^b f(x)\,dx$$

The symbol $\int_a^b f(x)dx$ is read as *the definite integral of* $f(x)$ *from a to b*. It is a *fixed number* for the given interval $[a, b]$ and it can change only if the interval is changed. In this notation, $f(x)$ is the integrand, "a" is the lower limit of integration and "b" is the upper limit. The symbol $\int$ is an integral sign. It resembles a capital S, which is appropriate because the definite integral is the limit of a sum. (We have already discussed about the acceptance of the same symbol $\int$ for the definite and indefinite integrals.)

7a.2   THE RULE FOR EVALUATING DEFINITE INTEGRALS

If $f(x)$ is a continuous function defined on some closed interval $[a, b]$, then the rule is as follows.
Find the *antiderivative* of $f(x)$, *by any method.*

Suppose, $\int f(x)dx = \phi(x)$. Then, by applying the *fundamental theorem of Calculus, we obtain,*

$$\int_a^b f(x)dx = \phi(b) - \phi(a).^{(1)}$$

We know that the *definite integral* $\int_a^b f(x)dx$ of the function $f(x)$ is a *fixed number* for the given interval $[a, b]$ and it can change only if the interval is changed. On the other hand, the *indefinite integral* of a function $f(x)$ is (in general) a *new function*. Thus, the definite integral $\int_a^b f(x)dx$, *is quite different from the indefinite integral* $\int f(x)dx$, although both look alike due to the symbol $\int$. It is when we wish to evaluate definite integrals that we make use of indefinite integrals.

By using the second fundamental theorem of integral Calculus we, in effect, get *a powerful tool for computing the limit of a sum in question,* which is the same as the value of the definite integral in question.

We now give the following definition of the definite integral of $f(x)$.

Definition: Let $f(x)$ be a function, *continuous on a closed interval* $[a, b]$, and *let $\phi(x)$ be the antiderivative of $f(x)$,* such that

$$\int f(x)dx = \phi(x).$$

Then, we write $\int_a^b f(x)dx$ to mean $[\phi(x)]_a^b = \phi(b) - \phi(a)$ where a and b are *real constants.* The symbol $\int_a^b f(x)dx$ is *called the definite integral of $f(x)$ from a to b.*

Note (1): *The second fundamental theorem of integral Calculus defines the definite integral* $\int_a^b f(x)dx$, *for $b > a$.*

[Later on, we will show that if $b < a$, then $\int_a^b f(x)dx = -\int_b^a f(x)dx$.]

In other words, the fundamental theorem can be applied for evaluating the definite integral even when the upper limit is smaller than the lower limit.

Note (2): Let $\int f(x)dx = \phi(x) + c$ (I)

$$\therefore \int_a^b f(x)dx = [\phi(x) + c]_a^b = [\phi(b) + c] - [\phi(a) + c] = \phi(b) - \phi(a) \qquad \text{(II)}$$

---

[1] Consider $\int f(x)dx = \phi(x) + c$, where c is an arbitrary constant, including zero. If we choose $c = 0$, then $\phi(x)$ is called (the) antiderivative of $f(x)$, since $\phi'(x) = f(x)$. Note that, $\phi(x) + c$ is called (an) antiderivative of $f(x)$. It is due to the arbitrary constant c [added to the antiderivative $\phi(x)$] that we call "$\phi(x) + c$", an antiderivative (or an indefinite integral), the indefiniteness being due to "c". For all practical purposes, we do not make a distinction between the antiderivative $[\phi(x)]$ and an antiderivative $[\phi(x)+c]$. However, for computing the definite integral $\int_a^b f(x)dx$, we shall always write the indefinite integral $\phi(x)$ without an arbitrary constant.

Observe that, *the value of the definite integral does not depend on the constant of integration.* This justifies the adjective *"definite"*. Therefore, *while evaluating the definite integral, we do not write the constant of integration with $\phi(x)$, which is the antiderivative of $f(x)$.*

Remark: Use of the sign $\int$ for an *indefinite integral* and for the *definite integral* does not help the beginner to keep the two ideas distinct from each other. From this point of view, *it is perhaps not the best notation to use.* However, once the distinction between the terms an *"indefinite integral"* and the *"definite integral"* is clearly understood, the integral sign is accepted conveniently and logically.

7a.3   SOME RULES (THEOREMS) FOR EVALUATION OF DEFINITE INTEGRALS

As in the case of indefinite integrals, *we can easily prove the following corresponding results for definite integrals.*

(1)   If f and g are *integrable functions*[2] then

$$\int_a^b [f(x) \pm g(x)]dx = \int_a^b f(x)dx \pm \int_a^b g(x)dx$$

(2)   If f is an *integrable function*[2] and k is a constant then

$$\int_a^b kf(x)dx = k \int_a^b f(x)dx$$

Corollary: If f and g are *integrable functions*[2] and k_1, k_2 are constants then

$$\int_a^b [k_1 f(x) \pm k_2 g(x)]dx = k_1 \int_a^b f(x)dx \pm k_2 \int_a^b g(x)dx$$

The above results can be extended to more than two functions involved in the sums and differences.

To prove these results, *we use the definition of the definite integral based on the fundamental theorem of integral Calculus.* For example, let us prove that,

$$\int_a^b [f(x) + g(x)]dx = \int_a^b f(x)dx + \int_a^b g(x)dx$$

---

[2] We have defined an "integrable function" in Chapter 5. (A function is said to be integrable on $[a, b]$ if it is bounded on $[a, b]$ and continuous there, except at a finite number of points.)

Proof: Let $\int f(x)\mathrm{d}x = \phi(x)$

Then,
$$\int_a^b f(x)\mathrm{d}x = [\phi(x)]_a^b = \phi(b) - \phi(a) \tag{1}$$

Again, let $\int g(x)Fx = F(x)$

Then,
$$\int_a^b g(x)\mathrm{d}x = F(b) - F(a) \tag{2}$$

Adding Equation (1) and Equation (2) we get

$$\int_a^b f(x)\mathrm{d}x + \int_a^b g(x) = [\phi(b) - \phi(a)] + [F(b) - F(a)]$$
$$= [\phi(b) + F(b)] - [\phi(a) + F(a)] \tag{3}$$

But for an indefinite integral, we know that,

$$\int [f(x) + g(x)]\mathrm{d}x = \int f(x)\mathrm{d}x + \int g(x)\mathrm{d}x = \phi(x) + F(x) + c$$
$$\therefore \quad \int_a^b [f(x) + g(x)]\mathrm{d}x = [\phi(x) + F(x) + c]_a^b$$
$$= [\phi(b) + F(b)] - [\phi(a) + F(a)] \tag{4}$$

Observe that the right-hand sides of Equation (3) and that of Equation (4) are identical. Therefore, equating their left-hand sides, we get

$$\int_a^b [f(x) + g(x)]\mathrm{d}x = \int_a^b f(x)\mathrm{d}x + \int_a^b g(x)\mathrm{d}x$$

On similar lines, we can prove the other results.

7a.3.1 Solved Examples

Example (1): Evaluate $\int_1^2 (5x^3 + x^2 - 4x)\mathrm{d}x$

Solution:
$$\int_1^2 [5x^3 + x^2 - x + 3]\mathrm{d}x = \left[\frac{5x^4}{4} + \frac{x^3}{3} - \frac{x^2}{2} + 3x\right]_1^2$$
$$= \left[5 \cdot \left(\frac{16}{4}\right) + \frac{8}{3} - \frac{4}{2} + 6\right] - \left[\frac{5}{4} + \frac{1}{3} - \frac{1}{2} + 3\right]$$
$$= \left[20 + \frac{8}{3} - 2 + 6\right] - \left[\frac{15 + 4 - 6 + 4}{12}\right]$$
$$= \frac{80}{3} - \frac{17}{12} = \frac{320 - 17}{12} = \frac{303}{12} = \frac{101}{4} \qquad \text{Ans.}$$

Example (2): Evaluate $\displaystyle\int_0^1 \frac{x^2 - 3x + 2}{\sqrt{x}}\, dx$

Solution: $\displaystyle\int_0^1 \frac{x^2 - 3x + 2}{\sqrt{x}}\, dx = \int_0^1 \left(x^{3/2} - 3x^{1/2} + 2x^{-1/2}\right) dx^{(3)}$

$$= \left[\frac{x^{5/2}}{5/2} - 3\frac{x^{3/2}}{3/2} + 2\frac{x^{1/2}}{1/2}\right]_0^1$$

$$= \left[\frac{2}{5}x^{5/2} - 2x^{3/2} + 4x^{1/2}\right]_0^1$$

$$= \left[\frac{2}{5} - 2 + 4\right] - 0 = \frac{12}{5} \qquad \text{Ans.}$$

Example (3): Evaluate $\int_0^{\pi/2} \sin^2 x\, dx$

Solution:

$$\int_0^{\pi/2} \sin^2 x\, dx = \int_0^{\pi/2} \frac{1 - \cos 2x}{2}\, dx, \qquad \begin{bmatrix} \because & \cos 2x = 1 - 2\sin^2 x \\ \\ \therefore & \sin^2 x = \dfrac{1 - \cos 2x}{2} \end{bmatrix}$$

$$= \frac{1}{2}\int_0^{\pi/2} [1 - \cos 2x]dx = \frac{1}{2}\left[\int_0^{\pi/2} 1dx - \int_0^{\pi/2} \cos 2x\, dx\right]$$

$$= \frac{1}{2}\left\{[x]_0^{\pi/2} - \left[\frac{\sin 2x}{2}\right]_0^{\pi/2}\right\} = \frac{1}{2}\left[\left(\frac{\pi}{2} - 0\right) - \frac{1}{2}(\sin \pi - \sin 0)\right]$$

$$= \frac{\pi}{4} - 0 = \frac{\pi}{4} \qquad \text{Ans.}$$

All the methods studied for evaluating indefinite integrals (such as *integration by substitution, integration by parts, and integration by partial fractions*) can be applied for computing definite integrals.

7a.3.2   Application of the Fundamental Theorem of Calculus

Evaluating definite integrals is generally a *two-step process*. First, we find the indefinite integral, then we apply the *fundamental theorem of Calculus*. If the indefinite integration is *easy, we can combine the two steps* as in Examples (1) and (2). However, if the computation is complicated enough to require a substitution, we typically separate the two steps. Thus, to calculate $\int_0^4 x\sqrt{x^2 + 9}\, dx$ we first write (using $x^2 + 9 = t$ and $2x\, dx = dt$).

$$\therefore \quad \int x\sqrt{x^2 + 9}\, dx = \frac{1}{2}\int \sqrt{x^2 + 9}(2x\, dx) = \frac{1}{2}\int t^{1/2}dt$$

$$= \frac{1}{2} \cdot \frac{2}{3}t^{3/2} + c = \frac{1}{3}\left(x^2 + 9\right)^{3/2} + c$$

---

[3] Observe that the integrand has been converted to the standard form. This is needed to write down an indefinite integral, before evaluating the definite integral.

Then, by the fundamental theorem, we write

$$\int_0^4 x\sqrt{x^2+9}\,dx = \left[\frac{1}{3}(x^2+9)^{3/2}\right]_0^4$$

$$= \frac{(4^2+9)^{3/2}}{3} - \frac{(0^2+9)^{3/2}}{3}$$

$$= \frac{(25)^{3/2}}{3} - \frac{(9)^{3/2}}{3}$$

$$= \frac{125}{3} - \frac{27}{3} = \frac{98}{3} \quad \text{Ans.}$$

7a.3.2.1 Simpler Way of Using Substitution Directly: From Two-Step Procedure to One-Step Procedure

Instead of going through a two-step process, *there is a simpler way of using substitution directly in a definite integral, as explained through the following example.*

Example (4): Evaluate $\displaystyle\int_0^{\sqrt{\pi/2}} x\sin^3(x^2)\cos(x^2)dx$

Solution: *First we will evaluate this using the two-step procedure* and then *evaluate it directly.*

Method (I): Two-step procedure.

$$\text{Put} \quad \sin(x^2) = t, \quad \text{so} \quad 2x\cos(x^2)dx = dt$$

$$\therefore \quad \int x\sin^3(x^2)\cos(x^2)dx = \frac{1}{2}\int \sin^3(x^2)\cdot 2x\cos(x^2)dx$$

$$= \frac{1}{2}\int t^3 dt = \frac{1}{2}\cdot\frac{t^4}{4} + c = \frac{1}{8}\sin^4(x^2) + c$$

Then, by the second fundamental theorem of Calculus,

$$\int_0^{\sqrt{\pi/2}} x\sin^3(x^2)\cos(x^2)dx = \left[\frac{1}{8}\sin^4(x^2)\right]_0^{\sqrt{\pi/2}}$$

$$= \frac{1}{8}\sin^4\left(\frac{\pi}{4}\right) - \frac{1}{8}\cdot 0$$

$$= \frac{1}{8}\left(\frac{1}{\sqrt{2}}\right)^4 - 0 = \frac{1}{32} \quad \text{Ans.}$$

Note: *In the two-step procedure* illustrated above, *it is necessary to express the indefinite integral in terms of the original variable x, before we apply the fundamental theorem.* This is because *the limits 0 and $\sqrt{\pi/2}$ apply to x, not to t.* But, in making the substitution $\sin(x^2) = t$, if we also make the corresponding changes in the limits of integration, to the new variable t, then we can complete the integration with t as variable. This is indicated in Method (II).

Method (II): Observe that for $x = 0$, $t = \sin(x^2) = \sin(0^2) = 0$, and for $x = \sqrt{\pi}/2$, $t = \sin\left[(\sqrt{\pi}/2)^2 \right] = \sin\left(\frac{\pi}{4}\right) = \frac{1}{\sqrt{2}} = \frac{\sqrt{2}}{2}$

$$\therefore \quad \int_0^{\sqrt{\pi}/2} x \sin^3(x^2) \cos(x^2) dx = \frac{1}{2} \int_0^{\sqrt{2}/2} t^3 dt$$

$$= \left[\frac{1}{2} \cdot \frac{t^4}{4} \right]_0^{\sqrt{2}/2} = \left[\frac{1}{8} \cdot \frac{4}{16} \right] = \frac{1}{32} \qquad \text{Ans.}$$

7a.3.2.2 Change of Limits in Definite Integrals During Substitution

Let us recall *the method of substitution for the indefinite integral*. There, we have seen that, by substituting $x = \phi(t)$ in the given integral $\int f(x) dx$, we get, $\int f(x) dx = \int f[\phi(t)] \phi'(t) dt$.

To evaluate a definite integral $\int_a^b f(x) dx$, where $x = \phi(t)$, we have the following form,

$$\int_a^b f(x) dx = \int_{t_1}^{t_2} f[\phi(t)] \phi'(t) dt$$

where t_1 and t_2 are numbers such that (a) $a = \phi(t_1)$ or $t_1 = \phi^{-1}$ and (b) $b = \phi(t_2)$ or $t_2 = \phi^{-1}$. *Thus, t_1 and t_2 are the limits of integration for the variable t, corresponding to the limits of integration a and b for the variable x.*

In practice, it is very simple to find t_1 and t_2 corresponding to the limits "a" and "b", respectively, as will be clearer from the solved examples given below. Thus, while changing the variable in a definite integral, we should change the limits of integration to reduce our work. Once this is done, we need not come back to the original variable x, for evaluating the definite integral.

Example (5): Evaluate $\int_0^1 \frac{x^3}{\sqrt{1+x^4}} dx$

Put $\quad 1 + x^4 = t \qquad \therefore \quad 4x^3 dx = dt, \qquad \therefore \quad x^3 dx = \frac{1}{4} dt$

for $x = 0$, $t = 1 + 0^4 = 1$, and for $x = 1$, $t = 1 + 1^4 = 2$

Solution: $\quad \therefore \quad \int_0^1 \frac{x^3}{\sqrt{1+x^4}} dx = \int_1^2 \frac{(1/4) dt}{\sqrt{t}} = \frac{1}{4} \int_1^2 t^{-1/2} dt$

$$= \frac{1}{4} \left[\frac{t^{1/2}}{1/2} \right]_1^2 = \frac{1}{2} \left[t^{1/2} \right]_1^2 = \frac{1}{2} \left[2^{1/2} - 1^{1/2} \right] = \frac{1}{2} \left[\sqrt{2} - 1 \right] \qquad \text{Ans.}$$

Example (6): Evaluate $\int_0^1 x^3 (1 - x^2)^{5/2} dx$[4]

---

[4] Note how the substitution converts the integrand to the standard form.

Solution: Put $1 - x^2 = t$ $\therefore$ $x^2 = 1 - t$ $\therefore$ $2x\,dx = -dt$ $\therefore$ $x\,dx = -\frac{1}{2}dt$ when $x = 0$, $t = 1 - 0 = 1$ and when $x = 1$, $t = 1 - 1^2 = 0$

$$\therefore \int_0^1 x^3(1 - x^2)^{5/2}dx = \int_0^1 x^2(1 - x^2)^{5/2} \cdot x\,dx$$

$$= \int_1^0 (1 - t)t^{5/2} \cdot \left(-\frac{1}{2}\right)dt$$

$$= -\frac{1}{2}\int_1^0 (1 - t)t^{5/2}dt = -\frac{1}{2}\int_1^0 \left(t^{5/2} - t^{7/2}\right)dt$$

$$= \frac{1}{2}\int_0^1 \left(t^{5/2} - t^{7/2}\right)dt \quad \text{[see Remark in Section 7a.2]}$$

$$= \frac{1}{2}\left[\frac{t^{7/2}}{7/2} - \frac{t^{9/2}}{9/2}\right]_0^1 = \frac{1}{2}\left[\frac{2}{7}t^{7/2} - \frac{2}{9}t^{9/2}\right]_0^1$$

$$= \frac{1}{2}\left[\left(\frac{2}{7} - \frac{2}{9}\right) - (0 - 0)\right] = \frac{9 - 7}{63} = \frac{2}{63} \quad \text{Ans.}$$

Example (7): Evaluate $\displaystyle\int_0^1 \frac{\tan^{-1}x}{1 + x^2}dx$

Solution:

$$\text{Put} \quad \tan^{-1}x = t \quad \therefore \quad \frac{1}{1 + x^2}dx = dt$$

$$\text{When } x = 0, \ t = \tan^{-1}0 = 0 \ (\because \ \tan 0 = 0)$$

$$\text{and when } x = 1, \ t = \tan^{-1}1 = \frac{\pi}{4} \quad \left(\because \ \tan\frac{\pi}{4} = 1\right)$$

$$\therefore \quad \int_0^1 \frac{\tan^{-1}x}{1 + x^2}dx = \int_0^{\pi/4} t\,dt = \left[\frac{t^2}{2}\right]_0^{\pi/4}$$

$$= \frac{1}{2}\left[t^2\right]_0^{\pi/4} = \frac{1}{2}\left[\left(\frac{\pi}{4}\right)^2 - 0^2\right] = \frac{\pi^2}{32} \quad \text{Ans.}$$

Example (8): Evaluate $\int_0^{\pi/2} \sin^2 x \cos x\,dx$

Solution:

$$\text{Put} \quad \sin x = t \quad \therefore \quad \cos x\,dx = dt$$

$$\text{When } x = 0, \ t = \sin 0 = 0$$

$$\text{and when } x = \frac{\pi}{2}, \ t = \sin\frac{\pi}{2} = 1$$

$$\therefore \quad \int_0^{\pi/2} \sin^2 x \cos x\,dx = \int_0^1 t^2 dt = \left[\frac{t^3}{3}\right]_0^1$$

$$= \frac{1}{3}\left[t^3\right]_0^1 = \frac{1}{3}\left[1^3 - 0^3\right] = \frac{1}{3} \quad \text{Ans.}$$

Example (9): Evaluate $\displaystyle\int_{-1}^{3} \frac{5x\,dx}{\sqrt{x^2+4}}$

Solution:

Put $x^2 + 4 = t^2$ $\therefore$ $2x\,dx = 2t\,dt$ $\therefore$ $x\,dx = t\,dt$

When $x = -1$, $t^2 = (-1)^2 + 4 = 5$ $\therefore$ $t = \sqrt{5}$

and when $x = 3$, $t^2 = (3)^2 + 4 = 13$ $\therefore$ $t = \sqrt{13}$

$$\therefore \quad \int_{-1}^{3} \frac{5x}{\sqrt{x^2+4}}\,dx = 5\int_{\sqrt{5}}^{\sqrt{13}} \frac{t\,dt}{t} = 5\int_{\sqrt{5}}^{\sqrt{13}} dt$$

$$= 5[t]_{\sqrt{5}}^{\sqrt{13}} = 5\left[\sqrt{13} - \sqrt{5}\right] \quad \text{Ans.}$$

Example (10): Evaluate $\displaystyle\int_{0}^{\pi} \frac{dx}{5 + 4\cos x}$

Solution: Let $\displaystyle\int \frac{dx}{5 + 4\cos x} = I \quad (\text{say})^{(5)}$

Consider $5 + 4\cos x$

$$= 5\left[\sin^2\frac{x}{2} + \cos^2\frac{x}{2}\right] + 4\left[\cos^2\frac{x}{2} - \sin^2\frac{x}{2}\right]$$

$$= \sin^2\frac{x}{2} + 9\cos^2\frac{x}{2}$$

$$\therefore \quad I = \int \frac{dx}{\sin^2(x/2) + 9\cos^2(x/2)}$$

Dividing N^r and D^r by $\cos^2\dfrac{x}{2}$, we get

$$I = \int \frac{\sec^2(x/2)}{\tan^2(x/2) + 9}\,dx$$

Put $\tan\dfrac{x}{2} = t$ $\therefore$ $\dfrac{1}{2}\sec^2\dfrac{x}{2}\,dx = dt$

$$\therefore \quad \sec^2\frac{x}{2}\,dx = 2dt$$

$$\therefore \quad I = 2\int \frac{dt}{t^2 + 9} = 2\int \frac{dt}{t^2 + (3)^2}$$

$^{(5)}$ The principal step in the evaluation of a definite integral is to find the related indefinite integral.

When $x = 0$, $t = \tan(0) = 0$ and when $x = \pi$, t tends to ∞ $\left(\because \quad \tan\dfrac{\pi}{2} = \infty \right)$.

Thus, as x varies from 0 to π, t varies from 0 to ∞.

$$\therefore \quad \int\limits_{0}^{\pi} \frac{dx}{5 + 4\cos x} = 2\int_{0}^{\infty} \frac{dt}{t^2 + (3)^2} = 2\left[\frac{1}{3}\tan^{-1}\frac{t}{3}\right]_{0}^{\infty}$$

$$= \frac{2}{3}\left[\tan^{-1}\infty - \tan^{-1}0\right]$$

$$= \frac{2}{3}\left(\frac{\pi}{2} - 0\right) = \frac{\pi}{3} \quad \text{Ans.}$$

Example (11): Evaluate $\displaystyle\int\limits_{0}^{\pi/2} \frac{\sin 2\theta \, d\theta}{\sin^4\theta + \cos^4\theta}$

Solution: Let $\displaystyle\int \frac{\sin 2\theta}{\sin^4\theta + \cos^4\theta}\, d\theta = I$

Put $\sin^2\theta = t$ $\quad \therefore \quad 2\sin\theta \cdot \cos\theta \, d\theta = dt$

or $\sin 2\theta \, d\theta = dt$

Now, $\quad \sin^4\theta + \cos^4\theta = t^2 + (1-t)^2 = t^2 + 1 - 2t + t^2$

$$= 2t^2 - 2t + 1 = 2\left[t^2 - t + \frac{1}{2}\right]$$

$$= 2\left[t^2 - 2(t)\left(\frac{1}{2}\right) + \left(\frac{1}{2}\right)^2 - \left(\frac{1}{2}\right)^2 + \frac{1}{2}\right]$$

$$= 2\left[\left(t - \frac{1}{2}\right)^2 + \left(\frac{1}{2}\right)^2\right]$$

$$\therefore \quad I = \frac{1}{2}\int \frac{dt}{(t - (1/2))^2 + (1/2)^2} = \frac{1}{2}\left[\frac{1}{1/2}\tan^{-1}\frac{(t - (1/2))}{1/2}\right] + c$$

$$= \tan^{-1}(2t - 1) + c$$

When $\theta = 0$, $t = \sin^2 0 = 0$

and when $\theta = \frac{\pi}{2}$, $t = \sin^2 \frac{\pi}{2} = 1$

Thus, when θ varies from 0 to $\dfrac{\pi}{2}$, t varies from 0 to 1.

$$\therefore \quad \int_0^{\pi/2} \frac{\sin 2\theta \, d\theta}{\sin^4\theta + \cos^4\theta} = \left[\tan^{-1}(2t-1)\right]_0^1 = \tan^{-1}(1) - \tan^{-1}(-1)$$

$$= \frac{\pi}{4} - \left(-\frac{\pi}{4}\right) = \frac{\pi}{2} \qquad \text{Ans.}$$

Note: Consider $\displaystyle\int \frac{\sin 2\theta \, d\theta}{\sin^4\theta + \cos^4\theta} = \int \frac{2\sin\theta \cdot \cos\theta}{\sin^4\theta + \cos^4\theta} d\theta = I$ (say)

Dividing N^r and D^r by $\cos^4\theta$, we get

$$I = \int \frac{2\tan\theta \cdot \sec^2\theta}{\tan^4\theta + 1} d\theta$$

Now, by putting $\tan^2\theta = t$ $\therefore$ $2\tan\sec^2\theta \, d\theta = dt$

It can also be easily shown that

$$\therefore \quad \int_0^{\pi/2} \frac{\sin 2\theta \, d\theta}{\sin^4\theta + \cos^4\theta} = \int_0^\infty \frac{dt}{1+t^2} = \left[\tan^{-1}t\right]_0^\infty = \frac{\pi}{2}$$

Example (12): Evaluate $\displaystyle\int_{\pi^2/9}^{\pi^2/4} \frac{\cos\sqrt{x}}{\sqrt{x}} dx$

Solution: Let $\sqrt{x} = t$ $\therefore$ $x = t^2$ $\therefore$ $dx = 2t \, dt$

$$\therefore \quad \int \frac{\cos\sqrt{x}}{\sqrt{x}} dx = \int \frac{\cos t}{t} \cdot 2t \, dt = 2\int \cos t \, dt$$

Now, for $x = \pi^2/9$, $t = \sqrt{x} = \dfrac{\pi}{3}$

and for $x = \pi^2/4$, $t = \sqrt{x} = \dfrac{\pi}{2}$

$$\therefore \quad \int_{\pi^2/9}^{\pi^2/4} \frac{\cos\sqrt{x}}{\sqrt{x}} = 2\int_{\pi/3}^{\pi/2} \cos t \, dt$$

$$= \left[2\sin t\right]_{\pi/2}^{\pi/3} = 2(1) - 2\left(\frac{\sqrt{3}}{2}\right)$$

$$= 2 - \sqrt{3} \qquad \text{Ans.}$$

Example (13): Evaluate $\displaystyle\int_0^{\pi/2} \frac{\cos x \, dx}{(1+\sin x)(2+\sin x)} dx$

Solution: Put $\sin x = t$ $\therefore$ $\cos x \, dx = dt$

When $x = 0$, $t = \sin 0 = 0$, and when $x = \dfrac{\pi}{2}$, $t = \sin \dfrac{\pi}{2} = 1$

$$\therefore \quad \int_0^{\pi/2} \frac{\cos x \, dx}{(1 + \sin x)(2 + \sin x)} \, dx = \int_0^1 \frac{dt}{(1 + t)(2 + t)}$$

Now resolving $\dfrac{1}{(1 + t)(2 + t)}$ into partial fractions, we get

$$\frac{1}{(1 + t)(2 + t)} = \frac{1}{(1 + t)} - \frac{1}{(2 + t)}$$

$$\therefore \quad \text{Given integral} = \int_0^1 \left[\frac{1}{(1 + t)} - \frac{1}{(2 + t)} \right] dt$$

$$= [\log(1 + t) - \log(2 + t)]_0^1 = \left[\log\left(\frac{1 + t}{2 + t}\right) \right]_1^0 = \log\frac{2}{3} - \log\frac{1}{2}$$

$$= \log\frac{2/3}{1/2} = \log\frac{4}{3} \qquad \text{Ans.}$$

7a.4   METHOD OF INTEGRATION BY PARTS IN DEFINITE INTEGRALS

We have proved the following result in connection with the evaluation of (certain) indefinite integrals, which involve product of two functions $u(x)$ and $v(x)$.

$$\int uv \, dx = u \int v \, dx - \int \left[\frac{d}{dx}(u) \int v \, dx \right] dx$$

(Obviously, it is clear that u represents the first function and v the second.)

In case of definite integrals this result is used in the following form

$$\int_a^b uv \, dx = \left[u \int v \, dx \right]_a^b - \int_a^b \left[\frac{d}{dx}(u) \int v \, dx \right] dx$$

The following solved examples illustrate how this result is applied to evaluate definite integrals.

Example (14): Evaluate $\int\limits_0^{\pi/4} x \cos x \, dx$

Solution: $\int\limits_0^{\pi/4} x \cos x \, dx$

$$= \left[x \cdot \int \cos x \, dx \right]_0^{\pi/4} - \int_0^{\pi/4} (1) \cdot \sin x \, dx$$

$$= [x \sin x]_0^{\pi/4} - [-\cos x]_0^{\pi/4}$$

$$= \left(\frac{\pi}{4} \cdot \frac{1}{\sqrt{2}} - 0 \right) + [\cos x]_0^{\pi/4}$$

$$= \frac{\pi}{4\sqrt{2}} + \frac{1}{\sqrt{2}} - 1 \qquad \text{Ans.}$$

Example (15): Evaluate $\int\limits_{0}^{\pi/2} x \sin x \, dx$

Solution:

$$\int\limits_{0}^{\pi/2} x \sin x \, dx = \left[x \int \sin x \, dx \right]_{\pi/2}^{0} - \int\limits_{0}^{\pi/2} (1).(-\cos x) dx$$

$$= [x(-\cos x)]_{\pi/2}^{0} + [\sin x]_{\pi/2}^{0}$$

$$= [-x \cos x]_{\pi/2}^{0} + [\sin x]_{\pi/2}^{0}$$

$$= \left[-\frac{\pi}{2} \left(\cos \frac{\pi}{2} \right) - 0(\cos 0) \right] + \left[\sin \frac{\pi}{2} - \sin 0 \right]$$

$$= = (0 - 0) + (1 - 0) = 1 \quad \text{Ans.}$$

Example (16): Evaluate $\int\limits_{0}^{1} x \, e^{2x} \, dx$

Solution:

$$\int\limits_{0}^{1} x \, e^{2x} dx = \left[x \int e^{2x} dx \right]_{0}^{1} - \int\limits_{0}^{1} (1) \frac{e^{2x}}{2} dx$$

$$= \left[x \cdot \frac{e^{2x}}{2} \right]_{0}^{1} - \left[\frac{1}{2} \cdot \frac{e^{2x}}{2} \right]_{0}^{1} = \left[1 \cdot \frac{e^2}{2} - 0 \cdot \frac{e^0}{2} \right] - \left[\frac{e^2}{4} - \frac{e^0}{4} \right]$$

$$= \left(\frac{e^2}{2} - 0 \right) - \left(\frac{e^2}{4} - \frac{1}{4} \right) = \frac{e^2}{4} + \frac{1}{4} = \frac{e^2 + 1}{4} \quad \text{Ans.}$$

Example (17): Evaluate $\int\limits_{2}^{3} x \log x \, dx$

Solution: $\int\limits_{2}^{3} x \log x \, dx = \int\limits_{2}^{3} (\log x) \cdot x \, dx$

$$\int\limits_{2}^{3} (\log x) \cdot x \, dx = \left[\log x \cdot \frac{x^2}{2} \right]_{2}^{3} - \int\limits_{2}^{3} \left(\frac{1}{x} \cdot \frac{x^2}{2} \right) dx$$

$$= \left[\frac{x^2}{2} \cdot \log x \right]_{2}^{3} - \frac{1}{2} \int\limits_{2}^{3} x \, dx = \left[\frac{x^2}{2} \cdot \log x \right]_{2}^{3} - \left[\frac{1}{2} \cdot \frac{x^2}{2} \right]_{2}^{3}$$

$$= \left[\frac{9}{2} \log 3 - \frac{4}{2} \log 2 \right] - \left[\frac{9}{4} - \frac{4}{4} \right] = \frac{9}{2} \log 3 - 2 \log 2 - \frac{5}{4} \quad \text{Ans.}$$

We emphasize that the principal step in the evaluation of a definite integral is to find the related indefinite integral. At times, finding the indefinite integral is lengthy. In such cases, we must first compute an indefinite integral and then apply the fundamental theorem to compute the definite integral in question, as clear from the following examples. Using certain indefinite integrals obtained earlier, we evaluate their definite integrals.

Example (18): $\int x \sec^2 x \, dx = x \tan x - \log(\sec x) + c$

$$\therefore \quad \int_0^{\pi/4} x \sec^2 x \, dx = [x \cdot \tan x - \log(\sec x)]_0^{\pi/4}$$

$$= \left[\frac{\pi}{4} \tan \frac{\pi}{4} - \log \left(\sec \frac{\pi}{4} \right) \right] - [0 - \log(\sec 0)]$$

$$= \frac{\pi}{4} - \log \sqrt{2} - 0$$

$$= \frac{\pi}{4} - \log 2^{1/2} = \frac{\pi}{4} - \frac{1}{2} \log 2 \qquad \text{Ans.}$$

Example (19): $\int x^2 a^x dx = \dfrac{x^2 a^x}{\log a} - \dfrac{2xa^x}{(\log a)^2} + \dfrac{2a^x}{(\log a)^3} + c$

$$\therefore \quad \int_0^1 x^2 a^x dx = \left[\frac{x^2 a^x}{\log a} - \frac{2xa^x}{(\log a)^2} + \frac{2a^x}{(\log a)^3} \right]_0^1$$

$$= \left[\frac{a}{\log a} - \frac{2a}{(\log a)^2} + \frac{2a}{(\log a)^3} \right] - \left[\frac{2}{(\log a)^3} \right] \qquad \text{Ans.}$$

Example (20): $\int_0^2 x^3 e^x dx = [e^x(x^3 - 3x^2 + 6x - 6)]_0^2$

$$= [e^2(8 - 12 + 12 - 6)] - (-6)$$

$$= 2e^2 + 6 \quad \text{Ans.}$$

Example (21): $\int_0^1 x \tan^{-1} x \, dx = \left[\dfrac{x^2}{2} \tan^{-1} x - \dfrac{x}{2} + \dfrac{1}{2} \tan^{-1} x \right]_2^1$

$$= \left[\frac{1}{2} \cdot \frac{\pi}{4} - \frac{1}{2} + \frac{1}{2} \cdot \frac{\pi}{4} \right] - \left[0 - 0 + \frac{1}{2} \cdot 0 \right]$$

$$= \frac{\pi}{4} - \frac{1}{2} = \frac{\pi - 2}{4} \qquad \text{Ans.}$$

Example (22): $\int_1^e \log x \, dx = [x \cdot \log_e x - x]_1^e$

$$= [e \log_e e - e] - [1 \log_e 1]$$

$$= [e - e] - [0 - 1] = 1 \qquad \text{Ans.}$$

7b Some Important Properties of Definite Integrals

7b.1 INTRODUCTION

In Chapter 7a, we have learnt to evaluate definite integrals, by first computing their antiderivatives and then applying the second fundamental theorem of Calculus. Also, we have discussed *how this two-step procedure can be reduced to a one-step procedure*, whenever substitution is involved in the process of integration.

But, there are many definite integrals involving certain complicated functions (integrands) whose antiderivatives cannot be obtained. In many such cases, it is possible to evaluate the definite integrals *by applying certain properties of definite integrals*. They are very useful in easily many integrals. Besides, they help us in evaluating (*certain*) definite integrals whose antiderivatives cannot be evaluated.[1]

We state and prove these special properties of definite integrals.

7b.2 SOME IMPORTANT PROPERTIES OF DEFINITE INTEGRALS

$$P_0: \int_a^b f(x)\mathrm{d}x = \int_a^b f(t)\mathrm{d}t$$

$$P_1: \int_a^b f(x)\mathrm{d}x = -\int_b^a f(x)\mathrm{d}x$$

(In particular, $\int_a^a f(x)\mathrm{d}x = 0$)

$$P_2: \int_a^b f(x)\mathrm{d}x = \int_a^c f(x)\mathrm{d}x + \int_c^b f(x)\mathrm{d}x$$

7b-Methods for evaluating definite integrals (continued) (Some important properties of definite integrals and their applications)

[1] This might seem a little strange, but it is true. The reason being, for certain functions, it is basically simpler to find the difference between two particular values of the antiderivative $\phi(x)$ [i.e., $\phi(b)-\phi(a)$], than it is to find $\phi(x)$ itself. *Further Calculus* by F.L. Westwater (Teach Yourself Books) p. 104.

Introduction to Integral Calculus: Systematic Studies with Engineering Applications for Beginners, First Edition.
Ulrich L. Rohde, G. C. Jain, Ajay K. Poddar, and A. K. Ghosh.
© 2012 John Wiley & Sons, Inc. Published 2012 by John Wiley & Sons, Inc.

$$P_3: \int_a^b f(x)dx = \int_a^b f(a+b-x)dx$$

$$P_4: \int_0^a f(x)dx = \int_0^a f(a-x)dx$$

[Note that (P_4) is a particular case of (P_3)]

$$P_5: \int_0^{2a} f(x)dx = \int_0^a [f(x) + f(2a-x)]dx \text{ that is,}$$

$$\int_0^{2a} f(x)dx = \int_0^a f(x)dx + \int_0^a f(2a-x)dx$$

$$P_6: \text{[Deductions from property } (P_5)]$$

(i) If $f(2a-x) = f(x)$,

then,
$$\int_0^{2a} f(x) = \int_0^a f(x)dx + \int_0^a f(x)dx = 2\int_0^a f(x)dx$$

(ii) If $f(2a-x) = -f(x)$,

then,
$$\int_0^{2a} f(x)dx = \int_0^a f(x)dx - \int_0^a f(x)dx = 0$$

$$P_7: \int_{-a}^a f(x)dx = \int_0^a [f(x) + f(-x)]dx$$

(i) $\int_{-a}^a f(x)dx = 2\int_0^a f(x)dx$, if f is an even function, that is, $f(-x) = f(x)$.

(ii) $\int_{-a}^a f(x)dx = 0$, if f is an odd function, that is, $f(-x) = -f(x)$.

7b.3   PROOF OF PROPERTY (P_0)

Now, we shall discuss the proofs of the properties (P_0) to (P_4).

Proof of Property (P_0)

Method (I): To prove $\int_a^b f(x)dx = \int_a^b f(t)dt$

Consider the left-hand side, $\int_a^b f(x)dx$

Making the substitution, $x = t$, we get $dx = dt$.

Also, for $x = a$, $t = a$, and for $x = b$, $t = b$.

$$\therefore \quad \int_a^b f(x)\mathrm{d}x = \int_a^b f(t)\mathrm{d}t \quad \text{(Proved)}$$

Method (II):

Proof: Let $\int f(x)\mathrm{d}x = \phi(x)$ (1)

$$\therefore \quad \int_a^b f(x)\mathrm{d}x = [\phi(x)]_b^a = \phi(b) - \phi(a) \tag{2}$$

Again from (1), we have,

$$\therefore \quad \int_a^b f(t)\mathrm{d}t = [\phi(t)]_b^a = \phi(b) - \phi(a) \tag{3}$$

From (2) and (3), we get

$$\therefore \quad \int_a^b f(x)\mathrm{d}x = \int_a^b f(t)\mathrm{d}t \quad \text{(Proved)}$$

This property implies that the value of a definite integral does not depend on the variable of integration as long as "the element of integration" is same.

Proof of Property (P_1)

To prove $\int_a^b f(x)dx = -\int_a^b f(x)dx$

Proof: Let $\int f(x)\mathrm{d}x = \phi(x)$ (4)

$$\therefore \quad \int_a^b f(x)\mathrm{d}x = [\phi(x)]_b^a = \phi(b) - \phi(a) \tag{5}$$

Now, $-\int_a^b f(x)\mathrm{d}x = -[\phi(x)]_a^b = -[\phi(a) - \phi(b)] = \phi(b) - \phi(a)$ (6)

From Equation (5) and Equation (6), we get $\therefore \quad \int_a^b f(x)\mathrm{d}x = -\int_b^a f(x)\mathrm{d}x$ (Proved)

Note (1): This property implies that if the limits of a definite integral are mutually interchanged, then its sign changes. In practice, *this property is used for absorbing the negative sign while solving problems and establishing other results.*

Note (2): We know that, $\int_a^b f(x)\mathrm{d}x$ is defined only when $a < b$. But, in practice, there are occasions to consider $\int_a^b f(x)\mathrm{d}x$ when $a > b$. In such situations, *we consider $\int_a^b f(x)\mathrm{d}x$ to mean* $-\int_b^a f(x)\mathrm{d}x$, *where b is less than a. This permits us to evaluate the definite integral, using the fundamental theorem of Calculus.*

Such situations may arise when we make substitutions to compute $\int_a^b f(x)\mathrm{d}x$. Suppose, we substitute $x = \phi(t)$, the given integral becomes $\int_{\phi^{-1}(a)}^{\phi^{-1}(b)} f[\phi(t)]\phi'(t)\mathrm{d}t$.

Now, there is no guarantee that $\phi^{-1}(a) < \phi^{-1}(b)$. (Refer to Chapter 7a, for the method of substitution in definite integrals.)

Proof of Property (P_2)

To prove
$$\int_a^b f(x)\mathrm{d}x = \int_a^c f(x)\mathrm{d}x + \int_c^b f(x)\mathrm{d}x$$

This property is known as *Interval Additive Property.*[2]

Proof: Let $\int f(x)\mathrm{d}x = \phi(x)$

$$\therefore \int_a^c f(x)\mathrm{d}x + \int_c^b f(x)\mathrm{d}x$$

$$= [\phi(x)]_c^a + [\phi(x)]_b^c = \phi(c) - \phi(a) + \phi(b) - \phi(c)$$

$$= \phi(b) - \phi(a) = [\phi(x)]_b^a = \int_a^b f(x)\mathrm{d}x \qquad \text{(Proved)}$$

$$\therefore \int_a^b f(x)\mathrm{d}x = \int_a^c f(x)\mathrm{d}x + \int_c^b f(x)\mathrm{d}x$$

Extension: $\int_a^b f(x)\mathrm{d}x = \int_a^c f(x)\mathrm{d}x + \int_c^d f(x)\mathrm{d}x + \int_d^b f(x)\mathrm{d}x$, where $f(x)$ is integrable on $[a, b]$ and $a, b, c, d \in [a, b]$.

Proof of Property (P_3)

To prove
$$\int_a^b f(x)\mathrm{d}x = \int_a^b f(a + b - x)\mathrm{d}x$$

---

[2] The property reads as follows: If f is integrable on an interval containing the three points a, b, and c, then $\int_a^b f(x)\mathrm{d}x = \int_a^c f(x)\mathrm{d}x + \int_c^b f(x)\mathrm{d}x$, no matter what the order of a, b, and c. For example, $\int_0^2 x^2\mathrm{d}x = \int_0^1 x^2\mathrm{d}x + \int_1^2 x^2\mathrm{d}x$, which most people readily believe. But, it is also true that $\int_0^2 x^2\mathrm{d}x = \int_0^3 x^2\mathrm{d}x + \int_3^2 x^2\mathrm{d}x$, which appears to be surprising. Note that, the function x^2 is integrable on any closed interval $[0, x]$. One may actually evaluate the above integrals to check that the equality holds. The reader will appreciate this property better after understanding the relation between a definite integral and the area under a curve $[y = f(x)]$, where $f(x)$ is integrable on an interval containing the points a, b, and c.

Proof: Consider right-hand side

Let $I = \displaystyle\int_b^b f(a+b-x)\mathrm{d}x$

Put $a+b-x = t$ $\therefore -\mathrm{d}x = \mathrm{d}t$ $\therefore \mathrm{d}x = -\mathrm{d}t$

When $x = a, t = a+b-a = b$, When $x = b, t = a+b-b = a$

$$\therefore \quad I = \int_a^b f(a+b-c)\mathrm{d}x = \int_b^a f(t)(-\mathrm{d}t) = -\int_b^a f(t)\mathrm{d}t$$

$$= \int_a^b f(t)\mathrm{d}t, \left[\quad \because \quad \int_b^a f(t)\mathrm{d}t = -\int_b^a f(t)\mathrm{d}t \right]$$

$$= \int_a^b f(x)\mathrm{d}x, \left[\quad \because \quad \int_a^b f(t)\mathrm{d}t = \int_a^b f(x)\mathrm{d}x \right]$$

$$\therefore \int_a^b f(x)\mathrm{d}x = \int_a^b f(a+b-x)\mathrm{d}x \quad \text{(Proved)}$$

Proof of Property (P_4)

To prove
$$\int_0^a f(x)\mathrm{d}x = \int_0^a f(a-x)\mathrm{d}x$$

Observe that this property involves only the change of argument x into $(a-x)$ and *it does not involve a change in the limits of integration.* From this point of view, *this property is really an exceptional one.* Let us prove P_4.

Proof: Consider $\int_0^a f(a-x)\mathrm{d}x$

Let $a-x = t$ $\therefore -\mathrm{d}x = \mathrm{d}t$ $\therefore \mathrm{d}x = -\mathrm{d}t$

When $x = 0, t = a-0 = a$

When $x = a, t = a-a = 0$

$$\therefore \int_0^a f(a-x)\mathrm{d}x = \int_a^0 f(t)(-\mathrm{d}t)$$

$$= -\int_a^0 f(t)\mathrm{d}t$$

$$= \int_0^a f(t)\mathrm{d}t = \int_0^a f(x)\mathrm{d}x,$$

$$\therefore \int_0^a f(x)\mathrm{d}x = \int_0^a f(a-x)\mathrm{d}x \quad \text{(Proved)}$$

Remark: Note that, the property P_4 is a particular case of P_3 (i.e., property P_3 is more general than property P_4). However, this observation does not help in any way in solving problems. In fact, for solving problems, it is important to remember the following.

If both the limits are of the form a to b (i.e., 3 to 4 or $\pi/6$ to $\pi/3$, etc.), but different from the limits of the form $-a$ to a, then we must apply the property (P_3). On the other hand, if the limits are of the form 0 to a (i.e., 0 to $\pi/2$ or 0 to 4, etc.), where the *lower limit* is always *zero* and *the upper limit is a positive number*, then we must apply the property (P_4). (The definite integrals involving limits of integration of the form $-a$ to a are discussed under property P_7.)

The following illustrative examples [(1)–(9)] will make the situation clear.

Illustrative Examples

Example (1): Evaluate $\displaystyle\int_5^4 \frac{\sqrt{x+5}}{\sqrt{x+5}+\sqrt{9-x}}\,dx$

Solution: Let $\displaystyle I = \int_0^4 \left(\frac{\sqrt{x+5}}{\sqrt{x+5}+\sqrt{9-x}}\,dx\right)$ $\qquad(7)$

$$\therefore\quad I = \int_0^4 \frac{\sqrt{(4-x)+5}}{\sqrt{(4-x)+5}+\sqrt{9-(4-x)}}\,dx, \quad \left[\because \int_0^a f(x)dx = \int_0^a f(a-x)dx\right]$$

$$I = \int_0^4 \frac{\sqrt{9-x}}{\sqrt{9-x}+\sqrt{5+x}} \qquad(8)$$

$$I = \int_0^4 \frac{\sqrt{9-x}}{\sqrt{x+5}+\sqrt{9-x}}$$

Adding Equation (7) and Equation (8), we get

$$2I = \int_0^4 \frac{\sqrt{x+5}+\sqrt{9-x}}{\sqrt{x+5}+\sqrt{9-x}}\,dx$$

$$= \int_5^4 1\cdot dx = [x]_0^4 = (4-0) = 4$$

$$\therefore\quad I = \frac{4}{2} \qquad \therefore\quad I = 2 \qquad \text{Ans.}$$

Example (2): Evaluate $\displaystyle\int_4^5 \frac{\sqrt{5-x}}{\sqrt{x-4}+\sqrt{5-x}}\,dx$

Solution: Let $I = \int\limits_{4}^{5} \dfrac{\sqrt{5-x}}{\sqrt{x-4}+\sqrt{5-x}}dx$
$\qquad\qquad$ (9)

$$\therefore \quad I = \int\limits_{4}^{5} \dfrac{\sqrt{5-(\overline{4+5}-x)}}{\sqrt{(\overline{4+5})-x-4}+\sqrt{5-(\overline{4+5}-x)}}dx,$$

$\qquad\qquad$ (10)

$$I = \int\limits_{4}^{5} \dfrac{\sqrt{x-4}}{\sqrt{5-x}+\sqrt{x-4}}dx$$

Adding Equation (9) and Equation (10), we get

$$2I = \int\limits_{4}^{5} \dfrac{\sqrt{5-x}+\sqrt{x-4}}{\sqrt{5-x}+\sqrt{x-4}}dx$$

$$= \int\limits_{4}^{5} 1 \cdot dx = [x]_4^5 = 5 - 4 = 1$$

$$\therefore \quad I = \dfrac{1}{2} \qquad \text{Ans.}$$

Example (3): Evaluate $\displaystyle\int\limits_{1}^{4} \dfrac{\sqrt[3]{x+6}}{\sqrt[3]{x+6}+\sqrt[3]{11-x}}dx$

Solution:

Let $\qquad\qquad I = \int\limits_{1}^{4} \dfrac{\sqrt[3]{x+6}}{\sqrt[3]{x+6}+\sqrt[3]{11-x}}dx$
$\qquad\qquad$ (11)

We know that $\int\limits_{b}^{b} f(x)dx = \int\limits_{b}^{b} f(\overline{a+b}-x)dx$

Applying this property, we get

$$I = \int\limits_{1}^{4} \dfrac{\sqrt[3]{(\overline{1+4}-x)+6}}{\sqrt[3]{(\overline{1+4}-x)+6}+\sqrt[3]{11-(\overline{1+4}-x)}}dx$$

$\qquad\qquad$ (12)

$$I = \int\limits_{1}^{4} \dfrac{\sqrt[3]{11-x}}{\sqrt[3]{11-x}+\sqrt[3]{x+6}}dx$$

Adding Equation (11) and Equation (12), we get

$$2I = \int\limits_1^4 \frac{\sqrt[3]{x+6}+\sqrt[3]{11-x}}{\sqrt[3]{11-x}+\sqrt[3]{x+6}}dx = \int\limits_1^4 1 \cdot dx$$

$$2I = [x]_1^4 = 4 - 1 = 3 \qquad \therefore I = 3/2 \qquad \text{Ans.}$$

Example (4): Evaluate $\displaystyle\int\limits_0^4 \frac{\sqrt[3]{x+5}}{\sqrt[3]{x+5}+\sqrt[3]{9-x}}dx = I$ (say)   (13)

Solution: We know that $\displaystyle\int\limits_0^a f(x)dx = \int\limits_0^a f(a-x)dx$

Applying this property, we get

$$I = \int\limits_0^4 \frac{\sqrt[3]{(4-x)+5}}{\sqrt[3]{(4-x)+5}+\sqrt[3]{9-(4-x)}}dx$$

(14)

$$I = \int\limits_0^4 \frac{\sqrt[3]{9-x}}{\sqrt[3]{9-x}+\sqrt[3]{x+5}}dx$$

Adding Equation (13) and Equation (14), we get

$$2I = \int\limits_0^4 \frac{\sqrt[3]{x+5}+\sqrt[3]{9-x}}{\sqrt[3]{9-x}+\sqrt[3]{x+5}}dx$$

$$2I = \int\limits_5^4 1 \cdot dx = [x]_0^4 = 4 - 0 = 4 \quad \therefore \quad I = 2 \qquad \text{Ans.}$$

Example (5): Evaluate $\displaystyle\int\limits_{10}^{15} \frac{\sqrt[4]{x}}{\sqrt[4]{25-x}+\sqrt[4]{x}}dx^{(3)}$

Solution: Let $\displaystyle I = \int\limits_{10}^{15} \frac{\sqrt[4]{x}}{\sqrt[4]{25-x}+\sqrt[4]{x}}dx$ (15)

---

[3] Look carefully at the integrand and also the limits of integration. Obviously, property (P_3) is applicable here. Observe that the number 25 which appears in the denominator of the integrand, equals the sum of the limits of integration.

$$\therefore \quad I = \int\limits_{10}^{15} \frac{\sqrt[4]{10+15-x}}{\sqrt[4]{25-(10+15-x)}+\sqrt[4]{10+15-x}}\,\mathrm{d}x$$

$$I = \int\limits_{10}^{15} \frac{\sqrt[4]{25-x}}{\sqrt[4]{x}+\sqrt[4]{25-x}}\,\mathrm{d}x \tag{16}$$

Adding Equation (15) and Equation (16), we get

$$2I = \int\limits_{10}^{15} \frac{\sqrt[4]{x}+\sqrt[4]{25-x}}{\sqrt[4]{x}+\sqrt[4]{25-x}}\,\mathrm{d}x$$

$$= \int\limits_{10}^{15} 1 \cdot \mathrm{d}x = [\,x\,]_{10}^{15} = 15-10 = 5$$

$$\therefore \quad I = \frac{5}{2} \quad \text{Ans.}$$

Example (6): Evaluate $\int\limits_{\pi/6}^{\pi/3} \frac{1}{1+\sqrt[3]{\tan x}}\,\mathrm{d}x = I \quad \text{(say)}$

Solution:

$$\therefore \quad I = \int\limits_{\pi/6}^{\pi/3} \frac{1}{1+\frac{\sqrt[3]{\sin x}}{\sqrt[3]{\cos x}}}\,\mathrm{d}x = \int\limits_{\pi/6}^{\pi/3} \frac{\sqrt[3]{\cos x}}{\sqrt[3]{\cos x}+\sqrt[3]{\sin x}}\,\mathrm{d}x \tag{17}$$

$$I = \int\limits_{\pi/6}^{\pi/3} \frac{\sqrt[3]{\cos\left(\frac{\pi}{6}+\frac{\pi}{3}-x\right)}}{\sqrt[3]{\cos\left(\frac{\pi}{6}+\frac{\pi}{3}-x\right)}+\sqrt[3]{\sin\left(\frac{\pi}{6}+\frac{\pi}{3}-x\right)}}\,\mathrm{d}x$$

$$\left[\because \int\limits_{a}^{b} f(x)\mathrm{d}x = \int\limits_{a}^{b} f(\overline{a+b}-x)\mathrm{d}x\right] \tag{18}$$

$$I = \int\limits_{\pi/6}^{\pi/3} \frac{\sqrt[3]{\cos\left(\frac{\pi}{2}-x\right)}}{\sqrt[3]{\cos\left(\frac{\pi}{2}-x\right)}+\sqrt[3]{\sin\left(\frac{\pi}{2}-x\right)}}\,\mathrm{d}x$$

$$I = \int\limits_{\pi/6}^{\pi/3} \frac{\sqrt[3]{\sin x}}{\sqrt[3]{\sin x}+\sqrt[3]{\cos x}}\,\mathrm{d}x$$

Adding Equation (17) and Equation (18), we get

$$2I = \int_{\pi/6}^{\pi/3} \frac{\sqrt[3]{\cos x} + \sqrt[3]{\sin x}}{\sqrt[3]{\sin x} + \sqrt[3]{\cos x}} dx$$

$$= \int_{\pi/6}^{\pi/3} 1 \cdot dx = [x]_{\pi/6}^{\pi/3} = \frac{\pi}{3} - \frac{\pi}{6} = \frac{\pi}{6} \quad \therefore \quad I = \frac{\pi}{12} \quad \text{Ans.}$$

Example (7): Evaluate $\int_0^a x^2 (a-x)^{3/2} dx$

Solution: Let $I = \int_0^a x^2 (a-x)^{3/2} dx$ [4]

$$\therefore \quad I = \int_0^a (a-x)^2 [a - (a-x)]^{3/2} dx \quad \left[\because \int_0^a f(x)dx = \int_0^a f(a-x)dx \right]$$

$$= \int_0^a (a-x)^2 x^{3/2} dx = \int_0^a (a^2 - 2ax + x^2) x^{3/2} dx$$

$$= \int_0^a (a^2 x^{3/2} - 2ax^{5/2} + x^{7/2}) dx$$

$$= \left[\frac{a^2 \cdot x^{5/2}}{5/2} - 2a \frac{x^{7/2}}{7/2} + \frac{x^{9/2}}{9/2} \right]_0^a$$

$$= \left[\frac{2}{5} a^2 \cdot a^{5/2} - \frac{4}{7} \cdot a \cdot a^{7/2} + \frac{2}{9} a^{9/2} \right] - 0$$

$$I = \frac{2}{5} a^{9/2} - \frac{4}{7} a^{9/2} + \frac{2}{9} a^{9/2}$$

$$I = \frac{(126 - 180 + 70)}{315} a^{9/2} = \frac{16}{315} a^{9/2} \quad \text{Ans.}$$

Example (8): Evaluate $\int_0^1 x^2 (1-x)^{5/2} dx$ [5]

---

[4] Look carefully at the integrand and also the limits of integration. Obviously, the property (P_4) has to be applied here.
[5] Look carefully at the integrand and also the limits of integration. As in the earlier example, the property (P_4) has to be applied here.

Solution: Let $I = \int\limits_0^1 x^2(1-x)^{5/2}dx$

$$\therefore \quad I = \int\limits_0^1 (1-x)^2[1-(1-x)]^{5/2}dx$$

$$\left[\because \int\limits_0^a f(x)dx = \int\limits_0^a f(a-x)dx \right]$$

$$= \int\limits_0^1 (1-x)^2 x^{5/2}dx = \int\limits_0^1 (1-2ax+x^2)x^{5/2}dx$$

$$= \int\limits_0^1 (x^{5/2} - 2x^{7/2} + x^{9/2})dx$$

$$= \left[\frac{x^{7/2}}{7/2} - 2\frac{x^{9/2}}{9/2} + \frac{x^{11/2}}{11/2} \right]_0^1$$

$$= \left[\frac{2}{7}x^{7/2} - \frac{4}{9}x^{9/2} + \frac{2}{11}x^{11/2} \right]_0^1 = \left[\frac{2}{7} - \frac{4}{9} + \frac{2}{11} \right] - 0$$

$$= \frac{198 - 308 + 126}{7 \times 9 \times 11} = \frac{16}{693} \qquad \text{Ans.}$$

Example (9): Evaluate $\int\limits_0^a \frac{dx}{x+\sqrt{a^2-x^2}}$

Let

$$I = \int\limits_0^a \frac{dx}{x + \sqrt{a^2 - x^2}}$$

Let $\quad x = a\sin t \qquad \therefore \quad dx = a\cos t\, dt$

When $x = 0, a\sin t = 0 \qquad \therefore \quad t = 0$

When $x = a, a\sin t = a \qquad \therefore \quad \sin t = 1 \qquad \therefore \quad t = \pi/2$

$$\therefore \quad I = \int\limits_0^{\pi/2} \frac{a\cos t\, dt}{a\sin t + \sqrt{a^2 - a^2\sin^2 t}}$$

$$= \int\limits_0^{\pi/2} \frac{a\cos t}{a\sin t + a\cos t}dt \quad \left[\because 1 - \sin^2 t = \cos^2 t \right] \qquad (19)$$

$$= \int\limits_0^{\pi/2} \frac{\cos t}{\sin t + \cos t}dt$$

$$I = \int\limits_{0}^{\pi/2} \frac{\cos(\pi/2 - t)}{\sin(\pi/2 - t) + \cos(\pi/2 - t)} dt$$

$$= \int\limits_{0}^{\pi/2} \frac{\sin t}{\cos t + \sin t}$$

(20)

Adding Equation (19) and Equation (20), we get

$$2I = \int\limits_{0}^{\pi/2} \frac{\cos t + \sin t}{\cos t + \sin t} dt = \int\limits_{0}^{\pi/2} 1 \cdot dt = [\, t\,]_{0}^{\pi/2}$$

$$2I = \pi/2 - 0 = \pi/2 \quad \therefore \quad I = \frac{\pi}{4} \qquad \text{Ans.}$$

Note that the properties P_3 and P_4 are very powerful, whenever they can be applied.

7b.3.1   Some More Definite Integrals Involving Complicated Integrands

Suppose it is convenient *to evaluate the definite integral* $\int_0^a f(x)dx$, whereas we have to evaluate $\int_0^a xf(x)dx$. *Here, the presence of the factor x is undesirable.* It is sometimes possible to remove the factor x, if *the condition* $f(x)=f(a–x)$ *is satisfied.* In particular, if $f(x)$ involves trigonometric functions and the limits of integration are 0 to π, then by *applying the property*

$$\int\limits_{0}^{a} f(x)dx = \int\limits_{0}^{a} f(a - x)dx,$$

we can remove the factor x, and evaluate the integral, *as will be clear from the following illustrative examples:*[6]

Illustrative Examples

Example (10):   Evaluate $\displaystyle\int\limits_{0}^{\pi} \frac{x \sin x}{1 + \cos^2 x} dx = I$ (say)   (21)

Solution: Here $f(x) = \sin x/(1 + \cos^2 x)$ and *the upper limit* $a = \pi$. Now,

$$f(\pi - x) = \frac{\sin(\pi - x)}{1 + \cos^2(\pi - x)} = \frac{\sin x}{1 + \cos^2 x} = f(x)$$

---

[6] The definite integral $\int_0^\pi x \cos^3 x dx$ cannot be evaluated by this method since the condition $f(x)=f(\pi-x)$ is not satisfied. Accordingly, x cannot be removed. [Observe that $f(\pi - x) = [\cos(\pi - x)]^3 = (-\cos x)^3 = -\cos^3 x \neq \cos^3 x$].

Note that $\cos^2(\pi - x) = [\cos(\pi - x)]^2 = (-\cos x)^2 = \cos^2 x$

$$\therefore \quad I = \int_0^\pi \frac{(\pi - x)\sin(\pi - x)}{1 + \cos^2(\pi - x)}\,dx = \int_0^\pi \frac{(\pi - x)\sin x}{1 + \cos^2 x}$$

$$\left[\because \int_0^a f(x)dx = \int_0^a f(a - x)dx \right] \tag{22}$$

$$\therefore \quad I = \int_0^\pi \frac{\pi \sin x}{1 + \cos^2 x}\,dx - \int_0^\pi \frac{x \sin x}{1 + \cos^2 x}\,dx$$

Adding Equation (21) and Equation (22), we get $2I = \pi \int_0^\pi \frac{\sin x}{1 + \cos^2 x}\,dx$

Put $\cos x = t$, $\therefore -\sin x\,dx = dt$ $\therefore \sin x\,dx = -dt$

When $x = 0$, $t = \cos 0 = 1$ and when $x = \pi$, $t = \cos \pi = -1$

$$\therefore \quad 2I = \pi \int_0^\pi \frac{\sin x}{1 + \cos^2 x}\,dx = \pi \int_1^{-1} \frac{-dt}{1 + t^2} = -\pi \int_1^{-1} \frac{-dt}{1 + t^2} = \pi \int_{-1}^1 \frac{dt}{1 + t^2}$$

$$= \pi[\tan^{-1} t]_{-1}^1 = \pi[\tan^{-1} 1 - \tan^{-1}(-1)] = \pi\left[\frac{\pi}{4} - \left(-\frac{\pi}{4}\right)\right] = \pi\left(\frac{\pi}{2}\right) = \frac{\pi^2}{2}$$

$$\therefore \quad I = \frac{1}{2} \cdot \frac{\pi^2}{2} = \frac{\pi^2}{4} \qquad \text{Ans.}$$

Example (11): Evaluate $\int_0^\pi \frac{x}{1 + \sin x}\,dx$

We know how to integrate $\int \frac{dx}{1 + \sin x}$, but we do not know how to evaluate the integral $\int \frac{x}{1 + \sin x}\,dx$.

Solution: Let $I = \int_0^\pi \frac{x}{1 + \sin x}\,dx$ $\qquad\qquad$ (23)

We note that $\sin(\pi - x) = \sin x$

$$\therefore \quad I = \int_0^\pi \frac{(\pi - x)}{1 + \sin(\pi - x)}\,dx, \qquad \left[\because \int_0^a f(x)dx = \int_0^a f(a - x)dx \right]$$

$$\tag{24}$$

$$\therefore \quad I = \int_0^\pi \frac{\pi - x}{1 + \sin x}\,dx = \int_0^\pi \frac{\pi}{1 + \sin x}\,dx - \int_0^\pi \frac{x}{1 + \sin x}\,dx$$

Adding Equation (23) and Equation (24), we get

$$2I = \pi \int_0^\pi \frac{1}{1+\sin x}dx \quad \therefore \quad I = \frac{\pi}{2}\int_0^\pi \frac{1}{1+\sin x}dx$$

Consider

$$\frac{1}{1+\sin x} = \frac{1}{1+\sin x} \cdot \frac{1-\sin x}{1-\sin x} = \frac{1-\sin x}{1-\sin^2 x} = \frac{1-\sin x}{\cos^2 x} = \sec^2 x - \sec x \cdot \tan x$$

$$\therefore \quad I = \frac{\pi}{2}\int_0^\pi (\sec^2 x - \sec x \cdot \tan x)dx = \frac{\pi}{2}[\tan x - \sec x]_0^\pi$$

$$= \frac{\pi}{2}[\{0-(-1)\} - \{0-1\}], \quad [\tan \pi = 0, \quad \sec \pi = -1, \quad \tan 0 = 0, \quad \sec 0 = 1]$$

$$= \frac{\pi}{2}[1+1] = \pi \quad \text{Ans.}$$

Example (12): Evaluate $\int_0^\pi x \sin x \cos^2 x \, dx$

Solution: Let $I = \int_0^\pi x \sin x \cos^2 x \, dx$ $\qquad\qquad\qquad\qquad\qquad$ (25)

Observe that $\sin x \cos^2 x = \sin(\pi-x).[\cos(\pi-x)]^2$ [i.e., $f(x)=f(a-x)$]

$$\text{Now,} \quad I = \int_0^\pi (\pi - x)\sin(\pi - x) \cdot \cos^2(\pi - x)dx \quad \left[\because \int_0^a f(x)dx = \int_0^a f(a-x)dx\right]$$

$$\therefore \quad I = \int_0^\pi (\pi - x)\sin(\pi - x) \cdot \cos^2(\pi - x)dx = \int_0^\pi (\pi - x)\sin x \cdot \cos^2 x dx \qquad (26)$$

$$= \int_0^\pi \sin x\cos^2 x dx - \int_0^\pi x\sin x \cdot \cos^2 x dx$$

Adding Equation (25) and Equation (26), we get

$$2I = \pi \int_0^\pi \sin x \cdot \cos^2 x dx$$

$$\therefore \quad I = \frac{\pi}{2}\int_0^\pi - \cos^2 x(-\sin x)dx = -\frac{\pi}{2}\int_0^\pi - \cos^2 x(-\sin x)dx$$

Put $\cos x = t \therefore -\sin x \, dx = dt$

When $x = 0$, $t = \cos 0 = 1$
   and when $x = \pi$, $t = \cos \pi = -1$

$$\therefore \quad I = -\frac{\pi}{2} \int_{1}^{-1} t^2 \mathrm{d}t = \frac{\pi}{2} \int_{-1}^{1} t^2 \mathrm{d}t$$

$$= \frac{\pi}{2}\left[\frac{t^3}{3}\right]_{-1}^{1} = \frac{\pi}{2}\left[\frac{(1)^3}{3} - \frac{(-1)^3}{3}\right]$$

$$= \frac{\pi}{2} \cdot \frac{2}{3} = \frac{\pi}{3} \qquad \text{Ans.}$$

Example (13): Evaluate $\displaystyle\int_{0}^{\pi} \frac{x\mathrm{d}x}{a^2 \cos^2 x + b^2 \sin^2} x\mathrm{d}x = I \quad$ (say) $\hfill(27)$

This problem is of the type $\int_0^a x \cdot f(x)\mathrm{d}x$, where $f(x)$ does not change when x is replaced by $(a-x)$. Hence the factor x can be removed by using the property $\int_0^a f(x)\mathrm{d}x = \int_0^a f(a - x)\mathrm{d}x$.

$$\therefore \quad I = \int_0^\pi \frac{\pi - x}{a^2 \cos^2 x + b^2 \sin^2 x}\mathrm{d}x \hfill(28)$$

Adding Equation (27) and Equation (28), we get

$$2I = \int_0^\pi \frac{\pi}{a^2 \cos^2 x + b^2 \sin^2 x}\mathrm{d}x \qquad [x \text{ is removed}]$$

$$\therefore \quad I = \frac{\pi}{2} \int_0^\pi \frac{\mathrm{d}x}{a^2 \cos^2 x + b^2 \sin^2 x} \quad^{(7)}$$

$$\therefore \quad I = 2\left[\frac{\pi}{2} \int_0^{\pi/2} \frac{\mathrm{d}x}{a^2 \cos^2 x + b^2 \sin^2 x}\right]$$

$$= \pi \int_0^{\pi/2} \frac{\mathrm{d}x}{a^2 \cos^2 x + b^2 \sin^2 x}$$

Dividing N^{r} and D^{r} by $\cos^2 x$, we get

$$I = \pi \int_0^{\pi/2} \frac{\sec^2 x\mathrm{d}x}{a^2 + b^2 \tan^2 x}\mathrm{d}x$$

---

[7] Recall that, $\int_0^{2a} f(x)\mathrm{d}x = 2\int_0^a f(x)\mathrm{d}x$ if $f(2a-x) = f(x)$. Here, the upper limit $\pi = 2 \times \pi/2$ $(=2a)$, and $f(2a - x) = a^2 \cos^2(\pi-x) + b^2 \sin^2(\pi-x) = a^2 \cos^2 x + b^2 \sin^2 x = f(x)$.

Put $\tan x = t \therefore \sec^2 x\, dx = dt$

When $x = 0$, $t = \tan 0 = 0$, and when $x = \pi/2$, $t = \tan \pi/2 = \infty$ (Thus, when $x \to \pi/2$, $t \to \infty$)

$$\therefore I = \pi \int_0^{\pi/2} \frac{\sec^2 x\, dx}{a^2 + b^2 \tan^2 x} = \pi \int_0^{\infty} \frac{dt}{a^2 + b^2 t^2} = \pi \int_0^{\infty} \frac{dt}{b^2 + (a^2/b^2) + t^2}$$

$$= \frac{\pi}{b^2} \int_0^{\infty} \frac{dt}{(a/b)^2 + t^2} = \frac{\pi}{b^2} \left[\frac{1}{(a/b)} \tan^{-1} \frac{t}{(a/b)} \right]_0^{\infty}$$

$$= \frac{\pi}{b^2} \left[\frac{b}{a} \tan^{-1} \frac{b}{a} t \right]_0^{\infty} = \frac{\pi}{ab} \left[\tan^{-1} \infty - \tan^{-1} 0 \right]$$

$$= \frac{\pi}{ab} \left[\frac{\pi}{2} - 0 \right] = \frac{\pi^2}{2ab} \qquad \text{Ans.}$$

Exercise

Evaluate the following integrals:

(1) $\int_0^{\pi} x \log \sin x\, dx$

Ans. $\frac{\pi^2}{2} \log \frac{1}{2}$

(2) $\int_0^{\pi} x \cdot \cos^2 x\, dx$

Ans. $\frac{\pi}{4}$

(3) $\int_0^{\pi} \frac{x \sec x \tan x}{2 + \tan^2 x}\, dx$

Ans. $-\frac{\pi^2}{4}$

(4) $\int_0^{\pi} \frac{x \sin x}{1 + \sin x}\, dx$

Ans. $\pi \left(\frac{\pi}{2} - 1 \right)$

7b.4   PROOF OF PROPERTY (P_5)

To prove $\int_0^{2a} f(x)dx = \int_0^a f(x)dx + \int_0^a f(2a - x)dx$

(Observe that the limits of integration on the left-hand side are 0 to $2a$, whereas those on the right-hand side are 0 to a.)

Proof:

$$\text{We have, } \int\limits_0^{2a} f(x)\mathrm{d}x = \int\limits_0^a f(x)\mathrm{d}x + \int\limits_a^{2a} f(x)\mathrm{d}x \quad (\text{Using } P_2) \tag{29}$$

[Now, we must show that, $\int\limits_a^{2a} f(x)\mathrm{d}x = \int\limits_0^a f(2a - x)\mathrm{d}x$]

Put $x = (2a-t)$, in the integral $\int\limits_a^{2a} f(x)\mathrm{d}x$ [Imp. Substitution]

$$\therefore \quad t = 2a - x, \text{ and } \mathrm{d}t = -\mathrm{d}x$$

when $x = a$, $t = a$, and when $x = 2a$, $t = 0$.

$$\therefore \int\limits_a^{2a} f(x)\mathrm{d}x = \int\limits_a^0 f(2a - t)(-\mathrm{d}t) = -\int\limits_a^0 f(2a - t)\mathrm{d}t$$

$$= \int\limits_0^a f(2a - t)\mathrm{d}t \quad (\text{using } P_1)$$

$$= \int\limits_0^a f(2a - x)\mathrm{d}x \quad (\text{using } P_0)$$

$\therefore$ Substituting in Equation (29), we get,

$$\therefore \quad \int\limits_0^{2a} f(x)\mathrm{d}x = \int\limits_0^a f(x)\mathrm{d}x + \int\limits_0^a f(2a - x)\mathrm{d}x \quad (\text{Proved})$$

Property (P_5) is particularly important in evaluating definite integrals of the form $\int_0^\pi \sin^n x \, \mathrm{d}x$, $\int_0^\pi \cos^n x \, \mathrm{d}x$, *and* $\int_0^\pi \cos^m x \cdot \sin^n x \, \mathrm{d}x$, *where* m, $n \in N$. In Section 7b.3.1, we have already evaluated some definite integrals of the form $\int_0^\pi x f(x)\mathrm{d}x$, by removing the undesirable factor x.

[Note that, in all these integrals, the lower limit is 0 and the upper limit is π, which is looked upon as $2 \cdot (\pi/2)$].

Observe that, under the property P_5, the given integral $\int_0^\pi f(x)\mathrm{d}x$ is expressed as a sum of two integrals in which the upper limit $2 \cdot (\pi/2)$ is halved to $(\pi/2)$. Besides, there is a change in the argument of one integral from x to $(\pi-x)$. The integrals on the right-hand side of the property (P_5) suggest two deductions from this property. These deductions are taken as the statement of property (P_6).

7b.4.1 Property (P_6) [Deductions from the Property (P_5)]

From the property (P_5) [i.e., $\int_0^{2a} f(x)\mathrm{d}x = \int_0^a f(x)\mathrm{d}x + \int_0^a f(2a - x)\mathrm{d}x$], we note that
If $f(2a-x) = f(x)$, then

$$\int\limits_0^{2a} f(x)\mathrm{d}x = \int\limits_0^a f(x)\mathrm{d}x + \int\limits_0^a f(x)\mathrm{d}x = 2\int\limits_0^a f(x)\mathrm{d}x \tag{I}$$

If $f(2a-x) = -f(x)$, then

$$\int_0^{2a} f(x)dx = \int_0^a f(x)dx - \int_0^a f(x)dx = 0 \tag{II}$$

Note: Let us examine, how the statements (I) and (II) are useful.

We know that, $\sin(\pi-x) = \sin x$. $\therefore \sin^n(\pi-x) = \sin^n x$
   Thus, *the condition $f(2a-x) = f(x)$ is satisfied.*

$$\therefore \int_0^{\pi} \sin^n x dx = 2 \int_0^{\pi/2} \sin^n x dx \tag{A}$$

Further, we know that, $\cos(\pi-x) = -\cos x$. Therefore, *if the power of* $\cos x$ *is even*, then we have,

$$\cos^{2n}(\pi - x) = (-\cos x)^{2n} = \cos^{2n} x,$$

which means that the condition $f(2a-x) = f(x)$ is satisfied.

$$\therefore \int_0^{\pi} \cos^{2n} x dx = 2 \int_0^{\pi/2} \cos^{2n} x dx \tag{B}$$

On the other hand, *if the power of* $\cos x$ *is odd*, we have,

$$\left[\cos^{2n+1}(\pi - x) = (-\cos x)^{2n+1} = -\cos^{2n+1} x \right]$$

which means that, *the condition $[f(2a-x) = -f(x)]$ is satisfied.*

$$\int_0^{\pi} \cos^{2n+1}(\pi - x)dx = 0 \tag{C}$$

The statements (I) and (II) (defining property P_6), can also be applied to evaluate integrals of the form

$$\int_0^{\pi} \cos^m x \cdot \sin^n x dx, \quad [m, n \in N]$$

If the power of $\cos x$ is an odd number, then we can *immediately* write $\int_0^{\pi} \cos^m x \cdot \sin^n x dx = 0$, *irrespective of whether power of* $\sin x$, *is odd or even*.
   This is simple. On the other hand, *if the power of* $\cos x$ *is an even number*, then we have

$$\int_0^{\pi} \cos^{2m} x \cdot \sin^n x dx = 2 \int_0^{\pi/2} \cos^{2m} x \cdot \sin^n x dx, \quad [m, n \in N]$$

Thus, our problem is reduced to evaluating integrals of the form.

$$\int_0^{\pi/2} \cos^{2m}x\,dx, \quad \int_0^{\pi/2} \sin^n x\,dx, \quad \int_0^{\pi/2} \cos^{2m}x \cdot \sin^n x\,dx, \quad [m, n \in N].$$

Of course, these integrals can be evaluated by using the *reduction formulae*, which are obtained by applying the *method of integration by parts*. Here, we shall not discuss the methods involving reduction formulae. *On the other hand, we shall be interested to check whether the given integral can be evaluated by the method of substitution.* Let us consider the following examples:

Example (14): To evaluate $\int_0^{\pi/2} \sin^5 x\,dx$ (note that the power of sin x is odd).

Solution: Let $I = \int_0^{\pi/2} \sin^5 x\,dx = \int_0^{\pi/2} \sin^4 x \cdot \sin x\,dx$

$$= \int_0^{\pi/2} (\sin^2 x)^2 \cdot \sin x\,dx$$

$$= \int_0^{\pi/2} (1 - \cos^2 x)^2 \cdot \sin x\,dx$$

Now we make the substitution, $\cos x = t$

$\therefore \quad -\sin x\,dx = dt$

$\therefore \quad \sin x\,dx = -dt$

$\therefore \quad$ when $x = 0$, $t = \cos 0 = 1$ and when $x = \pi/2$, $t = \cos(\pi/2) = 0$

$\therefore \quad I = \int_0^{\pi/2} (1 - \cos^2 x)^2 \cdot \sin x\,dx$

$$= \int_1^0 (1 - t^2)^2 \cdot (-dt) = -\int_1^0 (1 - 2t^2 + t^4) \cdot dt$$

$$= -\left[t - 2\frac{t^3}{3} + \frac{t^5}{5} \right]_1^0 = -\left[0 - \left(1 - \frac{2}{3} + \frac{1}{5} \right) \right]$$

$$= 1 - \frac{2}{3} + \frac{1}{5} = \frac{8}{15} \qquad \text{Ans.}$$

Similarly, we can compute $\int_0^{\pi/2} \cos^7 x\,dx$ (the power of cos x being odd) by substitution. $\int_0^{\pi/2} \sin^2 x \cdot \cos x\,dx$ can also be computed by substitution.

Of course, to compute integrals of the type given below, it will be convenient to use the reduction formulae.

$$\int\limits_{0}^{\pi/2} \sin^{10}x\,dx, \quad \int\limits_{0}^{\pi/2} \cos^{8}x\,dx, \quad \int\limits_{0}^{\pi/2} \sin^{2}x \cdot \cos^{4}x\,dx. \text{ (These integrals cannot be evaluated by}$$

substitution.)

Note that, $\int\limits_{0}^{\pi/2} \sin^{6}x \cdot \cos^{5}x\,dx$, $\int\limits_{0}^{\pi/2} \sin^{5}x \cdot \cos^{6}x\,dx$ and other similar integrals can be evaluated by substitution.

7b.5 DEFINITE INTEGRALS: TYPES OF FUNCTIONS

The next property of definite integrals is restricted to the types of functions called *even* and *odd* functions. Hence, before stating this property, *it is necessary to understand clearly the even* and *odd* functions.

This terminology is based on certain properties of even and odd integers, but even and odd functions do not have many properties of even and odd integers. For instance, whereas every integer is either even or odd, this is *not* true in the case of functions.

From the definitions of even and odd functions (being given below), it will also be observed that we can define certain functions that are neither even nor odd.

7b.5.1 Even Function

A function $f(x)$ *defined on an interval I* is called an *even* function if $f(-x) = f(x)$ for *all* $x \in I$.

Examples:

(1) Any polynomial function $p(x) = a_0 + a_1 x^2 + a_2 x^4 + \ldots + a_n x^{2n}$, in which there are *only even powers of x*, is an *even* function.

(2) A constant function $f(x) = a$, is an *even* function. Check this.

(3) We have seen that $\cos(-x) = \cos x$, for all x. Thus cosine function is an *even* function.

7b.5.2 Odd Function

A function $f(x)$ *defined on an interval I* is called an *odd* function if $f(-x) = -f(x)$ for all $x \in I$.

Examples:

(1) Any polynomial function in which there are *only odd powers of x* is an *odd* function. Thus, $f(x) = x, f(x) = x^3$ and $f(x) = x^5 + 4x^3 + x$, are *odd* functions.

(2) We have seen that $\sin(-x) = -\sin x$, and $\tan(-x) = -\tan x$, for *all* x. Thus, sine and tangent functions are *odd* functions.

Remark (1): Some functions may *be neither even nor odd*. For example, *sum or difference of one odd and one even function is neither* even *nor* odd.

Consider

$$f(x) = x^2 + x$$

$$\therefore -f(x) = -x^2 - x$$

Also, we have $f(-x) = (-x)^2 + (-x) = x^2 - x$.

Note that, $f(-x) \neq f(x)$, hence $f(x)$ is *not even*. Also, $f(-x) \neq -f(x)$, hence $f(x)$ is *not odd*. Similarly, $g(x) = \sin x + \cos x$, and $h(x) = x^2 + \sin x$, are neither *even* nor *odd*.

Remark (2): By definition, the sign of an odd function changes when the sign of its argument is changed.

The following are a *few rules, which decide whether a combination of even and odd functions will be even, odd or neither even nor odd.*

   (i) (Even function) $\pm$ (Even function) = Even function

  (ii) (Even function) $\times$ (Even function) = Even function

 (iii) $\frac{\text{Even function}}{\text{Even function}}$ = Even function

 (iv) (Odd function) $\pm$ (Odd function) = Odd function

  (v) (Odd function) $\times$ (Odd function) = Even function

 (vi) $\frac{\text{Odd function}}{\text{Odd function}}$ = Even function

(vii) (Even function) $\times$ (Odd function) = Odd function

(viii) $\begin{cases} \frac{\text{Even function}}{\text{Odd function}} = \text{Odd function} \\ \frac{\text{Odd function}}{\text{Even function}} = \text{Odd function} \end{cases}$

 (ix) (Even function) $\pm$ (Odd function) = *Neither even nor odd function*

Remark: Observe that in the rules (i) to (viii), an odd function behaves like a negative number. This observation is useful in deciding whether a given function is *even* or *odd*. If we simply remember that for an *odd* function $f(-x) = -f(x)$, then it is easy to remember all these rules.

It is easy to prove the above rules. For example, let us prove (i).

Let
$$h(x) = f(x) \pm g(x)$$

where $f(x)$ and $g(x)$ both are even functions.

Consider
$$h(-x) = f(-x) \pm g(-x)$$
$$= f(x) \pm g(x), \textit{by definition of an even function}$$
$$= h(x)$$
$$\therefore\ h(-x) = h(x)$$

Therefore, $h(x)$ is an even function. The rest of the rules can also be proved similarly. In solving problems, *the above rules may be directly used.*

We now prove the next property of definite integrals.

Property P_7:
$$\int_{-a}^{a} f(x)dx = \int_{0}^{a}[f(x)+f(-x)]dx = \int_{0}^{a} f(x)dx + \int_{0}^{a} f(-x)dx$$

Proof: We have
$$\int_{-a}^{a} f(x)dx = \int_{-a}^{0} f(x)dx + \int_{0}^{a} f(x)dx \ldots [\text{By } P_2] \tag{30}$$

Let $I = \int_{-a}^{0} f(x)dx$

[We will show that $\int_{-a}^{0} f(x)dx = \int_{0}^{a} f(-x)dx.$]

Put $x = -t \quad \therefore dx = -dt$

When $x = -a, -t = -a \quad \therefore t = a$

and when $x = 0, -t = 0 \quad \therefore t = 0$

$$\therefore \quad I = \int_{a}^{0} f(-t)(-dt) = -\int_{a}^{0} f(-t)dt$$

$$= \int_{0}^{a} f(-t)dt, \quad [\text{By using } P_1]$$

$$\therefore \int_{-a}^{0} f(x)dx = \int_{0}^{a} f(-x)dx$$

Substituting in Equation (30), we get

$$\int_{-a}^{a} f(x)dx = \int_{0}^{a} f(-x)dx + \int_{0}^{a} f(x)dx \tag{31}$$

$$\int_{-a}^{a} f(x)dx = \int_{0}^{a}[f(x)+f(-x)]dx \text{ (Proved)}$$

Deductions

(i) Let $f(x)$ be an *even* function.

$$\therefore \quad f(-x) = f(x) \forall x$$

∴ Substituting in Equation (31), we get

$$\int\limits_{-a}^{a} f(x)\mathrm{d}x = \int\limits_{0}^{a} f(x)\mathrm{d}x + \int\limits_{0}^{a} f(x)\mathrm{d}x$$

$$= 2\int\limits_{0}^{a} f(x)\mathrm{d}x, \quad \text{if } f(x) \text{ is even.}$$

(ii) Let $f(x)$ be odd

$$\therefore \quad f(-x) = -f(x), \forall x$$

Substituting in Equation (31), we get

$$\int\limits_{-a}^{a} f(x)\mathrm{d}x = \int\limits_{0}^{a} -f(x)\mathrm{d}x + \int\limits_{0}^{a} f(x)\mathrm{d}x$$

$$= 0, \quad \text{if } f(x) \text{ is odd.}$$

Note (1): *It is important to remember the statements of properties P_6 and P_7 and the limits of integration involved.*

If the limits of integration are of the form $-a$ to a [i.e., $(-\pi$ to $\pi)$ or $(-\pi/2$ to $\pi/2)$ or $(-5$ to $5)$, etc.], then we have to use property P_7 for evaluation of the integral, but if the limits of integration are of the form $(0$ to $a)$ or $(0$ to $\pi)$ [i.e., $(a$ to $b)$], then we must use a suitable property which may be P_3, P_4, P_5, or P_6.

Note (2): We emphasize that for deciding whether a given function is even or odd, *it is sufficient to use the rules stated above.* (In other words, it is not necessary to go step-by-step to find out whether the given function is even or odd.)

If the given function is odd, then its definite integral (from $-a$ to a) is zero, by the above property. On the other hand, if the given function $f(x)$ is even, then we have $\int_{-a}^{a} f(x)\mathrm{d}x = 2\int_0^a f(x)\mathrm{d}x$, which can be evaluated by applying the available methods and the properties of definite integrals.

Illustrative Examples

Example (15): Evaluate $\int\limits_{-\pi}^{\pi} \cos\left(\frac{x}{4}\right)\mathrm{d}x = I$ (say)

Solution: We know that $\cos(-t) = \cos t$

$$\therefore \quad f(x) = \cos\left(\frac{x}{4}\right) \text{ is an } even \text{ function.}$$

$$\therefore \quad I = \int\limits_{-\pi}^{\pi} \cos\left(\frac{x}{4}\right)\mathrm{d}x = 2\int\limits_{0}^{\pi} \cos\left(\frac{x}{4}\right)\mathrm{d}x, \quad [\text{Property (8)}]$$

Now, put $x/4 = t$ $\therefore dx = 4dt$

Also, for $x = 0$, $t = 0$ and for $x = \pi$, $t = \pi/4$

$$\therefore \quad I = 2 \int_0^{\pi/4} \cos t (4 \cdot dt) = 8 \int_0^{\pi/4} \cos t \, dt$$

$$= 8[\sin t]_0^{\pi/4} = 8 \left[\frac{\sqrt{2}}{2} - 0 \right] = 4\sqrt{2} \quad \text{Ans.}$$

Example (16): Evaluate $\int_{-3}^{3} \frac{x^3 + 5x}{2x^2 + 7} \, dx$

Solution: Here the integrand $f(x) = \frac{x^3 + 5x}{2x^2 + 7}$. Now, the N^r (i.e., $x^3 + 5x$) is an *odd* function and the D^r (i.e., $2x^2 + 7$) is an *even* function. $\therefore f(x)$ is an odd function.

$$\therefore \quad \int_{-3}^{3} \frac{x^3 + 5x}{2x^2 + 7} \, dx = 0 \quad \text{Ans.}$$

Example (17): Evaluate $\int_{-\pi/2}^{\pi/2} \sin^5 x \, dx$

Solution: Here $f(x) = \sin^5 x = (\sin x)^5$

$$f(-x) = [\sin(-x)]^5 = [-\sin x]^5 = -\sin^5 x = -f(x)$$
$$\therefore \quad f(x) \text{ is an odd function.}$$

$$\therefore \quad \int_{-\pi/2}^{\pi/2} \sin^5 x \, dx = 0 \quad \text{Ans.}$$

Example (18): Evaluate $\int_{-2}^{2} [x\sin^4 x + x^3 - x^4] dx$

Solution: *Here, the first two terms are odd functions and the last term is an even. Thus, we may write the integral as*

$$\int_{-2}^{2} (x\sin^4 x + x^3) dx - \int_{0}^{2} x^4 dx$$

$$= 0 - 2 \int_{0}^{2} x^4 dx = \left[-2\frac{x^5}{5} \right]_0^2 = \frac{-2}{5}(32 - 0) = \frac{-64}{5} \quad \text{Ans.}$$

Example (19): Evaluate $\int_{-\pi/2}^{\pi/2} \frac{\sin^4 x}{\sin^4 x + \cos^4 x} \, dx = I \quad \text{(say)}$

Here, the integrand $f(x) = \frac{\sin^4 x}{\sin^4 x + \cos^4 x}$ is an *even* function (since the $N^r = \sin^4 x$ is even and the $D^r = \sin^4 x + \cos^4 x$ is an even function).

$$\therefore \quad I = \int\limits_{-\pi/2}^{\pi/2} \frac{\sin^4 x}{\sin^4 x + \cos^4 x}\, dx = \int\limits_{0}^{\pi/2} \frac{\sin^4 x}{\sin^4 x + \cos^4 x}\, dx \qquad (32)$$

$$\therefore \quad I = 2 \int\limits_{0}^{\pi/2} \frac{\sin^4((\pi/2) - x)}{\sin^4((\pi/2) - x) + \cos^4((\pi/2) - x)}\, dx, \quad \left[\because \quad \int\limits_{0}^{a} f(x)\, dx = \int\limits_{0}^{a} f(a - x)\, dx \right]$$

$$= 2 \int\limits_{0}^{\pi/2} \frac{\cos^4 x}{\cos^4 x + \sin^4 x}\, dx \qquad (33)$$

Adding Equation (32) and Equation (33), we get

$$2I = 2 \int\limits_{0}^{\pi/2} \frac{\sin^4 x}{\sin^4 x + \cos^4 x}\, dx + 2 \int\limits_{0}^{\pi/2} \frac{\cos^4 x}{\sin^4 x + \cos^4 x}\, dx$$

$$= 2 \int\limits_{0}^{\pi/2} \frac{\sin^4 x + \cos^4 x}{\sin^4 x + \cos^4 x}\, dx = 2 \int\limits_{0}^{\pi/2} dx$$

$$= 2[x]_0^{\pi/2} = 2\left(\frac{\pi}{2} - 0\right) = \pi$$

$$\therefore \quad I = \frac{\pi}{2} \qquad \text{Ans.}$$

Exercise

Evaluate the following integrals:

(1) $\int\limits_{-a}^{a} \frac{1+x^3}{4-x^2}$

Ans. $\frac{1}{2} \log\left(\frac{2+a}{2-a}\right)$

$$\left[\text{Hint:} \frac{1 + x^3}{4 - x^2} = \frac{1}{4 - x^2} + \frac{x^3}{4 - x^2} \quad \text{(odd function)} \right]$$

(2) $\int\limits_{-\pi/4}^{\pi/4} \tan^2 x \sec x\, dx$

Ans. 0

Miscellaneous Solved Examples: An Important Definite Integral

Example (20): Prove that

(i) $\int_0^{\pi/2} \log_e(\sin x)dx - \frac{\pi}{2} \cdot \log 2 = \frac{\pi}{2} \cdot \log_e\left(\frac{1}{2}\right)$

(ii) $\int_0^{\pi/2} \log_e(\sin x)dx = \int_0^{\pi/2} \log_e(\cos x)dx$

Solution: Let $I = \int_0^{\pi/2} \log_e(\sin x)dx$ \hfill (34)

$$\therefore \quad I = \int_0^{\pi/2} \log_e\left(\sin\left(\frac{\pi}{2} - x\right)\right)dx, \quad \left[\text{sin ce } \int_0^a f(x)dx = \int_0^a f(a - x)dx\right]$$

$$= \int_0^{\pi/2} \log_e(\cos x)dx \hfill (35)$$

[Thus, the result (ii) is proved.]

To evaluate, $\int_0^{\pi/2} \log_e(\sin x)dx$, we add Equation (34) and Equation (35) and get

$$2I = \int_0^{\pi/2} \log_e(\sin x)dx + \int_0^{\pi/2} \log_e(\cos x)dx$$

$$= \int_0^{\pi/2} [\log_e(\sin x) + \log_e(\cos x)]dx = \int_0^{\pi/2} \log_e(\sin x \cdot \cos x)dx$$

$$= \int_0^{\pi/2} \log_e\left(\frac{\sin 2x}{2}\right)dx = \int_0^{\pi/2} [\log_e(\sin 2x) - \log_e 2]dx \quad \left[\because \sin x \cdot \cos x = \frac{1}{2}\sin 2x\right]$$

$$= \int_0^{\pi/2} \log_e(\sin 2x)dx - \log_e 2 \int_0^{\pi/2} dx$$

$$= \int_0^{\pi/2} \log_e(\sin 2x)dx - \log_e 2[x]_0^{\pi/2}$$

$$\therefore \quad 2I = \int_0^{\pi/2} \log_e(\sin 2x)dx - \frac{\pi}{2} \cdot \log_e 2 \hfill (36)$$

[Now we will show that $\int_0^{\pi/2} \log_e(\sin 2x)dx = I$.]

Consider $\int_0^{\pi/2} \log_e(\sin 2x)dx$

Put $2x = t$ $\therefore 2\,dx = dt$ $\therefore dx = \frac{1}{2}dt$

Limits, when $x = 0$, $t = 0$

and when $x = \frac{\pi}{2}$, $t = 2\left(\frac{\pi}{2}\right) = \pi$

$$\therefore \int_0^{\pi/2} \log_e(\sin 2x)dx = \int_0^{\pi} \log_e(\sin t) \cdot \frac{dt}{2}$$

$$= \frac{1}{2} \int_0^{\pi} \log_e(\sin t)dt$$

$$= \frac{1}{2} \int_0^{\pi/2} [\log_e(\sin t) + \log_e \sin(\pi - t)]dt \qquad \left[\because \int_0^{2a} f(x)dx = \int_0^a [f(x) + f(2a - x)]dx \right]$$

$$= \frac{1}{2} \int_0^{\pi/2} 2\log_e(\sin t)dt \quad [\because \sin(\pi - t) = \sin t]$$

$$= \int_0^{\pi/2} \log_e(\sin t)dt$$

$$= \int_0^{\pi/2} \log_e(\sin x)dx, \qquad \left[\because \int_a^b f(t)dt = \int_a^b f(x)dx \right]$$

$$= I$$

Therefore, from Equation (36), we get

$$2I = I - \frac{\pi}{2}\log_e 2 \qquad \therefore \quad I = -\frac{\pi}{2}\log_e 2$$

$$\therefore \int_0^{\pi/2} \log_e(\sin t)dt = -\frac{\pi}{2}\log_e \qquad \text{Ans.}$$

$$\text{Cor.} \int_0^{\pi/2} \log_e \cos x\, dx = -\pi/2 \log_e 2$$

Remark: It is important to note that $\int_0^{\pi/2} \log_e(\sin x)dx = \int_0^{\pi/2} \log_e(\sin 2x)dx$.

Example (21): Prove that $\displaystyle\int_0^{\pi/2} \log_e(\tan x)dx = 0$ $(37)^{(8)}$

Solution: Let $I = \displaystyle\int\limits_0^{\pi/2} \log_e(\tan x)dx$

$$\therefore \quad I = \int\limits_0^{\pi/2} \log_e\left[\tan\left(\frac{\pi}{2} - x\right)\right]dx = \int\limits_0^{\pi/2} \log_e(\cot x)dx \qquad (38)$$

Adding Equation (37) and Equation (38), we get

$$= \int\limits_0^{\pi/2} \log_e(\tan x \cdot \cot x)dx$$

$$= \int\limits_0^{\pi/2} \log_e(1)dx = \int\limits_0^{\pi/2} 0 \cdot dx \qquad [\because \quad \log_e 1 = 0]$$

$$= 0 \quad \text{Ans.}$$

Example (22): Evaluate $\displaystyle\int\limits_0^{\pi/4} \log_e(1 + \tan x)dx$

Solution: Let $I = \displaystyle\int\limits_0^{\pi/4} \log_e(1 + \tan x)dx$

Also $I = \displaystyle\int\limits_0^{\pi/4} \log_e[1 + \tan((\pi/4) - x)]dx$

$$= \int\limits_0^{\pi/4} \log_e\left[1 + \frac{\tan(\pi/4) - \tan x}{1 + \tan(\pi/4) \cdot \tan x}\right]dx$$

$$= \int\limits_0^{\pi/4} \log_e\left[1 + \frac{1 - \tan x}{1 + \tan x}\right], \quad [\because \quad \tan(\pi/4) = I] = \int\limits_0^{\pi/4} \log_e\left(\frac{2}{1 + \tan x}\right)dx$$

---

(8) Here, the lower limit is "0" and the upper limit is a positive number. Hence, the property P_4 [i.e., $\int_0^a f(x)dx = \int_0^a f(a - x)dx$] is applicable.

$$I = \int_0^{\pi/4} [\log_e 2 - \log_e(1 + \tan x)]dx$$

$$= \log_e 2 \int_0^{\pi/4} dx - \int_0^{\pi/4} \log_e(1 + \tan x)dx$$

$$\therefore \quad I = \log_e 2[x]_0^{\pi/4} - I \quad \therefore \quad 2I = \log_e 2\left[\frac{\pi}{4} - 0\right] = \frac{\pi}{4} \cdot \log_e 2$$

$$\therefore \quad I = \frac{\pi}{8}\log_e 2 \quad \text{Ans.}$$

Example (23): Evaluate $\int_0^1 \frac{\log_e(1+x)}{1+x^2}dx$

Solution: Let $I = \int_0^1 \frac{\log_e(1+x)}{1+x^2}dx$

Put $x = \tan t \quad \therefore \quad dx = \sec^2 t \, dt \quad \therefore \quad t = \tan^{-1}x$

When $x = 0$, $t = \tan^{-1} 0 = 0$ and when $x = 1$, $t = \tan^{-1} 1 = \pi/4$

$$\therefore \quad I = \int_0^{\pi/4} \frac{\log_e(1 + \tan t) \cdot \sec^2 t dt}{1 + \tan^2 t}$$

$$\therefore \quad I = \int_0^{\pi/4} \log_e(1 + \tan t)dt, \quad [\because 1 + \tan^2 t = \sec^2 t]$$

$$\therefore \quad I = \frac{\pi}{2}\log_e 2. \text{ [As already proved in Example (22)]} \quad \text{Ans.}$$

Example (24): Evaluate $\int_{-\pi/2}^{\pi/2} \sin^7 x \cdot dx$

Solution: Let $f(x) = \sin^7 x$

Consider $f(-x) = [\sin(-x)]^7 x = -\sin^7 x = -f(x)$

Thus, f is an odd function.

$$\therefore \quad \int_{-\pi/2}^{\pi/2} \sin^7 x \cdot dx = 0$$

Note: Whenever the limits of integration are of the form $-a$ to a [i.e., $\left(-\frac{\pi}{2} \text{ to } \frac{\pi}{2}\right)$ or $(-3 \text{ to } 3)$, etc.] then we must always check whether the integrand is an even or odd function, and proceed further.

Example (25): Prove that $\int\limits_{0}^{2a} \frac{f(x)}{f(x)+f(2a-x)} dx = a$

Solution: Let $I = \int\limits_{0}^{2a} \frac{f(x)}{f(x)+f(2a-x)} dx$ (39)

$$= \int\limits_{0}^{2a} \frac{f(2a-x)}{f(2a-x)+f[2a-(2a-x)]} dx \qquad \left[\because \int\limits_{0}^{a} f(x) \cdot dx = \int\limits_{0}^{a} f(a-x) \cdot dx \right]$$

$$\therefore \quad I = \int\limits_{0}^{2a} \frac{f(2a-x)}{f(2a-x)+f(x)} dx \qquad (40)$$

Adding Equation (39) and Equation (40), we get

$$2I = \int\limits_{0}^{2a} \frac{f(x)+f(2a-x)}{f(x)+f(2a-x)} dx = \int\limits_{0}^{2a} dx$$

$$= [x]_{0}^{2a} = 2a$$

$$\therefore \quad I = a \quad \text{(Proved)}$$

Example (26): Prove that $\int\limits_{a}^{b} \frac{f(x)}{f(x)+f(a+b-x)} dx = \frac{b-a}{2}$

Solution: Let $I = \int\limits_{a}^{b} \frac{f(x)}{f(x)+f(a+b-x)} dx$ (41)

$$= \int\limits_{a}^{b} \frac{f(a+b-x)}{f(a+b-x)+f[(a+b)-(a+b-x)]} dx \qquad [\text{by } P_3]$$

$$= \int\limits_{a}^{b} \frac{f(a+b-x)}{f(a+b-x)+f(x)} dx \qquad (42)$$

Adding Equation (41) and Equation (42), we get

$$2I = \int\limits_{a}^{b} \frac{f(x)+f(a+b-x)}{f(x)+f(a+b-x)} dx = \int\limits_{0}^{a} \cdot 1 dx$$

$$2I = [x]_{a}^{b} = b-a$$

$$\therefore I = \frac{b-a}{2} \quad \text{Ans.}$$

Example (27): Evaluate $\int\limits_{-1}^{2} |x^3 - x|\,dx$

Solution: Here *the limits of integration* are -1 to 2. By the definition of the *absolute value of a number*, we know that:

whenever $x^3 - x \geq 0$, $|x^3 - x| = x^3 - x$, and whenever $x^3 - x < 0$, $|x^3 - x| = -(x^3 - x) = x - x^3$.

To find the zeros of the function $|x^3 - x|$, we solve the following equation

$$x^3 - x = 0$$
$$\therefore \quad x(x^2 - 1) = 0 \Rightarrow x = 0,\ x = \pm 1$$

Thus, the function $(x^3 - x)$ has three zero, namely -1, 0, and 1. It follows that, the sign of this function must change in the intervals $[-1, 0]$, $[0, 1]$, and $[1, 2]$. Let us check how the sign changes.

(i) In $[-1, 0]$, $x^3 - x = x(x^2 - 1) = (-ve)\,(-ve) \geq 0$

(ii) In $[0, 1]$, $x^3 - x = x(x^2 - 1) = (+ve)\,(-ve) \leq 0$

(iii) In $[1, 2]$, $x^3 - x = x(x^2 - 1) = (+ve)\,(+ve) \geq 0$

Now, $\int\limits_{-1}^{2} |x^3 - x|\,dx = \int\limits_{-1}^{0} |x^3 - x|\,dx + \int\limits_{0}^{1} |x^3 - x|\,dx + \int\limits_{1}^{2} |x^3 - x|\,dx$

$$= \int\limits_{-1}^{0} x^3 - x\,dx + \int\limits_{0}^{1} x - x^3\,dx + \int\limits_{1}^{2} x^3 - x\,dx$$

$$= \left[\frac{x^4}{4} - \frac{x^2}{2}\right]_{-1}^{0} + \left[\frac{x^2}{2} - \frac{x^4}{4}\right]_{0}^{1} + \left[\frac{x^4}{4} - \frac{x^2}{2}\right]_{1}^{2}$$

$$= -\left[\frac{1}{4} - \frac{1}{2}\right] + \left[\frac{1}{2} - \frac{1}{4}\right] + \left[(4 - 2) - \left(\frac{1}{4} - \frac{1}{2}\right)\right]$$

$$= -\frac{1}{4} + \frac{1}{2} + \frac{1}{2} - \frac{1}{4} + 2 - \frac{1}{4} + \frac{1}{2} = \frac{3}{2} - \frac{3}{4} + 2$$

$$\frac{6 - 3 + 8}{4} = \frac{11}{4} \quad \text{Ans.}$$

Example (28): Prove that

(i) $\int\limits_{0}^{1} \sin^{-1}x\,dx = \frac{\pi}{2} - 1$

(ii) $\int\limits_{0}^{1} \cos^{-1}x\,dx = 1$

(iii) $\int\limits_{0}^{1} \tan^{-1}x\,dx = \frac{\pi}{4} - \frac{1}{2}\log_e 2$

(i) To prove $\int_0^1 \sin^{-1} x \, dx = \frac{\pi}{2} - 1$

Let $I = \int_0^1 \sin^{-1} x \, dx$

Put $x = \sin t$ $\therefore$ $t = \sin^{-1} x$

$$\Rightarrow dx = \cos t \, dt$$

When $x = 0$, $\sin t = 0 \Rightarrow t = 0$ and when $x = 1$, $\sin t = 1 \Rightarrow t = \frac{\pi}{2}$,

$$\therefore \quad I = \int_0^{\pi/2} \sin^{-1}(\sin t) = \cos t \, dt = \int_0^{\pi/2} t \cos t \, dt$$

Integrating by parts, we get

$$I = [t \sin t]_0^{\pi/2} - \int_0^{\pi/2} 1 \cdot \sin t \, dt$$

$$= \left[\frac{\pi}{2} \cdot \sin\frac{\pi}{2} - 0\right] - [-\cos t]_0^{\pi/2}$$

$$= \left[\frac{\pi}{2} \cdot 1\right] + [\cos t]_0^{\pi/2} = \frac{\pi}{2} + [0 - 1] = \frac{\pi}{2} - 1 \quad \text{(Proved)}$$

Similarly, Examples (ii) and (iii) can be solved.

Example (29): Evaluate $\int_{-3}^3 |x| \, dx$

Solution: Let $I = \int_{-3}^3 |x| \, dx$

By definition of the absolute value, we know that
For $x \geq 0$, $|x| = x$, and for $x < 0$, $|x| = -x$
(This suggests that we must break the interval of integration from -3 to 0 and from 0 to 3.)

$$\text{Thus, } I = \int_{-3}^0 |x| \, dx + \int_0^3 |x| \, dx \quad \left[\because \int_a^b f(x) \, dx + \int_b^c f(x) \, dx + \int_c^b f(x) \, dx\right]$$

$$= \int_{-3}^0 (-x) \, dx + \int_0^3 x \, dx = -\int_{-3}^0 x \, dx + \int_0^3 x \, dx$$

$$= \int_0^{-3} x \, dx + \int_0^3 x \, dx$$

$$= \left[\frac{x^2}{2}\right]_0^{-3} + \left[\frac{x^2}{2}\right]_0^3 = \left[\frac{(-3)^2 - 0}{2}\right] + \left[\frac{(3)^2 - 0}{2}\right]$$

$$= \frac{9}{2} + \frac{9}{2} = 9 \quad \text{Ans.}$$

Remark: Let $f(x) = |x|$. By definition of $|x|$, we have,

for $x \geq 0$, $|x| = x$, and for $x < 0$, $|x| = -x = -(-x) = x$

Thus, $f(x) = |x|$ is an even function.

$$\therefore \int_{-3}^{3} |x| dx = 2 \int_{0}^{3} x \, dx \text{ [By Property } (P_7)]$$

$$= 2 \left[\frac{x^2}{2} \right]_0^3 = 2 \left[\frac{(3)^2}{2} - 0 \right]_0^3 = 2 \cdot \frac{9}{2} = 9 \quad \text{Ans.}$$

Example (30): Prove that $\int_{0}^{\pi} |\cos x| dx = 2$

Solution: Let $I = \int_{0}^{\pi} |\cos x| dx = 2$

Note: We know that $-1 \leq \cos x \leq 1$

By definition of absolute value, we have,

$|\cos x| = \cos x$, if $\cos x \geq 0$ and $|\cos x| = -\cos x$, if $\cos x < 0$. Also, we know that

- $\cos x \geq 0$, when x varies from 0 to $\frac{\pi}{2}$.
- $\cos x \leq 0$, when x varies from $\frac{\pi}{2}$ to π.

(This suggests the way we should break the interval of integration.)

$$\therefore I = \int_{0}^{\pi/2} |\cos x| dx + \int_{\pi/2}^{\pi} |\cos x| dx$$

$$= \int_{0}^{\pi/2} \cos x \, dx + \int_{\pi/2}^{\pi} (-\cos x) dx$$

$$= [\sin x]_0^{\pi/2} - [\sin x]_{\pi/2}^{\pi}$$

$$= \left[\sin \frac{\pi}{2} - \sin 0 \right] - \left[\sin \pi - \sin \frac{\pi}{2} \right]$$

$$= (1 - 0) - (0 - 1) = 1 + 1 = 2 \quad \text{Ans.}$$

Miscellaneous Exercise

(1) $\int_{-\pi/2}^{\pi/2} \frac{x \sin x}{x^3 + \sin^3 x} dx$

Ans. 0

(2) Show that $\int_{0}^{\pi/2} (x \cot x) dx = \frac{\pi}{2} \log_e 2$

(3) $\int\limits_{0}^{\pi/2} \log_e(\sec x)dx$

Ans. $(\pi/2)\log 2$

(4) $\int\limits_{0}^{\pi} \log_e(1 + \cos x)dx$

Ans. $-\pi \log 2$

(5)   Show that $\int\limits_{0}^{\pi/2} f(\sin x)dx = \int\limits_{0}^{\pi/2} f(\cos x)dx$

(6)   Show that $\int\limits_{0}^{\pi/2} \log_e(\tan x)dx = 0$

(7)   Show that $\int\limits_{0}^{\pi/2} \log_e(\cot x)dx = 0$

(8)   Show that $\int\limits_{0}^{1} \cos^{-1}x dx = 1$

(9)   Show that $\int\limits_{0}^{1} \tan^{-1}x dx = \frac{\pi}{4} - \frac{1}{2}\log_e 2$

(10)   Show that $\int\limits_{0}^{\pi/2} \frac{dx}{1+\cot x} = \frac{\pi}{2}$

(11)   Show that $\int\limits_{-\pi/4}^{\pi/4} \sin^2 x dx = \frac{\pi}{4} - \frac{1}{2}$

(12)   Show that $\int\limits_{-1}^{3/2} |x\sin(\pi x)|dx = \frac{3}{\pi} + \frac{1}{\pi^2}$

Note: So far, we have defined $\int_a^b f(x)dx$ only for f continuous on $[a, b]$. *It follows from the maximum–minimum theorem that such a function is bounded on* $[a, b]$, in the sense that for some number M, $|f(x)| \le M$ for all x in $[a, b]$.

More generally, if I *is any interval (finite or infinite), then we say that* f *is bounded on* I, if there is a constant M such that $|f(x)| \le M$ for all x in I.

A function which is *not bounded* on a given interval "*I*", inside its domain is said to be *unbounded on I.*

We can extend the definition of definite integrals to include integrals of the form

$$\int\limits_{a}^{-\infty} f(x)dx \cdot \int\limits_{-\infty}^{-b} f(x)dx \cdot \int\limits_{-\infty}^{-\infty} f(x)dx^{(9)}$$

---

(9) We encounter such integrals while computing the potential of gravitational or electrostatic forces.

Such integrals (with infinite limits) are called *improper integrals*. By solving such integrals, we answer two questions:

(i) The geometric question: Whether an area can be defined for the region under the graph of a *non-negative function which is unbounded on a bounded interval*?

(ii) Whether it is possible to define the area of a region under the graph of a non-negative function on an unbounded interval?

The answer to (i) above is "NO", whereas the answer to (ii) is "YES". For necessary details, the reader may go thorough the detailed topic on improper integrals.

8a Applying the Definite Integral to Compute the Area of a Plane Figure

8a.1 INTRODUCTION

In elementary geometry, we have learnt to calculate areas of various geometrical figures like rectangles, triangles, trapezia, and so on. *These figures are enclosed by straight lines.* The formulae for calculating the areas of these figures are fundamental in the applications of mathematics to many real life problems. However, these formulae are *inadequate* for calculating the *areas enclosed by curves.*[1]

In Chapter 5, we have discussed the concept of area of a plane region and introduced a procedure to find the area bounded by the curve $y = f(x)$, the ordinates $x = a$, $x = b$, and the x-axis. There, we obtained the formula for the area in question and denoted it by

$$A = \int_a^b f(x)\mathrm{d}x = [\phi(x)]_a^b = \phi(b) - \phi(a), \quad \text{where } \int f(x)\mathrm{d}x = \phi(x)$$

8a.1.1 Some Applications of Integral Calculus

Many *applications of integral Calculus* involve measuring something, like

- the *area* of a plane region,
- the *volume* of a solid object,
- the *net distance* a moving object travels over an interval,
- the *work done against gravity* in raising a satellite into orbit,
- the *length* of a curve from one point to another, and so on.

Note: In a few simple cases, for example,

- measuring the area of a rectangular region,
- the length of a straight line segment, or
- the distance covered by an object moving at a constant speed, and so on,

Applications of the definite integral 8a-Applying the definite integral to find the area bounded by simple curves. Steps for constructing a rough sketch of the graph of a function to identify the region in question for computing its area.

[1] In elementary geometry, the formula for calculating the area of a circle is denoted by the formula $A = \pi r^2$, where "r" is the radius of the circle and π stands for the real number, represented by the ratio of circumference to the diameter of the circle. The approximate value of π is 3.14. In this chapter, we shall obtain this value with precision using definite integrals.

Introduction to Integral Calculus: Systematic Studies with Engineering Applications for Beginners, First Edition. Ulrich L. Rohde, G. C. Jain, Ajay K. Poddar, and A. K. Ghosh. © 2012 John Wiley & Sons, Inc. Published 2012 by John Wiley & Sons, Inc.

The quantities in question can be found by common sense alone, and *no big machinery from Calculus is needed.*

However, we know that, in practice, *most regions are not rectangular, most curves are not straight, and most speeds are not constant. In these more common and more interesting situations, Calculus tools,* in particular—*the definite integral, are indispensable.*

8a.1.2   To Get a Better Grip of the Subject, We Make the Following Simplifying Assumptions, and Revise Certain Technical Facts Related to Definite Integrals

8a.1.2.1   Assuming Good Behavior   Throughout this section, we assume that all the integrals $\int_a^b f(x)\mathrm{d}x$, that we meet, make good mathematical sense. *To guarantee this,* it is enough *to assume* that every integrand "$f(x)$" is continuous on $[a, b]$, as we do from now on.

(In fact, discontinuous integrands do sometimes arise in practical applications. Even in such cases, the basic ideas of this section often apply, although perhaps in slightly different forms.)

8a.1.2.2   Definite Integrals and Area: Revision   It is assumed that the reader has understood the concept of the definite integral geometrically in terms of area as discussed in Chapter 5. Thus, for any continuous function "$f(x)$" on $[a, b]$,

$$\int_a^b f(x)\mathrm{d}x = \text{Signed area bounded by } f\text{-graph for } a \leq x \leq b$$

In this connection, it must be remembered, that *any area below the x-axis counts as negative whereas that above the x-axis counts as positive.* To keep a track of positive and negative areas demands a little care, but the basic link between (definite) integrals and areas (as discussed in Chapter 5) is known to us and that they represent real numbers. The following easy example will make the issue clear to mind.

Example (1):   Consider the following two integrals.[2]

(I)
$$I_1 = \int_0^\pi \sin x \, \mathrm{d}x = 2$$

and

(II)
$$I_2 = \int_0^{2\pi} \sin x \, \mathrm{d}x = 0$$

Solution: (I) $I_1 = \int_0^\pi \sin x \, \mathrm{d}x = [-\cos x]_0^\pi$

$$= -[\cos x]_0^\pi = -[\cos \pi - \cos 0] = -[(-1) - (1)] = -[-2] = 2$$

(II) $I_2 = \int_0^{2\pi} \sin x \, \mathrm{d}x = [-\cos x]_0^{2\pi} = -[\cos x]_0^{2\pi}$

$$= -[\cos 2\pi - \cos 0] = -[1 - 1] = 0$$

The graph of $\sin x$ (Figure 8a.1) shows that on $[0, \pi]$, $\sin x \geq 0$. Thus, I_1 measures ordinary area: 2 square units *under one arch of the sine curve.* The value of I_2 means that the "*net*", or "*signed area*" on the interval $[0, 2\pi]$ is zero.

---

[2] As stated at the end of Chapter 5, we use the term integral to stand for both the definite integral and an indefinite integral. In fact, the context makes the meaning clear.

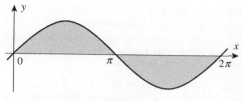

FIGURE 8a.1

(This makes good *geometric sense*: $\sin x \leq 0$ on the interval $[\pi, 2\pi]$ and the symmetry of the graph guarantees that area above and below the x-axis exactly cancel each other out.)

8a.1.2.3   Interpretation of Definite Integral in a Broader Sense [Not Just as Area]   *Every definite integral $\int_a^b f(x)dx$ can be interpreted as a signed area. In fact, definite integrals can be used to measure (or model) many quantities, other than areas.*

Volume, arc length, distance, work, mass, fluid pressure, and so on, can all be calculated *as definite integrals.* Choosing the right integral and interpreting the result appropriately depends on the problem at hand.

Remark: If definite integrals measured only areas, they would not deserve the fuss we make over them and the amount of attention we pay to study and interpret them.

Two Views of Definite Integrals
As mentioned above, definite integrals can be used to measure many disparate quantities. Usually, the key considerations are

- which function to integrate, and
- over what interval.

In applying the (definite) integral in varied settings, it is useful to remember two interpretations of the definite integral $\int_a^b f(x)dx$, both being *different but closely related* as discussed below.

(i) *A Limit of Approximating Sums:* The integral is defined formally *as a limit of approximating sums.* In Chapter 5, we have discussed and compared several kinds of *approximating sums* (by choosing each subinterval Δx_i *of equal length,* and the points $\bar{x}_1, \bar{x}_2, \ldots, \bar{x}_n$, at the left-hand and the right-hand end points). Recall that by using the right sums, we can write

$$\int_a^b f(x)dx = \lim_{n \to \infty} \sum_{i=1}^n f(x_i) \cdot \Delta x$$

where the inputs x_i are *the right end points* of n equal-length subintervals of $[a, b]$. From this point of view, *the (definite) integral* "adds up" *small contributions, each of the form* $f(x_i)\Delta x$.

(ii) *Accumulated Change in an Antiderivative: The second fundamental theorem of Calculus* states that

$$\int_a^b f(x)dx = [\phi(x)]_a^b = \phi(b) - \phi(a)$$

where the function ϕ (on the right-hand side) can be *any antiderivative* of "$f(x)$" on $[a, b]$. The difference, $\phi(b) - \phi(a)$, represents, in a natural way, *the accumulated change (or net change) in ϕ over the interval $[a, b]$*. Thus, to find the accumulated change in ϕ over $[a, b]$, we integrate the rate function "$f(x)$".[3]

Mathematically speaking, *these two approaches to the integral* [at (i) and (ii) above] *are equivalent, as the second fundamental theorem of Calculus* states. It guarantees that *both the methods give the same answer.* Having two different ways to think about the (definite) integral makes it more versatile in applications. *Which viewpoint is better depends on the situation.*[4]

8a.2 COMPUTING THE AREA OF A PLANE REGION

Recall that in Chapter 5 we have discussed *how to find the area of a right angled triangle formed by the lines:* $y = f(x) = 2x$, the x-axis and the ordinate $x = 1$. Even though *the area of any such triangle* can be easily found using *geometry and algebra*, we introduced *Archimedes' method of exhaustion* for computing the area in question as the *limit of a sum of areas of rectangles.* We have seen that this method applies to *more complex regions, and leads to the definition of the definite integral* $\int_a^b f(x)dx$. We also observed that the computation of definite integrals is very much simplified with the application of *the second fundamental theorem of Calculus.* We are now in a position to give *an easy, convenient, and intuitive method* of computing area(s) bounded by the curve $y = f(x)$, the x-axis and the ordinates $x = a$ and $x = b$, where $f(x) \geq 0$.

We may consider the area under the curve $y = f(x)$, as composed of *large number of very thin vertical strips* (each of which is a curvilinear trapezoid) (see Figure 8a.2a).

8a.2.1 Area of an Elementary Strip

Consider *an arbitrary strip of height y and width* dx. Then (*approximately*), the area of the elementary strip denoted by dA is given by

$$dA \approx y\,dx = f(x)dx \qquad [\because \quad y = f(x)]$$

This area is called *the elementary area*, located at *an arbitrary position* within the region from $x = a$ to $x = b$.

We can think of *the total area A of the region as the result of adding up the elementary areas of thin strips across the region* in question.

Symbolically, we express

$$A = \int_a^b dA = \int_a^b y\,dx = \int_a^b f(x)dx \tag{1}$$

---

[3] We know that to integrate the function "$f(x)$", means to find a function $\phi(x)$, which should satisfy the equation $\phi'(x) = f(x)$. Recall that, we have studied various methods in Chapters 1, 2, 3a, 3b, 4a, and 4b, for finding an antiderivative $\phi(x)$ of the integrand $f(x)$.

[4] When an integral presents itself in simple symbolic form, antidifferentiation is the obvious next step. But, for the data given graphically (or in tabular form), using approximating sums is a natural strategy. Here, we shall consider only the cases where antidifferentiation is sufficient for the purpose.

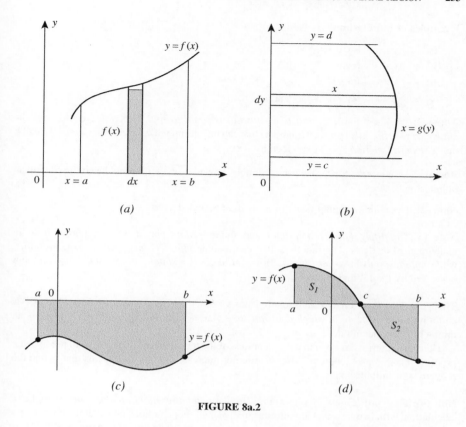

FIGURE 8a.2

Thus, a definite integral can be regarded as a sum, or, *more correctly* the limit of a sum of the areas of an infinite number of rectangles, one side of each of which (dx in the above) is *infinitesimally small*.[5]

At this stage, it is useful to clarify the meaning of the term "*infinitesimally small*".

8a.2.1.1   The Concept of Infinitesimal(s)   Infinitesimals are functions that approach zero, as the argument (say "x") varies in a certain manner.

Definition: A function $\alpha(x)$ *tending to zero*, as $x \to a$ (or as $x \to \infty$) is called an infinitesimal as $x \to a$.[6]

---

[5] Now, the use of the term "integral" will be clear. The word "integrate" means "to give the total sum". The first letter of the word sum appears in the symbol "$\int$", which is the old fashioned elongated "s". It is also evident why the infinitesimal, dx, must necessarily appear as a factor in an integral (see Chapter 1, Section 1.2).

[6] From the definition of a limit, it follows that (if $\lim_{x \to a} \alpha(x) = 0$, then) for any preassigned, arbitrarily small positive (number) ε, there will be positive (number) δ, such that for all x satisfying the condition $|x - a| < \delta$, the condition $|\alpha(x)| < \varepsilon$, will be satisfied.

Examples of infinitesimal(s)

(i) $y = x^2$ for $x \to 0$,

(ii) $y = x - 1$ for $x \to 1$,

(iii) $y = \frac{1}{x}$ for $x \to \infty$,

(iv) $y = 2^x$ for $x \to -\infty$, and so on.

Note (1): We know that *the limit of a constant is equal to that constant*. Accordingly, the number zero is *the only constant* (number) that can be interpreted *as an infinitesimal*. [We call it the *"constant function zero"* expressed by $f(x) = 0$.]

Remark: A concrete nonzero constant number, however small, must not be confused with an infinitesimal.

Note (2): Every infinitesimal (say, for $x \to a$) is bounded as $x \to a$.

Note (3): *Infinitely large magnitudes* and *infinitesimals* play a very important role in mathematical analysis. The functions that can assume *infinitely large values* have *no limits*, while *infinitesimals* have *zero limits*. However, there is a *simple relationship* between them expressed by the following theorem.

Theorem: If a function $f(x)$ tends to infinity as $x \to a$, then $1/f(x)$ is an *infinitesimal*. If $\alpha(x)$ is an infinitesimal as $x \to a$, *which does not take a zero value for $x \neq a$*, then $1/(\alpha(x))$ is an infinitely large magnitude.[7]

Here, our interest has been to get a clear idea of the term infinitesimal. We shall not give the proof of this (simple) theorem. Also, we are not interested in other theorems that define the properties of infinitesimals.

Note (4): The word *"Calculus"* is an abbreviation for *"infinitesimal Calculus"*, which implies a calculation, with numbers that are infinitesimally small. For example, consider the growth of a small plant. In the ordinary way, we know that it grows gradually and continuously. *If it is examined after an interval of a few days, the growth will be obvious and readily measured.*

On the other hand, if it is observed after an interval of a *few minutes*, although some growth has taken place, the amount is *too small* to be distinguished. If an observation takes place after a *still smaller interval* of time, say a *few seconds*, although *no change can be detected*, we know that *there has been growth*.

To express such a (small) growth, we use a *mathematical term*: we call it an *infinitesimally small growth* or an *infinitesimal growth*.[8]

Note (5): The process of gradual and continuous growth may be observed in innumerable instances. What is of real importance in most cases is not necessarily the actual amount of growth, *but the rate of growth* [i.e., the rate at which a quantity *"$f(x)$"* increases or decreases with the infinitesimal increases in x]. This is the subject of that part of *infinitesimal Calculus* that we call *differential Calculus*. (The meaning of the term "differential" is discussed in Chapter 16 of Part I.)

8a.2.1.2 *Area Under a Curve*   In considering the total area under a curve [as the limit of a sum of an *infinite number of products of the form "$f(x) \cdot \Delta x$"*], one must keep in mind that the narrower the rectangles, the *closer will be the approximation to the true area under the curve*. The process of narrowing the rectangles may be continued indefinitely (i.e., to any extent). *This*

---

[7] Note that the infinitesimal $\alpha(x)$ is not allowed to assume the value zero, for any $x \neq a$.

[8] *Teach Yourself Calculus* by P Abbott, B.A, ELBS Publication by Hodder and Stoughton Ltd, London.

is our infinite process to which the limit concept may be applied. In the (sum of an) *infinite number of algebraic products* of the form $f(x) \cdot \Delta x$, *one factor* (namely Δx), *in the limit, becomes infinitely small.* As Δx approaches zero, the sum of rectangular areas approaches the true total area under the curve. Thus, the expression, $\lim_{\Delta x \to 0} \sum_{x=a}^{x=b} f(x) \cdot \Delta x$ is identified with $\int_a^b f(x) dx$. We know that the latter expression represents the definite integral of the function $f(x)$.

Further, we know that the differential "dx" stands for an *arbitrary increment* in the independent variable "x", and so "dx" can be as small as we wish. Thus, in the above (definite) integral, "dx" *does not have any independent meaning; it refers to the values of independent variable "x".* For this reason, we say that "x" in "dx" stands for the variable of integration in the interval [a, b] (see Chapter 1, Table 1.1).

Note: In fact, *the definite integral is used as a device, for illustrating the process that can be applied to the summation of any such series,* subject to the conditions necessary for the integration of "$f(x)$", as discussed in Chapter 5. This is of great practical importance, since it enables us to calculate not only areas but also volumes, lengths of curves, moments of inertia, and so on, *which are capable of being expressed in the form* $\lim_{\Delta x \to 0} \sum_{x=a}^{x=b} f(x) \cdot \Delta x$. They can then be represented by the definite integral $\int_a^b f(x) dx$.

Since the concept of the definite integral is developed in connection with the computation of the area under a curve, *we shall first consider different situations under which we may have to compute the area(s) under a curve,* and then consider some examples for computation of areas.

Sometimes, the area can be easily determined by integrating with respect to y, rather than x. This situation arises when the curve is given in the form $x = g(y)$, where $g(y) \geq 0$ and $c \leq y \leq d$. In this case, it is more convenient to consider the elementary strip(s) that are adjoining the y-axis. Then, the area A bounded by the curve $x = g(y)$, the y-axis and the lines $y = c$ and $y = d$ is given by

$$A = \int_c^d x \, dy = \int_c^d g(y) dy \quad [\because \quad x = g(y)] \tag{2}$$

Note: In both the cases as shown in Figure 8a.2a and b, we have considered the situations, wherein the curves are *above the x-axis.* Now we shall consider the cases in which the given curve is below the x-axis.

If the position of the curve under consideration is below the x-axis, then we have $f(x) \leq 0$ from $x = a$ to $x = b$ (Figure 8a.2c). The area bounded by the curve $y = f(x)$, the x-axis and the straight lines $x = a$ and $x = b$, *comes out to be negative.*

But, it is only *the numerical value of the area* that is taken into consideration. Thus, if the area is negative we take its *absolute value* and write

$$A = \left| \int_a^b f(x) dx \right| \tag{3}$$

Generally, it may happen that some portion of the curve is above the x-axis and some is below x-axis. Then, *the figure bounded by the curve* $y = f(x)$, the x-axis and the straight lines $x = a$ and $x = b$ *is situated on both sides of* the x-axis (Figure 8a.2d), and its area "S" is given by

$$S = S_1 + S_2 = \int_a^c f(x) dx + \left| \int_c^b f(x) dx \right| \tag{4}$$

Note carefully, the limits of integrals on the right-hand side, and the expression of the last integral in equation (4) given above.

8a.2.1   Area Between Two Curves

Now, we are in a position to compute the area bounded by two curves $y = \varphi(x)$ and $y = f(x)$ where $\varphi(x) \geq f(x)$ in $[a, b]$. In this situation, as shown in Figure 8a.3a, we have

$$\int_a^b \varphi(x)dx \geq \int_a^b f(x)dx$$

If the length of the ordinate of the upper curve is denoted by y_{upper} and the ordinate of the lower curve is denoted by y_{lower}, then we can express the area (between the two curves) by

$$S = \int_a^b (y_{upper})dx - \int_a^b (y_{lower})dx = \int_a^b \varphi(x)dx - \int_a^b f(x)dx = \int_a^b [\varphi(x) - f(x)]dx$$

Thus, the *positive difference* $[\varphi(x) - f(x)]$ can be treated as a *single function* for computing the area "S". Accordingly, we write

$$S = \int_a^b [\varphi(x) - f(x)]dx \tag{5}$$

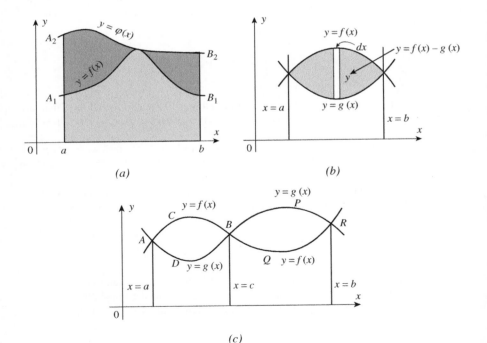

(a)   *(b)*

(c)

FIGURE 8a.3

If the two curves in question *intersect at one or more points*, then the Equation (5) can be *suitably modified* and applied. In fact, the point(s) of intersection indicate the limits for the definite integrals in question.

If the equations of the curves are in the form $y = f(x)$ and $y = g(x)$, then the points of intersection (of the curves) are given by $x = a$ and $x = b$ $(a < b)$ at which the curves have *common values* of y.[9]

In such cases, it is convenient to take an elementary area in the form of vertical strip(s) and the (correct) intervals of integration (see Figure 8a.3b and c).

In Figure 8a.3b, the area of an arbitrary elementary strip is given by $dA = [f(x) - g(x)]dx$, so that the total area can be taken as

$$A = \int_a^b [f(x) - g(x)]dx$$

Further, in view of the earlier discussion, the area of the shaded region in Figure 8a.3c is given by

$$A = \int_a^c [f(x) - g(x)]dx + \int_c^b [g(x) - f(x)]dx$$

Irrespective of whether we take elementary area in the form of *vertical* or *horizontal strips*, the measure of the area in question (i.e., the value of the definite integral) will be same. In fact, wherever possible, *vertical strips* are preferred for practical convenience.

The cases discussed above (with Figures 8a.2a–d and 8a.3a–c) suggest that to compute the desired area(s), *we should have a rough sketch of the region in question*. Obviously, this will help us in identifying the limits of the (definite) integrals involved.

The importance of a *rough sketch* [of the curve $y = f(x)$] will be realized when the graph of the curve encloses any region below the x-axis (for using vertical strips) or to the left of y-axis (for using horizontal strips). This requirement is easily met if we remember certain properties of curves using coordinate geometry and differential Calculus.[10]

We give below some important points that should be useful in constructing a rough sketch of the curve(s).

8a.3   CONSTRUCTING THE ROUGH SKETCH [CARTESIAN CURVES]

For the purpose of graphing a curve, we consider the *equation of the given curve* and check:

(I) *Whether the Curve Passes Through the Origin*

Example (2):   In the equation of the curve, $y^2 = 4ax$, if we put $x = 0$, we get $y = 0$. Thus, the point $(0, 0)$, lies on the curve. In other words, *the curve passes through the origin*. Similarly, the curves $y^2 = 4x$, $y^2 = -x$, $y = x^2$, $x^2 = -y$, $y = x^3$, and $y = -x^3$ all pass through the origin.

---

[9] If the equations of the curves are in the form $x = f(y)$ and $x = g(y)$, then the points of intersection (of the curves) are given by $y = a$ and $y = b$ $(a < b)$, at which the curves have *common values* of x.

[10] In fact, differential Calculus offers a general scheme for constructing graphs of functions. However, for drawing a "rough sketch of the region", we do not require many of the concepts like finding the asymptotes to the graph, intervals of monotonicity and its extremum, intervals of convexity of the graph and the points of inflection, and so on. This reduces our work, since we need only a "rough sketch" of the curve to visualize the region under consideration.

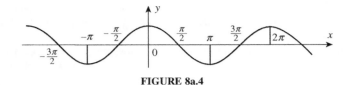

FIGURE 8a.4

Note: The best way to decide whether the origin lies on the curve is to see that the equation *does not contain any constant term.*

Remark: Note that, the circle $x^2 + y^2 = a^2$ and the ellipse $((x^2/a^2) + (y^2/b^2)) = 1$, do not pass through the origin.

(II) *Symmetry*

The most important point for tracing the curves is *to judge its symmetry*, which we do as follows:

(a) *Symmetry About y-Axis [Even Functions]*:   If the equation of the curve involves *even and only even powers of x,* then there is *symmetry about y-axis.*

Note: For a curve to have symmetry about y-axis, its equation should not have any odd power term of x.

Example (3): The parabolas $x^2 = 4ay$ and $x^2 = -9y$, the circle $x^2 + y^2 = a^2$, and ellipse $((x^2/a^2) + (y^2/b^2)) = 1$ all are *symmetrical* about y-axis.

Definition: A function whose graph is *symmetric with respect to y-axis is called an even function.*

An even function $f(x)$ has the property: $f(x) = f(-x)$. Thus, $f(x) = \cos x$ *is another example of an even function* (see Figure 8a.4).

(b) *Symmetry About x-Axis* If the equation of the curve involves *even and only even powers of y,* then there is *symmetry about x-axis.*[11]

Example (4): The *parabolas* $y^2 = 4ax$, $y^2 = -x$, *the circle* $x^2 + y^2 = a^2$, *and the ellipse* $((x^2/a^2) + (y^2/b^2)) = 1$, *all are symmetrical about x-axis.*

(c) *Symmetrical about Both the Axes*:   If the equation of the curve involves *even and only even powers of both x and y,* then there is *symmetry about both the axes.*

Example (5): The circle $x^2 + y^2 = a^2$ and the ellipse $((x^2/a^2) + (y^2/b^2)) = 1$, *both are symmetrical about both the axes.*

(d) *Symmetry With Respect to the Origin [Odd Functions]*:   In the given equation $y = f(x)$, if replacing x by $(-x)$ and y by $(-y)$ *gives an equivalent equation*, then the graph of such a function is symmetric with respect to the origin.

In other words, the graph of a function is symmetric with respect to the origin, if the point $(-x, -y)$ is on the graph whenever a point (x, y) is. It follows that *each point on the graph is*

---

[11] A function symmetric about the x-axis does not have any qualifying name. The definition of an odd function will follow shortly.

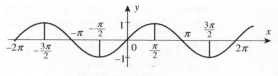

FIGURE 8a.5

matched by another point on the graph, which is on the other side of the origin, such that the values $f(-x)$ and $[-f(x)]$ are same.

Thus, for a function $f(x)$ to be *odd*, $(-x)$ must be in the domain of "f" whenever x is, and the relation $f(-x) = -f(x)$ must hold.

An example of an odd function is $f(x) = \sin x$, since $\sin(-x) = -\sin x$ (see Figure 8a.5).

A good example of an odd function is $y = x^3$. Its graph is symmetrical with respect to the origin. Note that from $f(x) = x^3$, we get $f(-x) = (-x)^3 = -x^3$.

Thus, $f(-x) = -f(x)$. Similarly, the graph of $y = -x^3$ is symmetrical with respect to the origin. Graphs of both these functions are given below (Figure 8a.6a and b).

Definition: A function whose graph is *symmetrical with respect to the origin* is called an *odd function*.

(We have discussed at length about even and odd functions, in Chapter 7b of this volume.)

Important Note: If a curve has symmetry about the coordinate axis and about the origin then it is helpful in tracing the "rough sketch" of the curve. Further, it becomes easier to find symmetrical parts of the curve that enclose equal areas.

In particular, the idea of symmetry is found very useful in computing areas enclosed by parabolas, circles, ellipses, and trigonometric functions, namely, $y = \sin x$ and $y = \cos x$. This will become clearer when we solve the problems which follow.

(III) *Points of Intersection [With the Axes and Between Two Given Curves]*
We can find *the points where the given curve intersects the axes*. For this purpose, we proceed as follows: Put $y = 0$, in the equation of the given curve and solve it for x. Thus, we get the

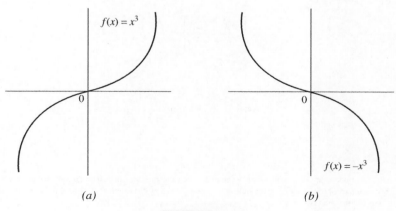

(a)   *(b)*

FIGURE 8a.6

x coordinate(s) of the *point(s)* where the *curve cuts the x-axis*. Similarly, if we put $x = 0$, in the equation of the curve, and solve it y, we get the y-coordinate(s) of the point(s) where the curve cuts the y-axis.

Thus, it is easy to see that the ellipse $((x^2/a^2) + (y^2/b^2)) = 1$, cuts the x-axis at the points $(a, 0)$ and $(-a, 0)$, and the y-axis at the points $(0, b)$ and $(0, -b)$. Also, the circle $x^2 + y^2 = 9$ cuts the x-axis at the points $(3, 0)$ and $(-3, 0)$, and the y-axis at the points $(0, 3)$ and $(0, -3)$.

To find the *points of intersection between the given two curves*, we solve the *system of equations representing the curves*. (It means we may use any of the given equations in the other one to obtain an equation in a single variable and then solve it.) Further, using these values (in the given equations), we obtain the *points of intersection between the two curves*.

Note: In this chapter, we shall be computing the areas enclosed by simple curves when

- The region is bounded by a curve, the x-axis (or the y-axis) and two lines perpendicular to the coordinate axis meeting the curve, as shown in Figure 8a.2a and b.
- The region is bounded by the circles and the ellipses (standard forms only). These curves are symmetric to both the axes.
- The region is bounded between two curves both of which may be arcs of simple curves like circles, parabolas, and ellipses (standard forms) or one of them may be an straight line intersecting the given curve. Now we start with simple examples.

Illustrative Examples

Example (6): Compute the area of the figure bounded by the parabola $y = x^2$, the x-axis, and the lines $x = 2$ and $x = 3$

Solutions: The given curve is $y = x^2$. It passes through the origin and is symmetrical with respect to the y-axis. Also, the curve is above the x-axis (Figure 8a.7).[12]

The area in question is given by

$$S = \int_2^3 x^2 \, dx = \left[\frac{x^3}{3}\right]_2^3$$

$$= \left[\frac{(3)^3}{3} - \frac{(2)^3}{3}\right]$$

$$= \left[\frac{27}{3} - \frac{8}{3}\right]$$

$$= \left[6\frac{1}{3}\right] \text{ square units} \qquad \text{Ans.}$$

---

[12] From the equation $y = x^2$, we note that for any value of x, the value of y is positive. Thus, the given curve is above the x-axis. This is a matter of observation and the student is supposed to know this simple fact. To save time such details need not be included in the solution.

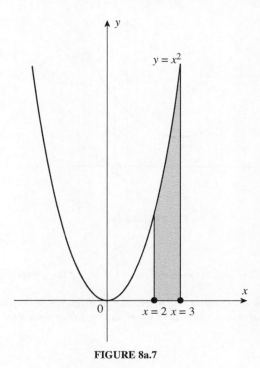

FIGURE 8a.7

Example (7): Find the area of the region bounded by the curve $y = x^2 + 2x + 2$, the lines $x = -2$ and $x = 1$, and the x-axis

Solution: The given curve is

$$y = x^2 + 2x + 2 \tag{6}$$

Note that equation (6) *does not have real roots*. (It means that the curve does not intersect the x-axis.) The rough sketch of the curve is given on right side (Figure 8a.8). [13]
The area of the region in question is given by

$$S = \int_{-2}^{1} (x^2 + 2x + 2)\,dx$$

$$S = \left[\frac{x^3}{3} + \frac{2x^2}{2} + 2x \right]_{-2}^{1}$$

$$= \left(\frac{1}{3} + 1 + 2 \right) - \left(\frac{-8}{3} + 4 - 4 \right)$$

$$= \frac{10}{3} + \frac{8}{3} = 6 \text{ square units} \qquad \text{Ans.}$$

---

[13] We have, for $x = -2$, $y = 2$, and for $x = 1$, $y = 5$. From these values of y, and the fact that the curve does not intersect x-axis, it follows that the curve from Equation (6) lies above the x-axis. Again, these details need not be given in the solution, to save time.

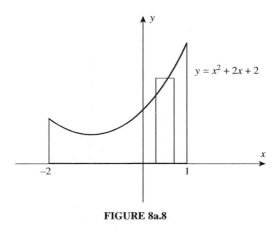

FIGURE 8a.8

Example (8):   Find the area between the curve $y = e^x$ and the x-axis from $x = 1$ to $x = 2$.

Solution: *Rough sketch of the region* in question is given below (Figure 8a.9). (The upper part of the vertical strip lies on the curve $y = e^x$ and the lower part on the line $y = 0$, which is the x-axis.)

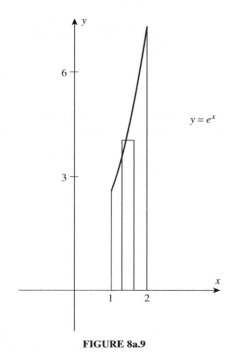

FIGURE 8a.9

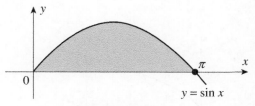

FIGURE 8a.10

The area of the region in question is given by

$$S = \int_1^2 e^x \, dx$$

$$= [e^x]_1^2$$

$$= e^2 - e$$

$$= e(e - 1) \text{ square units} \qquad \text{Ans.}$$

Example (9): Compute the area of the figure bounded by the curve $y = \sin x$ and the x-axis from $x = 0$ to $x = \pi$

Solution: For $x = 0$, we have $\sin x = 0 \Rightarrow y = 0$, and for $x = \pi$, we have $\sin \pi = 0 \Rightarrow y = 0$. Thus, *the desired area is bounded by a half-wave of the sine curve*, and the x-axis, from $x = 0$ to $x = \pi$ (Figure 8a.10). The area in question is given by

$$S = \int_0^\pi \sin x \, dx = [-\cos x]_0^\pi$$

$$= -[\cos \pi - \cos 0]$$

$$= -\cos \pi + \cos 0$$

$$= -(-1) + 1 = 1 + 1$$

$$= 2 \text{ square units} \qquad \text{Ans.}$$

Remark: In evaluating definite integrals, *if the antiderivative involved has a negative sign*, then it is useful to take the negative sign outside the bracket, before evaluating the integral. This helps us in avoiding possible mistakes in computation.

Note: In this chapter, we have previously shown that

$$\int_0^{2\pi} \sin x \, dx = 0 \text{ (Section 8a.1.2, Figure 8a.1)}$$

We interpreted this result as follows:

On the interval $[0, 2\pi]$ the sine curve makes two half waves, the first of which lies above the x-axis and the other lies below it. *The areas bounded by them are equal and the corresponding terms that appear in the geometric representation of the definite integral, cancel each other.* If we consider *this integral in terms of area of the region enclosed*, then its value must equal 4 square units.

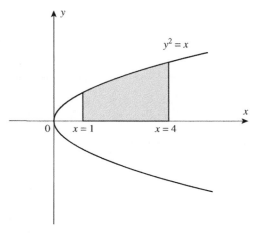

FIGURE 8a.11

Example (10): Compute the area bounded by parabola $y^2 = x$, $y \geq 0$, $x = 1$, and $x = 4$

Solution: The parabola $y^2 = x$ passes through the origin $(0, 0)$, and it is symmetrical to the x-axis.

(The given condition $y \geq 0$ suggests that the area in question is bounded by the upper part of the parabola $y^2 = x$, the x-axis, and the straight lines $x = 1$ and $x = 4$.)[14]

A *rough sketch* of the region in question is given above (Figure 8a.11).

In this case, we can write $y = \sqrt{x}$, $a = 1$, and $b = 4$. (Note that *the function $y = \sqrt{x}$ is defined only for nonnegative values of x.*) The desired area is given by

$$S = \int_1^4 \sqrt{x}\,dx = \int_1^4 x^{1/2}\,dx$$

$$= \left[\frac{x^{3/2}}{3/2}\right]_1^4 = \left[\frac{2}{3}x^{\frac{3}{2}}\right]_1^4 = \frac{2}{3}\left[(4)^{3/2} - (1)^{3/2}\right]$$

$$= \frac{2}{3}(8 - 1) = \frac{14}{3} \text{ square units} \qquad \text{Ans.}$$

Note: *Conditions of symmetry should be made use of where they exist, to shorten computations. If the area in question is symmetric with respect to the x-axis (or the y-axis), [as may happen in case of parabola(s), circles, ellipses, and trigonometric functions like $\sin x$ and $\cos x$], then we must choose one symmetrical part of the region (preferably the one which is above the x-axis or to the right of the y-axis) and then multiply the result suitably.*

Example (11): Compute the area bounded by the curve $y = -x^2 + 4$ and $y = 0$ (i.e., x-axis).

---

[14] These details are reflected here for better understanding of the problem. However, they need not be reflected in the proof. This will save time.

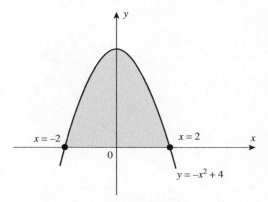

FIGURE 8a.12

Solution: The given curve is $y = -x^2 + 4$, which is *symmetrical with respect to the y-axis*. For $y = 0$, we get $x^2 = 4 \Rightarrow x = \pm 2$, which are the points at which *the curve intersects the x-axis*. This observation is helpful in drawing a rough sketch of the region in question as shown below. We have to compute the *area of the shaded region* that is symmetrical with respect to the y-axis (Figure 8a.12).

We shall compute the area of the region situated in the first quadrant, that is, half of the total area in question. If the total area is denoted by "S" then, we have

$$\frac{1}{2}S = \int_0^2 (-x^2 + 4)\mathrm{d}x = \left[\frac{-x^3}{3} + 4x\right]_0^2 = \left[-\frac{8}{3} + 8\right] = \frac{16}{3}$$

$$S = \frac{32}{3} = 10\frac{2}{3} \text{ square units} \qquad \text{Ans.}$$

Note: In this example, we have chosen to compute half of the area in question (which is above the x-axis) taking a vertical strip as the elementary area. This approach is definitely convenient. However, we may as well choose the elementary *horizontal* strip to compute the area of the region, which is on the right-hand side of the y-axis. With this approach, the area in question will be given by the definite integral $\int_a^b x \, \mathrm{d}y$, (where x is to be replaced in terms of y and the limits a and b, must be found out).

The given curve is $y = -x^2 + 4$.

$$\therefore \quad x^2 = 4 - y$$
$$\therefore \quad x = (4 - y)^{1/2}$$

To find the limits of the integral, we have, for $x = 0$, $y = 4$, and for $x = 2$, $y = 0$. (Now we integrate from lower value of y to upper value of y.) Thus, half of the area A is given by

$$\frac{1}{2}A = \int_0^4 (4 - y)^{1/2} \, \mathrm{d}y = -\int_0^4 (4 - y)^{1/2} \, \mathrm{d}y = -\left[\frac{(4 - y)^{3/2}}{3/2}\right]_0^4$$

$$= -\left[\frac{2}{3}(-8)\right] = \frac{16}{3}$$

$$\therefore \quad A = \frac{32}{3} \qquad \text{Ans.}$$

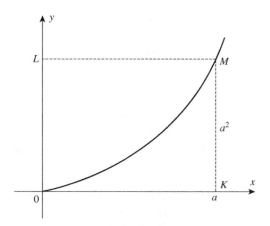

FIGURE 8a.13

Example (12): Find the area of a curvilinear triangle bounded by the x-axis, the parabola $y = x^2$ and the straight line $x = a$

Solution: The given curve (parabola) is

$$y = x^2 \qquad (7)$$

It is symmetrical to the y-axis.[15]

The required area is given by

$$S = \int_0^a x^2 \, dx = \left[\frac{x^3}{3}\right]_0^a = \frac{a^3}{3} - 0 = \frac{a^3}{3} \text{ square units} \qquad \text{Ans.}$$

Remark: Note that the area of the *curvilinear triangle* OKM, shown above is *one-third of the area of the rectangle* OKML, as explained below (Figure 8a.13).

Observe that $\dfrac{a^3}{3} = \dfrac{1}{3}(a) \cdot (a^2) = \dfrac{1}{3}$ (base) $\cdot$ (height)

Example (13): Find the area bounded by the curve $y = \cos x$ between $x = 0$ and $x = 2\pi$

Solution: A rough sketch of the curve $y = \cos x$ from $x = 0$ to $x = 2\pi$ is shown in Figure 8a.14. (Note that a part of the region that is below the x-axis will have a "negative area".)

The given curve is $y = \cos x$

$$\text{For } x \in \left[0, \frac{\pi}{2}\right], \quad \cos x \geq 0$$

$$x \in \left[\frac{\pi}{2}, \frac{3\pi}{2}\right], \quad \cos x \leq 0$$

[15] Note that for any nonzero value of x, the value of y is always positive, we consider the line $x = a$, for $a > 0$.

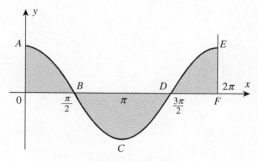

FIGURE 8a.14

and

$$x \in \left[\frac{3\pi}{2}, 2\pi\right], \quad \cos x \geq 0,$$

∴ Required area of the shaded region

$$= \int_{0}^{\pi/2} \cos x \, dx - \int_{\pi/2}^{3\pi/2} \cos x \, dx + \int_{3\pi/2}^{2\pi} \cos x \, dx^{\infty(16)}$$

$$= [\sin x]_{0}^{\pi/2} - [\sin x]_{\pi/2}^{3\pi/2} + [\sin x]_{3\pi/2}^{2\pi}$$

$$= \left[\sin \frac{\pi}{2} - \sin 0\right] - \left[\sin \frac{3\pi}{2} - \sin \frac{\pi}{2}\right] + \left[\sin 2\pi - \sin \frac{3\pi}{2}\right]$$

$$= [(1 - 0) - (-1 - 1) + (0 - (-1))] = 1 + 2 + 1 = 4 \text{ square units} \qquad \text{Ans.}$$

Example (14): Find the area enclosed by the circle of radius "a" units

Solution: Consider the circle represented by the equation

$$x^2 + y^2 = a^2 \tag{8}$$

which has the *center at the origin* and the *radius* "a" units as indicated in the *rough sketch*. As the circle is *symmetrical about both the coordinate axes,* we choose to compute the area of the region AOBA, *which is in the first quadrant and above the x-axis* (Figure 8a.15). (This area is *one fourth of the area* "S" of the circle.)

From Equation (8), we get

$$y = \pm\sqrt{a^2 - x^2} \, dx$$

---

[16] It is important to remember that area of the regions below the *x*-axis is always a negative number. (This can be seen only from the sketch of the function.) In such cases, we must express the area of the region by the expression $\left[-\int_{a}^{b} f(x) dx\right]$, or by the expression $\int_{a}^{b} |f(x)| dx$ to count it positive.

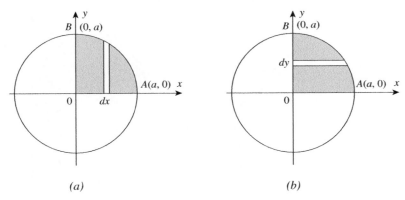

(a)   (b)

FIGURE 8a.15

Since, the portion of the curve from Equation (8) is considered in the first quadrant, we have

$$\frac{1}{4}S = \int\limits_{0}^{a} y\,dx = \int\limits_{0}^{a} \sqrt{a^2 - x^2}\,dx$$

$$\therefore \quad S = 4\int\limits_{0}^{a} \sqrt{a^2 - x^2}\,dx \qquad \text{(Taking vertical strips)}$$

$$= 4\left[\frac{x}{2}\sqrt{a^2 - x^2} + \frac{a^2}{2}\sin^{-1}\frac{x}{a}\right]_0^a$$

$$= 4\left[\left(\frac{a}{2}\cdot 0\right) + \frac{a^2}{2}\sin^{-1} 1\right] = 4\left[\frac{a^2}{2}\cdot\frac{\pi}{2}\right]$$

$$= \pi a^2 \text{ square units} \qquad \text{Ans.}$$

Remark: In the above example, we have considered a *vertical strip to represent an elementary area*. If we consider *horizontal strips*, the area of the circle will still be the same. This is quite natural. In that case, we write [using Equation (8)],

$$S = 4\int\limits_{0}^{a} \sqrt{a^2 - y^2}\,dy$$

$$= 4\left[\frac{y}{2}\sqrt{a^2 - y^2} + \frac{a^2}{2}\sin^{-1}\frac{y}{a}\right]_0^a$$

$$= 4\left[\left(\frac{a}{2}\cdot 0\right) + \frac{a^2}{2}\sin^{-1} 1\right] = 4\left[\frac{a^2}{2}\cdot\frac{\pi}{2}\right] = \pi a^2 \text{ square units} \qquad \text{Ans.}$$

Similarly, it can be shown that the area enclosed by the ellipse $\dfrac{x^2}{a^2} + \dfrac{y^2}{b^2} = 1$ is πab square units.

Remark: In the above example, we have used the *integration formula obtained in* Chapter 4b. There, we never had an idea that *this formula will be useful in computing the area enclosed by a*

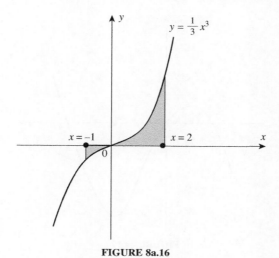

FIGURE 8a.16

circle. In fact, *this is a difficult integral.* However, there are *other simpler techniques of representing the area of a circle.* We shall discuss these techniques shortly. First we consider the following thought provoking examples.

Example (15): Compute the area of the region bounded by the curve $y = \dfrac{1}{3}x^3$, $y = 0$, $x = -1$, and $x = 2$

Solution: A rough sketch of the curve $y = \dfrac{1}{3}x^3$ is shown in Figure 8a.16 and the region in question is shaded.

(Observe that *the region in question is situated on both sides* of the x-axis.)

Let S_1 is the area of the region below the x-axis and S_2 is the area of the region above the x-axis. Thus, the desired area is given by

$$S = -S_1 + S_2 = -\int\limits_{-1}^{0}\left(\frac{1}{3}x^3\right)dx + \int\limits_{0}^{2}\left(\frac{1}{3}x^3\right)dx$$

$$= -\left[\frac{x^4}{12}\right]_{-1}^{0} + \left[\frac{x^4}{12}\right]_{0}^{2} = -\left[0 - \frac{1}{12}\right] + \left[\frac{16}{12} - 0\right]$$

$$= \frac{1}{12} + \frac{16}{12} = \frac{17}{12} = 1\frac{5}{12} \text{ square units} \qquad \text{Ans.}$$

Here, the area of the region, *below the x-axis is given by the expression* "$-\int_{-1}^{0}((1/3)x^3)dx$" or it may be given by $\int_{-1}^{0}|(1/3)x^3|dx$, both being same. The important point to be noted is that in the expression $\int_{-1}^{0}|(1/3)x^3|dx$, we do not put the negative sign outside the integral symbol. The definition of absolute value takes care that the area in question adds in a positive manner. However, the expression $-\int_{-1}^{0}((1/3)x^3)dx$ is more convenient from practical point of view.

Example (16): Find the total area computed between the x-axis and the curve $y = x(x+1)(x-2)$

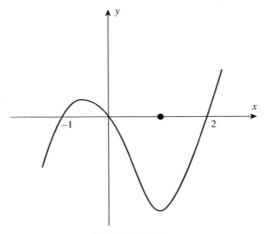

FIGURE 8a.17

Solution: By solving the equation, $x(x+1)(x-2)=0$, we get that the curve crosses the x-axis at $x=-1, x=0$, and $x=2$, as shown in the rough sketch of Figure 8a.17. *Note that a part of the region below the x-axis has "negative area."* The desired area of the shaded region is given by

$$A = \int_{-1}^{0} (x^3 - x^2 - 2x)dx - \int_{0}^{2} (x^3 - x^2 - 2x)dx$$

$$= \left[\frac{x^4}{4} - \frac{x^3}{3} - 2\frac{x^2}{2} \right]_{-1}^{0} - \left[\frac{x^4}{4} - \frac{x^3}{3} - 2\frac{x^2}{2} \right]_{0}^{2}$$

$$= \left[0 - \left(\frac{1}{4} - \frac{(-1)}{3} - 1 \right) \right] - \left[\left(4 - \frac{8}{3} - 4 \right) - 0 \right]$$

$$= \left[-\frac{1}{4} - \frac{1}{3} + 1 \right] - \left[\frac{-8}{3} \right]$$

$$= \left[\frac{5}{12} \right] + \left[\frac{8}{3} \right] = \frac{37}{12} \text{ square units} \qquad \text{Ans.}$$

Example (17): Find the area of the region bounded by the curve $y = x^2 - x - 2$ and $y = 0$ (the x-axis) from $x = -2$ to $x = 2$

Solution: The given curve is

$$y = x^2 - x - 2 = f(x)$$

Let us find *the zeros of this function, by solving the equation.*

$$x^2 - x - 2 = 0$$
$$x^2 - 2x + x - 2 = 0$$
$$x(x-2) + 1(x-2) = 0$$

or $\qquad\qquad (x-2)(x+1) = 0$

∴ $x = -1$ and $x = 2$ are the zeros of $f(x)$.

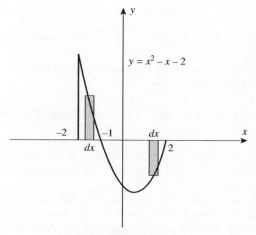

FIGURE 8a.18

It follows that the given curve crosses the x-axis at $x = -1$ and $x = 2$. We have to find the area of the region from $x = -2$ to $x = 2$. The rough sketch is given above.

Note that, *a part of the curve* (from -2 to -1) *is above the x-axis and the other part* (from -1 to 2) *is below the x-axis.*

This situation points out the importance of sketching the region (Figure 8a.18).

On $[-2, -1]$, *the area of the region is given by*

$$S_1 = \int_{-2}^{-1} (x^2 - x - 2)dx$$

$$= \left[\frac{x^3}{3} - \frac{x^2}{2} - 2x \right]_{-2}^{-1}$$

$$= \left[\left(-\frac{1}{3} - \frac{1}{2} + 2 \right) - \left(-\frac{8}{3} - \frac{4}{2} + 4 \right) \right]$$

$$= \left[-\frac{1}{3} - \frac{1}{2} + \frac{8}{3} + 2 + 2 - 4 \right]$$

$$= \frac{-2 - 3 + 16}{6} = \frac{11}{6} \text{ square units}$$

Again, on $[-1, 2]$ *the (positive) area of the region is given by*

$$S_2 = -\int_{-1}^{2} (x^2 - x - 2)dx = -\left[\frac{x^3}{3} - \frac{x^2}{2} - 2x \right]_{-1}^{2}$$

$$S_2 = -\left[\left(\frac{8}{3} - \frac{4}{2} - 4 \right) - \left(-\frac{1}{3} - \frac{1}{2} + 2 \right) \right] = -\left[\left(\frac{8}{3} - \frac{6}{1} + \frac{1}{3} + \frac{1}{2} - 2 \right) \right]$$

$$= -\left[\frac{16 + 2 + 3 - 48}{6} \right] = -\frac{21}{6} + \frac{48}{6} = \frac{27}{6} = 4\frac{1}{2} \text{ square units}$$

∴ Total area

$$S = S_1 + S_2 = \frac{11}{6} + \frac{27}{6} = \frac{38}{6} = \frac{19}{3} \text{ square units} \qquad \text{Ans.}$$

Note: *In such problems it is always better to evaluate S_1 and S_2 separately, as done above*, thus avoiding possible mistakes in calculations.

8a.4   COMPUTING THE AREA OF A CIRCLE (DEVELOPING SIMPLER TECHNIQUES)

In a circular measure of an angle, the *unit angle* is called the *radian*. *One radian* is the angle subtended at the center of a circle by "*an arc of the circle*", whose length is equal to the radius of the circle.

In fact, no part of a curve however small can be superimposed on any portion of a straight line, so that it coincides with it. In other words, the length of a curve cannot be found by comparison with a straight line of a known length. However, under the application of definite integrals we have a method of determining the lengths of arcs of plane curves whose equations are known. Though *it is not possible to measure the length of an arc of a circle*, this difficulty is (initially) overcome by the assumption that the length of an arc of a semicircle equals π radians, where π is a constant. Accordingly, the length of the circumference of a circle is 2π radians.

Since, the (entire) length of the circumference of a circle is consumed in measuring the (total) angle subtended by the circle at the center, we can write

$$\frac{\text{circumference(of the circle)}}{\text{radius("r" of the circle)}} = 2\pi \text{ radians}$$

Therefore, the circumference $= 2\pi r$ units of length. Even at this stage, the difficulty faced by the student remains unchanged, since he does not have any method of finding the value of π. Of course, the ancient Greeks estimated the value of π (through practical methods to be approximately 3.14159... or some less accurate approximation), but otherwise it is undetermined.[17]

The method involved in measuring the length of an arc, is similar to that used for computing areas. An expression is found for "*an element of length*" of the curve, and the sum of all such elements is obtained by computing its definite integral. (Of course, we shall be discussing these matters in the next chapter.) *For the time being, we agree that the length of the circumference of a circle is $2\pi r$, where "r" is the radius of the circle, and π is approximately equal to* 3.1415...Thus, we have a practical method of computing the length of circumference of a circle. The following method will be found very useful in applications.

8a.4.1   Area of a Circle: A Detailed Discussion Involving the Ideas of Limits

The *area of a circle can be thought of* as the *area of a plane figure* that is *traced out by a finite straight line* OA ($= r$ units) as it rotates around one of its ends (say, "O"), and makes a complete rotation. In the figure below, the line OP ($= r$ units) starting from *the fixed position* OA (on OX) makes a *complete rotation* around a *fixed point* "O". Thus, the point "P" *describes the circumference of a circle.*

---

[17] The determination of the value of π occupied mathematicians through the centuries. Various devices (with which we are not concerned here) were applied and approximate values of π were found. Fortunately, modern mathematics, with the help of *Calculus*, has solved the problem. It can now be proved that the above ratio (representing π) is incommensurable, and that its value to any required degree of accuracy can be calculated with accuracy. Now, it is also known that "π" is not restricted to the circle, and that it appears in many other contexts. *Teach Yourself Calculus* by P. Abbott.

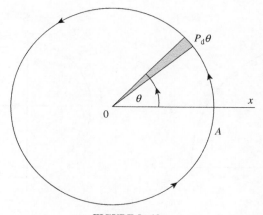

FIGURE 8a.19

Our interest is to find *the area marked out by* OA *in one rotation* (i.e., *area of the circle of radius "r"*). Suppose the point P, has rotated from OA, such that it has described an angle "θ". Then, AOP is a *sector of a circle* of radius "r". Now, suppose OP is rotated further through an *infinitesimally small angle* dθ. The *infinitely small sector* so described would be *an element of area* and the sum of all such sectors when OP makes a complete rotation from OX back again to its original position, will be the area of the circle (Figure 8a.19).

Note that the limit concept is applicable in this case. Hence, the *infinitely small arc* subtending *an infinitely small angle* dθ at the center of the circle, *"in the limit"*, can be regarded *as a "straight line"* and the infinitely small sector as a "triangle". Further, the *altitude of the triangle* can be regarded, *"in the limit"*, as the *radius "r"* of the circle.

Thus, the triangle in question (which is *a sector of the circle, in this case*) can be regarded *"in the limit"* as a *"right triangle"*. Using the formula for the *area of a triangle*, we have

Element of area (of the *triangular sector* under consideration)

$$= \frac{1}{2}(\text{base}) \cdot (\text{altitude})$$

[Now, we explain below (in the footnote) that the base of the triangular sector (i.e., arc length), in the limit equals $r\,d\theta$.][18]

$$= \frac{1}{2}(r \cdot d\theta) \cdot (r) = \frac{1}{2}r^2 \cdot d\theta$$

Now, the angle corresponding to a *complete rotation* is 2π radians.

$$\therefore \quad \text{Area} = \int_0^{2\pi} \frac{1}{2}r^2 \cdot d\theta = \left(\frac{1}{2}r^2\right) \cdot \int_0^{2\pi} d\theta = \left(\frac{1}{2}r^2\right)[\theta]_0^{2\pi}$$

$$= \left(\frac{1}{2}r^2\right) \cdot [2\pi - 0] = \pi r^2 \text{ square units}$$

---

[18] The result: Base of the triangular sector of a circle $= r \cdot d\theta$, follows from the trigonometric limit $\lim_{\theta \to 0}(\sin\theta/\theta) = 1$ (where θ is expressed in radians). Now, we may write, $\lim_{d\theta \to 0}(\sin d\theta/d\theta) = 1$, where the small angle dθ is expressed in radians.

Once it is accepted that the (small) arc in question (in the limit) can be treated as a line segment, we can define the ratio $\sin d\theta = (\text{length of the arc}/r) = d\theta$ (since the angle dθ is expressed in radians). $\therefore$ Length of the arc $= r \cdot d\theta$ (we emphasize that this statement is valid "in the limit").

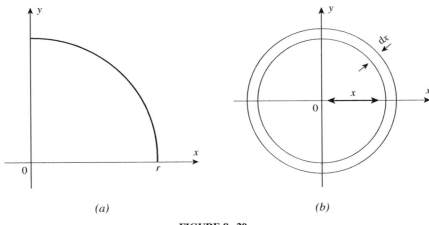

(a)   (b)

FIGURE 8a.20

Now, we give below *two simple techniques* for computing the area of a circle. In both the methods, we identify an element of (variable) area, which can generate the circle in a simple process. Let us discuss.

Method (1): Consider a circle of a *variable radius* "x". Suppose the area of the circle is $A(x)$. When x increases by an infinitely small increment dx, let the corresponding small increment in the area of the circle be denoted by $dA(x)$. This increment in the area is equal to the area of a narrow strip (in the form of a ring) between the circle of radius x and that of radius $x + dx$ (Figure 8a.20). (In this process, the *independent variable* is the radius "x" and the *dependent variable* is the "area" of the circular ring that can grow to a circle.)

The width of the strip is dx and we can take $2\pi x$ as its length. (Here again the important role is played by the length of the circumference of a circle that we have assumed to be $2\pi x$.)
    Thus,

$$dA(x) = 2\pi x \, dx$$

[This expression represents an element of (variable) area, which can grow to the area of the given circle.]
    It follows that area of a circle of radius "r" is

$$A = \int_0^r 2\pi x \, dx = 2\pi \left[\frac{x^2}{2} \right]_0^r = 2\pi \frac{r^2}{2} = \pi r^2 \text{ square units}$$

Method (2): Now, consider a sector of the circle with a *variable angle* "φ". The increment to the area $A(\varphi)$ of the sector, is the area of the triangle with base $r \, d\varphi$, and altitude "r". It is denoted by

$$dA(\varphi) = \frac{1}{2} r d\varphi r = \frac{1}{2} r^2 d\varphi$$

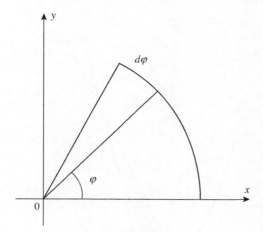

FIGURE 8a.21

Therefore, the area of the circle of radius "r" is given by

$$\frac{1}{2}\int_0^{2\pi} r^2 \, d\varphi$$

$$A = = \frac{1}{2}r^2[2\pi - 0]$$

$$= \pi r^2 \text{ square units} \qquad \text{Ans.}$$

Note: In this process, the *independent variable* is the angle φ and the *dependent variable* is the *area of the sector*. Here again, the final product is the *area of the circle*, but the basic concept involved is that the circular base of the sector is treated in the limit as a line segment. This permits us to define (an element of) the variable area of a circle (Figure 8a.21).

Remark: Both the methods discussed above are traditional elementary methods that are found to be more *efficient* than the method of calculating areas by using the *standard result(s) of integral(s)*.

8a.4.2   Area Between Two Curves

Geometrically, the *concept of the definite integral in terms of area implies* (roughly speaking) that the area of any region is calculated by considering the region in the form of *large number of thin strips*, [whose width must be *indefinitely small* (i.e., as small as we wish)] and then *adding up these elementary strips*.

In the process of computing a definite integral, the *most important role is played by the concept of limit*, which permits us to obtain the *actual area of the region, even bounded by curves*. This is considered as *the greatest achievement in the field of mathematics*, as we know it, today.

Having obtained formulae for computing areas of such regions, we now proceed to give some solved examples.

Example (18):   Compute the area of the region bounded by the parabola $y = x^2$ and the straight line $y = 2x$

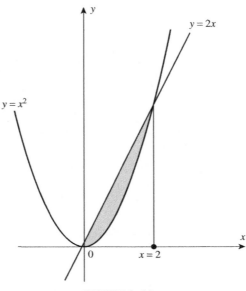

FIGURE 8a.22

Solution: The parabola $y = x^2$ passes through the origin and it is symmetrical to the y-axis. Its rough sketch is given below (Figure 8a.22).

To find the points of intersection, we solve the system of equations $y = x^2$ and $y = 2x$, we get $x^2 = 2x$, so that $x^2 - 2x = 0$ or $x(x - 2) = 0$. Thus, the points of intersection are $x = 0$ and $x = 2$. Now, it is important to note carefully that *the upper portion of the shaded region is bounded by the straight line $y = 2x$ and that its lower portion is bounded by the arc of the parabola $y = x^2$.* Hence, the desired area of the shaded region is given by

$$S = \int_0^2 (y_{\text{upper}} - y_{\text{lower}}) dx$$

$$= \int_0^2 (2x - x^2) dx$$

$$= \left[x^2 - \frac{x^3}{3} \right]_0^2 = \left[4 - \frac{8}{3} \right] - 0 = \frac{4}{3} \text{ square units} \qquad \text{Ans.}$$

Example (19): Find the area enclosed between the parabola $y^2 = 4ax$ $(a > 0)$ and the line $y = mx$ $(m > 0)$

Solution: The two curves are

$$y^2 = 4ax \qquad \qquad (9)$$

$$y = mx \qquad \qquad (10)$$

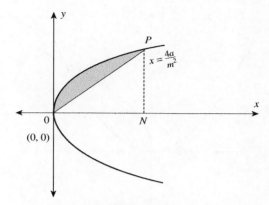

FIGURE 8a.23

To find the points of intersection, we solve *the system of above equations*, and get

$$m^2 x^2 = 4ax$$
$$\Rightarrow x(m^2 x - 4a) = 0$$
$$\therefore x = 0 \quad \text{and} \quad x = \frac{4a}{m^2}$$

Note that the parabola using Equation (9) is symmetrical to the x-axis.

[Taking the slope of the line having Equation (10), as *positive*, the rough sketch of the shaded region in question, is indicated in Figure 8a.23.][19]

The desired area is given by

$$S = \int\limits_{0}^{4a/m^2} (y_{\text{upper}} - y_{\text{lower}})\mathrm{d}x\theta = \int\limits_{0}^{4a/m^2} (\sqrt{4ax} - mx)\mathrm{d}x$$

$$= \int\limits_{0}^{4a/m^2} \left[2\sqrt{a}x^{1/2} - mx \right]\mathrm{d}x$$

$$= \left[2\sqrt{a}\frac{x^{3/2}}{3/2} - m\frac{x^2}{2} \right]_{0}^{4a/m^2} = \left[\frac{4\sqrt{a}}{3}x^{3/2} - m\frac{x^2}{2} \right]_{0}^{4a/m^2}$$

$$= \left[\frac{4\sqrt{a}}{3}\left(\frac{4a}{m^2}\right)^{3/2} - \frac{m}{2}\left(\frac{4a}{m^2}\right)^{2} \right] - 0$$

$$S = \left[\frac{4\sqrt{a}}{3}\cdot\frac{8a^{3/2}}{m^3} - \frac{m}{2}\frac{16a^2}{m^4} \right]$$

$$= \frac{32a^2}{3m^3} - \frac{8a^2}{m^3} = \frac{8a^2}{3m^3} \text{ square units} \qquad \text{Ans.}$$

[19] If the slope of the line $y = mx$ is taken as negative, then the enclosed region will be below the x-axis. Of course, its area will remain unchanged.

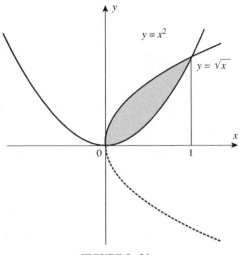

FIGURE 8a.24

Example (20): Compute the area bounded by the curves $y = \sqrt{x}$ and $y = x^2$

Solution: The given curves are

$$y = \sqrt{x}, \quad x \geq 0$$

or

$$y^2 = x, \quad x \geq 0 \tag{11}$$

and

$$y = x^2 \tag{12}$$

Note that, the curve Equation (11) is a parabola that is symmetrical to the x-axis and we have to consider only the *upper half* for $x \geq 0$.

The curve $y = x^2$ is the parabola, which is *symmetrical to the y-axis*.

The points of intersection of the curves are given by $(x^2)^2 = x$.

$$\therefore \quad x^4 - x = 0 \quad \text{or} \quad x(x^3 - 1) = 0$$
$$\therefore \quad x = 0 \quad \text{and} \quad x = 1^{(20)}$$

Note carefully that the upper curve of the shaded region is $y = \sqrt{x}$ and the lower curve is $y = x^2$ (Figure 8a.24).

$$\therefore \quad \text{Area in question} = \int_0^1 \left(\sqrt{x} - x^2\right) dx$$

$$= \left[\frac{x^{3/2}}{3/2} - \frac{x^3}{3}\right]_0^1$$

$$= \frac{2}{3} - \frac{1}{3} = \frac{1}{3} \text{ square units} \qquad \text{Ans.}$$

---

[20] Observe that for $x = 0$, $y = 0$ and for $x = 1$, $y = 1$. It follows the region in question is above the x-axis.

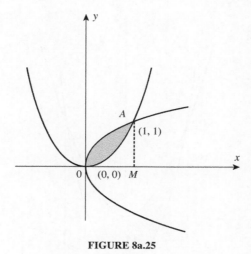

FIGURE 8a.25

Remark: If we are required to find the area bounded by the parabolas $y = x^2$ and $x = y^2$, then the *answer will remain the same* but there will be a slight change in the rough sketch, as will be clear from the sketch (Figure 8a.25). (In solving this problem, the figures do not make any difference. However, the correct understanding of the problem is important.)

Example (21): Compute the area of the region bounded between the curves (parabolas) given by $7x^2 - 9y + 9 = 0$ and $5x^2 - 9y + 27 = 0$

Solution: Let us rewrite the equations of the parabolas in the form

$$y = \frac{7}{9}x^2 + 1 \qquad\qquad (13)$$

$$y = \frac{5}{9}x^2 + 3 \qquad\qquad (14)$$

Both the curves represent parabolas that are symmetric to the y-axis. To find the points of intersection, we solve the system of these equations.
Thus, we get

$$7x^2 + 9 = 5x^2 + 27 \qquad \text{(Note that each side equals 9y)}$$
$$\therefore \quad 2x^2 = 18 \quad \text{or} \quad x^2 = 9$$

so that we get

$$x_1 = -3, \quad x_2 = 3$$

It is easy to draw a rough sketch of the curves having Equation (13) and Equation (14) (Figure 8a.26).

Note: Observe that for $x = 0$, Equation (13) gives $y = 1$ and Equation (14) gives $y = 3$. *From these values of y, it follows that the upper curve is given by* Equation (14) *and the lower curve is given by* Equation (13).
Further, observe that the shaded area is symmetric with respect to the y-axis.

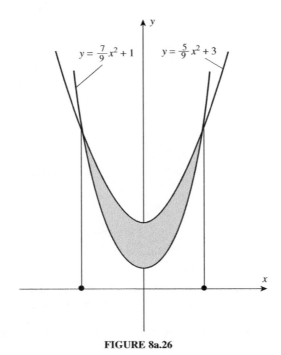

FIGURE 8a.26

Hence, we compute one half of the area in question, taking the limits of integration from 0 to 3, by

$$\frac{1}{2}S = \int_0^3 \left[\left(\frac{5}{9}x^2 + 3 \right) - \left(\frac{7}{9}x^2 + 1 \right) \right] dx$$

$$\int_0^3 \left[\left(2 - \frac{2}{9}x^2 \right) \right] dx$$

$$\left[2x - \frac{2}{9}\frac{x^3}{3} \right]_0^3 = \left[6 - \frac{2}{9}\frac{27}{3} \right] = 4 \text{ square units}$$

$$\therefore \quad S = 8 \text{ square units} \qquad \text{Ans.}$$

Example (22): Find the area common to the curves $2(y-1)^2 = x$ and $(y-1)^2 = x - 1$

Solution: The equations of the curves are

$$2(y-1)^2 = x \qquad (15)$$
$$(y-1)^2 = x - 1 \qquad (16)^{[21]}$$

Let us find *the points of intersection of the above curves.*

---

[21] Consider Equation (15) in the form $2t^2 = x$, where $t = (y-1)$. This equation represents a parabola that is symmetric to the x-axis. [In our case the curve is symmetric to the line $y = 1$, which is parallel to the x-axis.] Similarly, the parabola having Equation (16) is symmetric to the line $y = 1$. This observation is useful in drawing a rough sketch of the area in question.

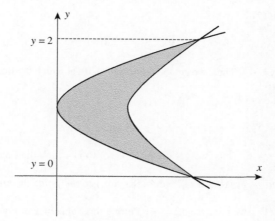

FIGURE 8a.27

Using Equation (16) in Equation (15), we have

$$2(x - 1) = x \qquad \therefore \quad x = 2$$

Putting this value of x in Equation (16), we get

$$(y - 1)^2 = 1 \qquad \therefore \quad y - 1 = \pm 1$$
$$\therefore \quad y = 2 \quad \text{or} \quad 0.$$

Thus, the curves having Equations (15) and (16) intersect in the points (2, 0) and (2, 2) (Figure 8a.27).

The area in question is given by

$$A = \int_0^2 \left[\left(1 + (y - 1)^2 \right) \right] dy - \int_0^2 2(y - 1)^2 dy$$

$$= \int_0^2 \left[1 - (y - 1)^2 \right] dy$$

$$= \left[y - \frac{(y - 1)^3}{3} \right]_0^2 = \left(2 - \frac{1}{3} \right) - \left(\frac{1}{3} \right)$$

$$= 2 - \frac{2}{3} = \frac{6 - 2}{3} = \frac{4}{3} \quad \text{square units} \quad \text{Ans.}$$

Example (23): Compute the area of the figure bounded by the inclined lines $x - 2y + 4 = 0$, $x + y - 5 = 0$, and $y = 0$

Solution: The given lines are

$$x - 2y + 4 = 0 \tag{17}$$

and

$$x + y - 5 = 0 \tag{18}$$

From the Equation (7)

$$x - 2y + 4 = 0$$

we get, for $y = 0$, $x = -4$. Thus, we get the point A $(-4, 0)$, at which the line of Equation (17) intersects the x-axis.

From the Equation (18)

$$x + y - 5 = 0$$

we get, for $y = 0$, $x = 5$. Thus, we get the point C $(5, 0)$ at which the line of Equation (18) intersect the x-axis.

Solving the *system of Equations* (17) and (18), we get *the point of intersection of these straight lines.*

From Equation (17), we have $x = (2y - 4)$. Putting this value of x in Equation (18), we get

$$(2y - 4) + y - 5 = 0$$
$$\therefore \quad 3y = 9, \text{ so that } y = 3$$

Therefore, by putting this value of y in Equation (18), we get $x = 2$. Thus, we get the point M $(2, 3)$ *at which the lines* having Equations (17) and (18) *intersect.* Now, we are in a position to construct the figure as given below (Figure 8a.28).

To compute the desired area, it is necessary to partition the triangle *AMC* into *two triangles AMN* and *NMC.* Because, *as x changes from A to N*, the area is *bounded by the straight line* having Equation (17) and when x varies from N to C the *area is bounded by the straight line* having Equation (18).

For the triangle *AMN, the role of the line* $x - 2y + 4 = 0$ (bounding the area) is expressed by the equation

$$y = \frac{1}{2}x + 2, \quad \text{with the limits} \quad a = -4 \quad \text{and} \quad b = 2$$

For the triangle *NMC*, the role of the line $x + y - 5 = 0$ (bounding the area) is expressed by the equation $y = -x + 5$ with the limits $a = 2$ and $b = 5$.

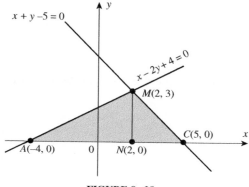

FIGURE 8a.28

Computing the area of each of the triangles and adding together the results so obtained, we get

$$S_{\Delta AMN} = \int_{-4}^{2} \left(\frac{1}{2}x + 2\right) dx = \left[\frac{1}{2}\left(\frac{x^2}{2}\right) + 2x\right]_{-4}^{2} \infty$$

$$= \left[\frac{1}{4}x^2 + 2x\right]_{-4}^{2}$$

$$= [(1 + 4) - (4 - 8)] = 9 \text{ square units}$$

$$S_{\Delta NMC} = \int_{2}^{5} (-x + 5) dx = \left[-\left(\frac{x^2}{2}\right) + 5x\right]_{2}^{5}$$

$$= \left[\left(-\frac{25}{2} + 25\right) - \left(\frac{-4}{2} + 10\right)\right]$$

$$= [15 - 10.5] = 4.5 \text{ square units}$$

$\therefore$ Total area in question $= 9 + 4.5 = 13.5$ square units   Ans.

Remark: From the above figure, it is easy to check the area of the triangle AMC. We have

$$S_{\Delta AMC} = \frac{1}{2} (\text{length } AC) \cdot (\text{length } NM)$$

$$= \frac{1}{2}[5 - (-4)] \cdot (3 - 0) = \frac{1}{2}(9)(3) = \frac{27}{2} = 13.5 \text{ square units.}$$

Let us recall some useful definitions pertaining to the standard equation of parabola

$$y^2 = 4ax, \qquad a > 0 \tag{19}$$

and the related important points.

[In the equation of parabola having Equation (19), the coordinates of the focus "S" are $(a, 0)$.][22]

Definition: Axis—The line through the focus and perpendicular to the directrix of a parabola is called the axis of the parabola. [The axis of the parabola having Equation (19) is the x-axis itself.]

Definition: Vertex—The point of intersection of a parabola with its axis is called vertex of the parabola. [The vertex of the parabola from Equation (19) is the origin O (0, 0).]

- The parabola having Equation (19) is symmetrical to the x-axis. As y^2 is always positive, therefore for $a > 0$ (as given above), x cannot be negative. Therefore, the curve entirely lies in the first and the fourth quadrant.

  It is useful to remember the following figures pertaining to various forms of parabola (Figure 8a.29).

- As x takes larger values, y becomes larger. Hence, the curve extends to infinity. For the standard form of Equation (19), we say that the parabola opens out in the positive direction of x-axis.

---

[22] For the parabola $y^2 = -4ax = 4(-ax)$, $a > 0$ the coordinates of the focus are $(-a, 0)$. Similarly, for the parabola $x^2 = 4ay$ ($a > 0$), the focus is at $(0, a)$, and for the parabola $x^2 = -4ay = x^2 = 4(-a)y$ ($a > 0$), the focus is $(0, -a)$.

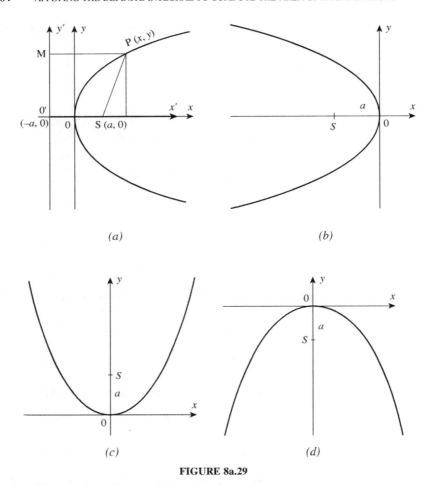

FIGURE 8a.29

Definition: Focal-Chord—A chord passing through the focus "S" is called the focal-chord of the parabola.

Definition: Latus-Rectum—The *focal-chord* of the parabola *perpendicular* to the axis is called the *latus-rectum* of the parabola.
For the parabola $y^2 = 4ax$, $a > 0$, the equation of the latus-rectum is $x = a$.

Note: For the parabola $y^2 = 24x$, the equation of the latus-rectum is $x = 6$.

Example (24): Find the area bounded between the parabola $y^2 = 4ax$ and its latus-rectum

Solution: The given curve (parabola) is

$$y^2 = 4ax \tag{20}$$

It is symmetrical about the x-axis and passes through the origin O (0, 0) (Figure 8a.30).
    The latus-rectum is the line perpendicular to the x-axis, which passes through the focus S $(a, 0)$ (Figure 8a.31).

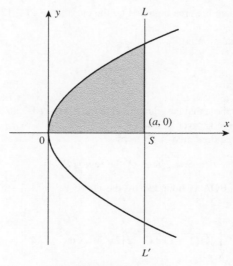

FIGURE 8a.30

The required area of the region is given by

$$A = 2 \,(\text{area of the shaded region OLSO})$$

$$= 2\int_0^a y \, dx = 2\int_0^a \sqrt{4ax} \, dx$$

$$= 4\sqrt{a}\int_0^a \sqrt{x} \, dx$$

$$= 4\sqrt{a}\frac{2}{3}\left[x^{3/2}\right]_0^a = \frac{8}{3}\sqrt{a} \cdot a^{3/2} = \frac{8}{3}a^2 \text{ square units} \qquad \text{Ans.}$$

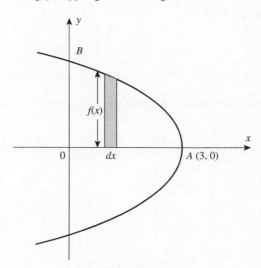

FIGURE 8a.31

Example (25): Calculate the area bounded by the curve $y^2 + 4x - 12 = 0$ and the y-axis

Solution: The given equation is

$$\begin{aligned} y^2 &= -4(x - 3) \\ &= 4(3 - x) \end{aligned} \tag{21}$$

It represents the parabola that is symmetrical to the x-axis. Also, its vertex is A $(3, 0)$. (The rough sketch of the curve is given below.)

By symmetry, the required area is given by

$$S = 2(\text{area of the region } BOA)$$

[Note that, the region BOA is bounded by the curve $y = \sqrt{4(3 - x)}$, the x-axis and the ordinates $x = 0$ and $x = 3$.]

$$\text{Hence, } S = 2\int_0^3 \sqrt{4(3 - x)}\, dx = 2\int_0^3 2\sqrt{3 - x}\, dx$$

$$= 4\int_0^3 \sqrt{3 - x}\, dx = 4\left[\frac{(3 - x)^{3/2}}{3/2}(-1)\right]_0^3$$

$$= \frac{-8}{3}\left[(3 - x)^{3/2}\right]_0^3 = \frac{-8}{3}\left[(0) - (3)^{3/2}\right]$$

$$S = \frac{-8}{3}\left[0 - \sqrt{27}\right] = \frac{-8}{3}\left[-3\sqrt{3}\right] = 8\sqrt{3} \text{ square units} \qquad \text{Ans.}$$

Example (26): Find the area enclosed between the curve $x^2 = 4y$ and the line $x = 4y - 2$

Solution: The given equations are

$$x^2 = 4y \tag{22}$$

$$x + 2 = 4y \tag{23}$$

The parabola having Equation (22) is symmetrical to the y-axis and passes through the origin (Figure 8a.32).

The points of intersection between the two curves are obtained by solving the system of Equations (22) and (23). We get

$$x^2 = x + 2 \quad [\text{using Equation}(23) \text{ in Equation}(22)]$$

$$\text{or} \quad x^2 - x - 2 = 0$$

$$x^2 - 2x + x - 2 = 0$$

$$x(x - 2) + 1(x - 2) = 0$$

$$\therefore \quad (x - 2)(x + 1) = 0$$

$$\therefore \quad x = 2 \quad \text{and} \quad x = 1$$

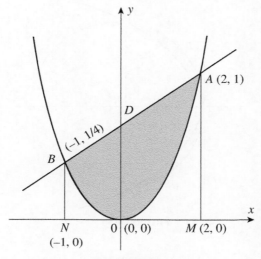

FIGURE 8a.32

The area of the shaded region is given by

$$A = \int_{-1}^{2} (y_{\text{upper}} - y_{\text{lower}}) dx$$

$$= \int_{-1}^{2} \left[\left(\frac{x}{4} + \frac{1}{2} \right) - \left(\frac{x^2}{4} \right) \right] dx$$

$$= \frac{1}{4} \int_{-1}^{2} (x + 2 - x^2) dx$$

$$= \frac{1}{4} \left[\frac{x^2}{2} + 2x - \frac{x^3}{3} \right]_{-1}^{2}$$

$$= \frac{1}{4} \left[\left(\frac{4}{2} + \frac{4}{1} - \frac{8}{3} \right) - \left(\frac{1}{2} - \frac{2}{1} + \frac{1}{3} \right) \right]$$

$$= \frac{1}{4} \left[\frac{20}{6} + \frac{7}{6} \right] = \frac{27}{24} = \frac{9}{8} \text{ square units} \qquad \text{Ans.}$$

Example (27): Find the area of the region enclosed between the circle $x^2 + y^2 = 2ax$ and parabola $y^2 = ax$ $(a > 0)$

Solution: The given curves are

$$x^2 + y^2 = 2ax \tag{24}$$

and

$$y^2 = ax \tag{25}$$

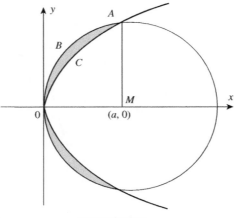

FIGURE 8a.33

Note that both the curves are *symmetrical to the x-axis*. Also, both the curves *pass through the origin*.

To find the points of intersection of the above curves, we solve the system of Equations (24) and (25). We get

$$x^2 + ax = 2ax \qquad [\text{using Equation}(25) \text{ in Equation}(24)]$$
$$\therefore \quad x^2 - ax = 0 \qquad \text{or} \quad x(x-a) = 0$$
$$\therefore \quad x = 0 \quad \text{or} \quad x = a. \quad \text{Now, for } x = 0, \ y = 0 \text{ and for } x = a, \ y = \pm a$$

Points of intersection are (a, a) and $(a, -a)$. It follows that the center of the circle is at $(a, 0)$ and the radius "a" units. The rough sketch of the intersecting curves is given below (Figure 8a.33).

We have to find the area of the region enclosed between the curves on *both the sides of the x-axis*

(For convenience, we shall consider the shaded area above the x-axis, and then multiply it by two.)

Let the shaded area above the x-axis be S square units.

$$\therefore \quad S = \int_0^a \left[y_{\text{upper}} - y_{\text{lower}} \right] dx$$

$$\therefore \quad = \int_0^a y_{\text{upper}} \, dx - \int_0^a y_{\text{lower}} \, dx$$

$$= \int_0^a \sqrt{2ax - x^2} \, dx - \int_0^a \sqrt{ax} \, dx \qquad (26)$$

Now consider, $2ax - x^2 = -[x^2 - 2ax]$
$$= -[x^2 - 2ax + a^2 - a^2]$$
$$= -[(x - a)^2 - a^2]$$
$$= a^2 - (x - a)^2$$

Using this expression in (26), we get

$$S = \int_0^a \sqrt{a^2 - (x-a)^2}\, dx - \sqrt{a} \int_0^a x^{1/2}\, dx$$

$$= \frac{1}{2}\left[(x-a)\sqrt{a^2 - (x-a)^2} + a^2 \sin^{-1}\frac{x-a}{a}\right]_0^a - \sqrt{a}\left[\frac{x^{3/2}}{3/2}\right]_0^a$$

$$S = \frac{1}{2}\left[a^2(\sin^{-1} 0 - \sin^{-1}(-1))\right] - \sqrt{a}\left[\frac{2}{3}a^{3/2}\right]$$

$$= \frac{1}{2}a^2\left[0 - \left(-\frac{\pi}{2}\right)\right] - \frac{2}{3}a^2$$

$$= \frac{a^2}{2}\cdot\frac{\pi}{2} - \frac{2}{3}a^2$$

$$= a^2\left[\frac{\pi}{4} - \frac{2}{3}\right]$$

$\therefore$ The total area between the curves having Equations (27) and (28) is given by

$$2S = a^2\left[\frac{\pi}{2} - \frac{4}{3}\right] \qquad \text{Ans.}$$

Example (28): Find the area lying above the x-axis and inclined between the circle $x^2 + y^2 = 8x$ and inside the parabola $y^2 = 4x$

Solution: The given curve are

$$x^2 + y^2 = 8x \tag{27}$$

$$y^2 = 4x \tag{28}$$

Note that both the curves are symmetrical to the x-axis, and both of them pass through the origin.

To find the points of intersection between Equations (27) and (28), we write

$$x^2 + 4x = 8x \qquad [\text{using Equation(28) in Equation(27)}]$$
$$\therefore \quad x^2 - 4x = 0$$
$$\therefore \quad x^2(x-4) = 0$$
$$\therefore \quad x = 0 \quad \text{or} \quad x = 4$$

Now, for $x = 0$, $y = 0$ and for $x = 4$, $y = \pm 4$.

It follows that the point $(4, 0)$ is the center of the circle.[23]

The shaded area indicated in the sketch has to be computed (Figure 8a.34).

[23] The coordinates of the center of the circle can also be obtained from Equation (27). We express Equation (27) as $x^2 - 2(4)x + 16 - 16 + y^2 = 0$.

$\therefore$ $(x-4)^2 + y^2 = 16$. This is the equation of the circle with the center $(4, 0)$ and radius 4 units.

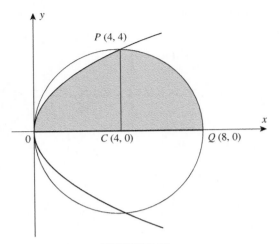

FIGURE 8a.34

The shaded area enclosed by the parabola above the x-axis is

$$S_1 = \int_0^4 \sqrt{x}\, dx$$

and the shaded area enclosed by the circle above the x-axis is

$$S_2 = \int_4^8 \sqrt{8x - x^2}\, dx = \int_4^8 \sqrt{4^2 - (x - 4)^2}\, dx$$

Now, $$S_1 = \left[\frac{x^{3/2}}{3/2}\right]_0^4 = \frac{2 \cdot 8}{3} = \frac{16}{3} \text{ square units}$$

The value of S_2 is *computed by substitution.*

Put $x - 4 = t$ $\therefore$ $dx = dt$

Also, when $x = 4$, $t = 0$, and when $x = 8$, $t = 4$.

$$\therefore \quad S_2 = \int_0^4 \sqrt{4^2 - (x - 4)^2}\, dt$$

$$= \left[\frac{t}{2}\sqrt{4^2 - t^2} + \frac{1}{2}(4)^2 \sin^{-1}\frac{t}{4}\right]_0^{4\,(24)}$$

$$S_2 = \left[0 + 8 \cdot (\sin^{-1} 1 - \sin^{-1} 0)\right]$$

$$= 8 \cdot \left(\frac{\pi}{2} - 0\right) = 4\pi$$

[24] Recall that $\int \sqrt{a^2 - x^2}\, dx = (1/2)x\sqrt{a^2 - x^2} + (1/2)a^2 \cdot \sin^{-1}(x/a) + c.$

$\therefore$ The required area is given by

$$S_1 + S_2 = \frac{16}{3} + 4\pi = \frac{4}{3}(4 + 3\pi) \qquad \text{Ans.}$$

Example (29):   Find the area of the loop of the curve $y^2 = x^2(1 - x)$

Solution: The given curve is

$$y^2 = x^2(1 - x)$$

Obviously, the given curve is symmetrical about x-axis.

To find the point(s) of intersection of the curve with the x-axis, we put $y = 0$, we get

$$0 = x^2(1 - x) \quad \therefore \quad x = 0 \quad \text{or} \quad x = 1$$

The rough sketch of the curve is given below (Figure 8a.35).
Observe that the curve has a loop between $x = 0$ and $x = 1$.
$\therefore$ Area "A", of the loop

$$= 2 \int_0^1 y \, dx = 2 \int_0^1 \sqrt{x^2(1 - x)} \, dx$$

$$A = \int_0^1 x\sqrt{1 - x} \, dx$$

Now, using the property

$$\int_0^1 f(x) dx = \int_0^1 f(a - x) dx$$

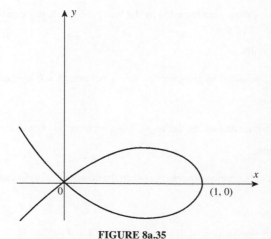

FIGURE 8a.35

We get

$$A = 2\int_0^1 (1-x)\sqrt{1-(1-x)}dx$$

$$= 2\int_0^1 (1-x)\sqrt{x}\,dx = 2\int_0^1 \left(\sqrt{x} - x^{3/2}\right)dx$$

$$A = 2\left[\frac{2}{3}x^{3/2} - \frac{2}{5}x^{5/2}\right]_0^1 = 2\left[\frac{2}{3} - \frac{2}{5}\right] = \frac{8}{15} \text{ square units} \qquad \text{Ans.}$$

Miscellaneous Exercise

Q(1) Calculate the area bounded by the curve $y^2 - 4x - 12 = 0$ and the y-axis.

Ans. $4\sqrt{3}$ square units.

Q(2) Find the area of the region enclosed between the line $y = x$ and the parabola $y = x^2$.

Ans.   1/3 square units.

Q(3) Find the area of the ellipse $\dfrac{x^2}{a^2} + \dfrac{y^2}{b^2} = 1$

Ans. πab square units.

Q(4) Find the area of the region in the first quadrant enclosed by the line $y = x$ and the circle $x^2 + y^2 = 32$.

Ans. 4π square units.

Q(5) Find the area of the region bounded by the triangle with vertices $(1, 0)$, $(2, 2)$, and $(3, 1)$, using integration.

Ans.   3/2 square units.

Q(6) Find the area of the region bounded by the line $y = 3x + 2$, the x-axis and the ordinates $x = -1$ and $x = 1$.

Ans.   13/3 square units.

Q(7) Find the area bounded by the curve $y = \cos x$ between $x = 0$ and $x = \pi$.

Ans.   2 square units.

Q(8) Find the area bounded by the curve $y = \sin x$ between $x = 0$ and $x = 2\pi$.

Ans.   4 square units.

Q(9) Find the area of the region bounded between the parabolas $y^2 = 4ax$ and $x^2 = 4ay$, where $a > 0$.

Ans. $16a^2/3$ square units.

[Hint: The points of intersection of the curves are $(0, 0)$ and $(4a, 4a)$.]

Q(10) Prove that the curves $y^2 = 4x$ and $x^2 = 4y$ divide the area of the square bounded by $x = 0$, $x = 4$, $y = 4$ and $y = 0$ into three equal parts.
[Hint: Refer to Example (8).]

Q(11) Find the area of the region enclosed between the two circles $x^2 + y^2 = 4$ and $(x - 2)^2 + y^2 = 4$.

Ans. $\dfrac{8\pi}{3} - 2\sqrt{3}$.

Q(12) Find the area contained between the curve $y = x^3$ and the straight line $y = 2x$.

Ans. 1 square unit.

Q(13) Using the method of integration find the area of the triangle ABC bounded by the lines $2x + y = 4$, $3x - 2y = 6$, and $x - 3y + 5 = 0$.

Ans. $\dfrac{7}{2}$ square units.

Q(14) Find the area of the region enclosed by the parabola $x^2 = y$ and the line $y = x + 2$.

Ans. $\dfrac{9}{2}$ square units.

8b   To Find Length(s) of Arc(s) of Curve(s), the Volume(s) of Solid(s) of Revolution, and the Area(s) of Surface(s) of Solid(s) of Revolution

8b.1   INTRODUCTION

In the previous chapter, Chapter 8a, we have seen how the methods of integration enabled us *to find areas of plane figures* by applying the definite integral. Now we shall consider certain fields of mathematics and those of engineering in which the *ideas of definite integrals are applied* for obtaining useful formulas and results, which cannot be obtained otherwise.

8b.2   METHODS OF INTEGRATION

The *methods of integration may be applied* to compute

- *the lengths of arcs of plane curves* (whose equations are known),
- the *volume(s) of the solid(s) of revolution* (which are marked out in space, when a *plane area is rotated about an axis*),
- the *surface areas of solids of revolution* and many other quantities (like *center of gravity, moment(s) of inertia*, etc., to be studied later in higher classes).

In fact, the convenient approach to learn these applications is to study them in the sequence they are mentioned above. The lengths of arcs of regular curves will be required in calculating the surface area of the solids of revolution. Let us discuss.

8b.2.1   The Measurement of the Length of a Curve

As already mentioned in the previous chapter, *no part of a curve*, however small, *can be superimposed on any portion* of *a straight line*. It means that the *length of any arc of a curve* (whether a circle or any other curve) *cannot be found by comparison with a straight line of a*

Applications of the definite integral 8b-Applying the definite integral to find the length(s) of arc(s) of curve(s), the volume(s) of solid(s) of revolution, and the area(s) of surface(s) of solid(s) of revolution.

Introduction to Integral Calculus: Systematic Studies with Engineering Applications for Beginners, First Edition.
Ulrich L. Rohde, G. C. Jain, Ajay K. Poddar, and A. K. Ghosh.
© 2012 John Wiley & Sons, Inc. Published 2012 by John Wiley & Sons, Inc.

known length. However, with *the applications of definite integral(s)*, it is possible to determine *the length(s) of arc(s) of plane curve(s)* whose equations are given in the *Cartesian, parametric Cartesian, or polar form.* The method involved is similar to that used for computing areas. An expression is found for "*an element of length*" of the curve, and the *sum of all such elements* is obtained by *integration* (i.e., by the application of definite integral). The process of finding the length (of an arc) of a curve is called *rectification of the curve.*

To understand the process (and the approach), it is important *to be very clear about the concept of limit* (discussed in Chapters 7a and 7b of Part I), and *the concept of infinitesimal(s)* (discussed in Chapter 8a). The student is advised that *with any vague ideas of these concepts,* they should not proceed to learn the applications in question. Also, one should not compromise with these ideas for any purpose (like getting marks in the examination). These ideas are very *simple, interesting,* and *paying in the long run.*

In this connection, the reader may go through Section 8a.5.1, wherein we have discussed the process of computing *area of a circle involving these ideas.* (The relevant footnote is very important.) There, we have treated (and accepted) *a small sector* of the circle, *in the limit,* as a "*right triangle*". Finally, using the *formula for the area of a triangle,* we obtained *the element of area of the triangular sector* as $(1/2)r^2\ d\theta$, where *the small angle $d\theta$ is expressed in radians.* This *expression,* for the elementary area (of the sector), *was then used in computing the area of the circle.* If the reader feels that, in obtaining the above expression (for the element of area of the triangular sector), he had to compromise at any stage, then it is advisable to revise the basic concept of limit, or even *better, approach a good teacher for guidance.*

Again, it is useful to go through Section 8a.5.2, wherein we have discussed another *method of computing the area of a circle, with a variable radius "x". In this case, the independent variable* is the radius "x", and the *dependent variable* is "*an element of area*", in the form of a *circular ring,* which can grow into a circle. The *width* of the ring is taken as "dx" and we take its *length* as $2\pi x$. Here again, the important role is played by the *small width* "dx" of the ring, which is taken to be *infinitesimally thin.*

From the definition of the *radian measure of an angle,* the circumference is regarded as consisting of 2π arcs, each of which is equal in length to the radius. It follows that (circumference/radius(r)) $= 2\pi$ radians or the circumference $= 2\pi r$. Thus, the length of the circumference (which is a curve) is expressed in terms of the length of the radius (which is a line segment). However, there is still a difficulty because we do not have an exact value of π. Of course, we have the *approximate value of π* as $3.1415\ldots$, obtained by practical methods. This permits us to express the elementary area of the ring to be $2\pi x\ dx$.[1]

Now, we shall establish the formula for computing the length of an arc of a curve *whose equation is known.*

8b.2.1.1   General Formula for the Length of a Curve in Cartesian Coordinates
Let AB represent a portion of the *plane curve* of a function $y = f(x)$, between the points A and B (Figure 8b.1).

E and F are two points on the curve and they have been joined by a straight line EF, *which is the chord* of the arc between them. In the *usual notation,* EG $= \Delta x$ and FG $= \Delta y$. Similarly, the part of the curve between E and F is equal to ΔS, where S measures length of curve.

The triangle EGF is a right-angled triangle, and therefore,

$$(\Delta y)^2 + (\Delta x)^2 = (\text{EF})^2$$

---

[1] Note carefully, how the concept of limit is involved in the above discussion.

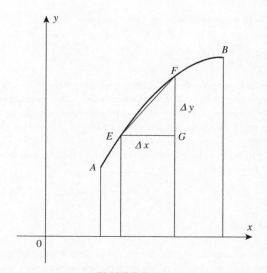

FIGURE 8b.1

If the point F is moved very close to E, then the length of chord EF will be very nearly equal to *the length of the arc* EF. As F is made *indefinitely close* to E, (i.e., as close to E as we wish) then, as Δx approaches zero, *the length of the chord approaches the length of the arc*. Then, the equation,

$$(\Delta y)^2 + (\Delta x)^2 = (\text{arc } EF)^2 \text{ is approximately true}$$

or

$$(\Delta y)^2 + (\Delta x)^2 = (\Delta s)^2 \text{ is approximately true}$$

and from this, we obtain the equivalent of Δs in two forms:

$$\frac{(\Delta s)^2}{(\Delta x)^2} = \frac{(\Delta y)^2}{(\Delta x)^2} + 1$$

$$\therefore \quad \Delta s = \sqrt{1 + \left(\frac{\Delta y}{\Delta x}\right)^2} \cdot \Delta x \tag{1}$$

and similarly,

$$\Delta s = \sqrt{1 + \left(\frac{\Delta x}{\Delta y}\right)^2} \cdot \Delta y \tag{2}$$

Note that Equations (1) and (2) are approximately true.

Now, it must also be clear (and it is easy to show) that

$$ds = \sqrt{1 + \left(\frac{dy}{dx}\right)^2} \cdot dx \quad \text{or} \quad ds = \sqrt{1 + \left(\frac{dx}{dy}\right)^2} \cdot dy \quad \text{(exactly)}$$

This follows because as Δx approaches zero, the ratio between *chord* EF and arc EF approaches 1. The length of the curve is s, and *this value may be found by integrating* either of the equivalents of ds. *In practice, it will be found that depending on the equation of the curve* [i.e., $y = f(x)$ or $x = g(y)$], *one form usually gives an easier calculation than the other.*

Let us consider the *first form of the integrand.*

Then, if we have to find the length of the arc from $x = a$ to $x = b$, we write,

$$s = \int_a^b \sqrt{1 + \left(\frac{dy}{dx}\right)^2} \cdot dx \tag{3}$$

This is an important result.

Note (1): The purpose of introducing this section is to demonstrate the usefulness of definite integrals and the power of Calculus. Here, we shall be dealing with very simple problems only, without going into the complicated situations. Let us demonstrate its application in calculating the length of circumference of a circle.

Example (1): Find the length of the circumference of a circle $x^2 + y^2 = r^2$

Solution: The equation of the circle is $x^2 + y^2 = r^2$. Differentiating the given equation with respect to x, we get

$$2x + 2y\frac{dy}{dx} = 0$$

$$\therefore \quad \frac{dy}{dx} = -\frac{x}{y}$$

Using Equation (3) above, we compute the arc length of one-fourth of the circle, taking the limit from 0 to r (Figure 8b.2):

We have,

$$s = \int_a^b \sqrt{1 + \left(\frac{dy}{dx}\right)^2} \cdot dx$$

$$\frac{s}{4} = \int_0^r \sqrt{1 + \left(\frac{-x}{y}\right)^2} \cdot dx = \int_0^r \sqrt{\frac{x^2 + y^2}{y^2}} \cdot dx$$

$$\therefore \quad \frac{s}{4} = \int_0^r \sqrt{\frac{r^2}{y^2}} \cdot dx = \int_0^r \frac{r}{y} \cdot dx = r\int_0^r \frac{dx}{\sqrt{r^2 - x^2}}$$

$$= r\left[\sin^{-1}\frac{x}{r}\right]_0^r = r\left[\frac{\pi}{2} - 0\right] = \frac{\pi r}{2}$$

$\therefore$ Circumference of the circle $= 4 \cdot \dfrac{\pi r}{2} = 2\pi r$ Ans.

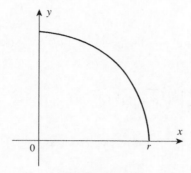

FIGURE 8b.2

Example (2): Find the length of the arc of the parabola $x^2 = 4y$ from vertex to the point where $x = 2$

Solution: The equation of the parabola is $x^2 = 4y$.
We can write the above equation in the form:

$$y = \frac{x^2}{4} \qquad \therefore \quad \frac{dy}{dx} = \frac{x}{2}$$

The sketch of the curve is shown in Figure 8b.3, where OQ represents the part of the curve of which the length is required.

The limits of x are from 0 to 2.

Using, $S = \int_a^b \sqrt{1 + \left(\frac{dy}{dx}\right)^2}\,dx$, on substitution,

$$S = \int_0^2 \sqrt{1 + \left(\frac{x^2}{4}\right)}\,dx = \frac{1}{2}\int_0^2 \sqrt{x^2 + 4}\,dx$$

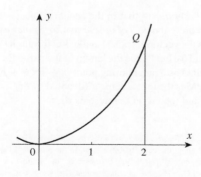

FIGURE 8b.3

Using the formula,

$$\int \sqrt{x^2 + a^2}\, dx = \frac{1}{2}x\sqrt{x^2 + a^2} + \frac{a^2}{2}\log_e(x + \sqrt{x^2 + a^2}) + c \quad \text{(see Chapter 4b)}$$

We get

$$S = \frac{1}{2}\int_0^2 x\sqrt{x^2 + (2)^2}\, dx$$

$$= \frac{1}{2}\left[\frac{1}{2}\cdot 2\cdot\sqrt{8} + 2\{\log_e(2+\sqrt{8}) - \log_e 2\}\right]$$

$$= \frac{1}{2}\left[2\sqrt{2} + 2\{\log_e(2+2\sqrt{2}) - \log_e 2\}\right]$$

$$= \left[\sqrt{2} + \log_e\left(\frac{2+2\sqrt{2}}{2}\right)\right]$$

$$= \sqrt{2} + \log_e(1 + \sqrt{2}) = 2.295 \text{ (approx)} \qquad \text{Ans.}$$

(Here, the log table to the base "*e*" has to be used.)

Remark: The formula giving the *length of an arc of a curve* can be applied to any other curve, whose equation is given in Cartesian coordinates.

Note (2): The calculation of the length of an ellipse reduces to the calculation of an integral *that cannot be expressed in terms of elementary functions*. This integral can be computed only by approximate methods (by Simpson's rule, for example). We will not discuss such methods here.[2]

The same situation occurs in the calculation of *length of the arc of the hyperbola* $y = \frac{1}{x}$, and the *length of the arc of a sine curve*. The length of an arc of a parabola can be reduced to an integral that is rather complicated, although it can be expressed in terms of elementary functions.

8b.3   EQUATION FOR THE LENGTH OF A CURVE IN POLAR COORDINATES

(The *method, in general, is similar* to that in the *rectangular coordinates*.)

In Figure 8b.4, let AB represent *a part of the curve whose polar equation* is $r = f(\theta)$, where r is the radius vector and θ is the vectorial (polar) angle. Let the angles made by OA and OB with (the polar axis) OX be θ_1 and θ_2. Let "s" be the length of the part AB (of the curve). Let P be any point (r, θ) on the curve and Q be a neighboring point $(r + \delta r, \theta + \delta\theta)$, such that $\angle$QOM is the increase in θ (say $\delta\theta$) and PM is the increase in "r"(denoted by δr). Let PQ be the *chord* joining P to Q. Then, QM $= r\,\delta\theta$ and the arc PQ represents δs.[3]

---

[2] For details, refer to *Differential and Integral Calculus* (Vol. I) by N. Piskunov (pp. 446–447), Mir Publishers, Moscow.
[3] The reader must be convinced about the equality QM $= r\,\delta\theta$. In the previous Chapter 8a (Section 8a.5.1) we have discussed at length that a (small) *sector of a circle*, can be regarded, "in the limit", as a right triangle. The same thing happens here when we treat the arc PQ, *in the limit* as a line segment. [In the entire approach, the concept of limit plays the most important role.]

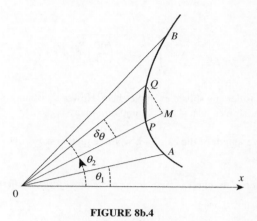

FIGURE 8b.4

Then $PQ^2 = QM^2 + PM^2$
or $PQ^2 = (r\,\delta\theta)^2 + (\delta r)^2$ (with the construction, $PM = \delta r$).
When Q is taken *indefinitely close to P* (i.e., $\delta\theta \to 0$), then, *in the limit*

$$(ds)^2 = (r\,d\theta)^2 + (dr)^2$$

Therefore $$ds = \sqrt{r^2(d\theta)^2 + (dr)^2}$$

$$= \sqrt{r^2 + \left(\frac{dr}{d\theta}\right)^2} \cdot d\theta \tag{E}$$

(Note that s is a function of two variables, namely r and θ.)

For now, we regard "s" as a function of "θ". Then, to find the *length of the curve from A to B*, we must integrate both sides of (E) from θ_1 to θ_2.

Thus, on integrating, we get

$$s = \int_{\theta_1}^{\theta_2} \sqrt{r^2 + \left(\frac{dr}{d\theta}\right)^2}\,d\theta \tag{I}$$

We may also write Equation (E), in the form

$$ds = \sqrt{1 + r^2\left(\frac{d\theta}{dr}\right)^2}\,dr$$

Now, we regard "s" as a function of "r". Hence, (to find the length of the curve from A to B) the limits of integration will be from r_1 to r_2.

We write,

$$s = \int_{r_1}^{r_2} \sqrt{1 + r^2 \left(\frac{d\theta}{dr}\right)^2} \, dr \qquad\qquad \text{(II)}$$

Note: We do not consider any solved examples here. However, as an exercise, we give below one problem, which is given as a solved example in many books.

Q. Find the complete length of the cardioid, whose equation is $r = a(1 - \cos\theta)$.
Ans. $s = 8a$.

Further, if the equation of the curve is given in the *parametric form,* $x = f(t), y = g(t)$, then it can be shown that

$$ds = \sqrt{\left(\frac{dx}{dt}\right)^2 + \left(\frac{dy}{dt}\right)^2} \cdot dt$$

Accordingly, the length "s" of the curve can be computed by integrating both sides, with respect to "t" (the parameter), from t_1 to t_2. (We will not discuss anything more about the length of curves, since it is not needed for beginners.)

8b.4   SOLIDS OF REVOLUTION

The solids with which we shall be concerned are those that are marked out in space, when a *continuous curve or an area is rotated about some axis*. These are termed *solids of revolution*. For example,

(a) If a *semicircle* is rotated *about its diameter*, the solid of revolution so obtained is a *sphere*.
(b) If a *rectangle* is rotated *about one of its sides*, we get a *right circular cylinder in a complete rotation*.
(c) If a *right-angled triangle rotates completely about one of the sides containing the right angle as an axis*, it will generate *a right circular cone*.

"Solid of revolution" is a *mathematical term*, whose meaning is clearly reflected in the "*solids of revolution*", marked out in space by the rotation of a curve, about an axis. A circle represents the *cross section* of a sphere, and we may be interested to calculate its volume.

If we rotate a quadrant (of the circle) about the x-axis, it will sweep out the volume of half the sphere. If we can calculate this volume, then we must double it to calculate the volume of the whole sphere. This again suggests to us that we should concentrate our knowledge (of Mathematics) on just one quadrant. Once a solid of revolution is generated, as mentioned above, one may like to calculate its volume or the surface area generated in the process.

Remark: When we revolve a curve about an axis, it implies that the area under the curve is revolved about the axis.

Now, we start finding the formula for volume of solid(s) of revolution.

8b.5   FORMULA FOR THE VOLUME OF A "SOLID OF REVOLUTION"

(a) *Rotation about the x-Axis*: Let the *area under a curve* $y = f(x)$, namely, $\int_a^b f(x)dx$, be revolved about the x-axis, thus generating a volume. The area of a cross section of this solid by a plane perpendicular to the x-axis is πy^2. The volume of a thin slice, "dx" thick would be $\pi y^2\,dx$.[4]

The *total volume of solid of revolution* between the two parallel planes $x = a$ and $x = b$ would therefore *be the sum of all such slices*

$$\text{or}\qquad V = \pi \int_a^b y^2\,dx = \pi \int_a^b [f(x)]^2\,dx$$

Now, it must be clear that to obtain the formula for computing *volume of a solid of revolution is quite simple.* (Let us see how the concept of limit is deeply involved in the process of obtaining the above formula.)

Let $y = f(x)$ be a *continuous nonnegative function* defined on the interval $[a, b]$. Imagine a solid resulting from the rotation of a *curvilinear trapezoid* about the x-axis, bounded by the function $y = f(x)$, the x-axis and the straight lines $x = a$ and $x = b$ (Figure 8b.5a).

The volume of this solid [say $V(x)$] is a function of x. For *an infinitesimal increment* dx (i.e., dx can be though of to be as small as we please) (or we may call it infinitely decreasing increment), the volume $V(x)$ increases by the volume of an infinitely thin layer of width dx, with the base area $\pi[f(x)]^2$. Thus, the element of increase in the volume is given by the expression,

$$dV(x) = \pi[f(x)]^2\,dx$$

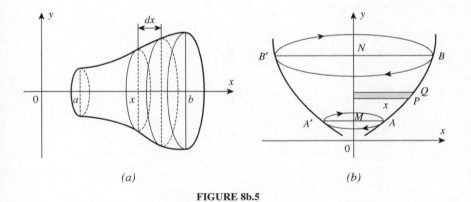

(a)   *(b)*

FIGURE 8b.5

---

[4] The disc of volume swept out by the element of area under the curve will always be circular whatever the shape of the curve. This may at first seem a little odd, *yet it is implied in the rotation of the curve about its axis.* [To rotate the curve means to turn it about its axis so that the whole curve describes a lateral circular motion.]

(To calculate the volume of this solid, we shall consider a similar solid with a variable right side, which cuts the *x*-axis at the point *x*.)

We then get, $V = V(x) = \int_a^b \pi y^2 \, dx = \int_a^b \pi [f(x)]^2 \, dx$

(b) *Rotation about the y-Axis*: Let AB be a potion of the curve $y = f(x)$ that is rotated about OY, so that A and B describe circles as indicated with centers M and N, on the *y*-axis (Figure 8b.5b).

Let OM = *a*, ON = *b*.

Let P(*x*, *y*) be any point on the curve and Q be another near by point on the curve with coordinates $(x + \delta x, y + \delta y)$.

Then, the volume of the slab generated by PQ becomes, *in the limit*, $\pi x^2 \, dy$.

∴ The volume of the whole solid is the sum of all such slabs between the limits $y = a$ to $y = b$.

$$\therefore \quad V = \pi \int_a^b x^2 \, dy$$

[Note that, here the variable of integration is "*y*". Therefore, from the given equation $y = f(x)$, we must replace "*x*" in terms of y.]

Solved Examples

Example (3): Calculate the volume of a sphere of radius *r*

Solution: The equation of a circle with the center at the origin and radius "*r*" is given by

$$x^2 + y^2 = r^2 \tag{4}$$

If the *quadrant* AOB *is rotated about* OX the volume described will be that of a *hemisphere*.

Observe that, the volume of the slab (generated by PQ on rotation about *x*-axis) becomes, in the limit, $\pi y^2 \, dx$ (Figure 8b.6).

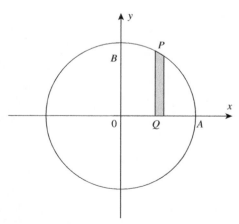

FIGURE 8b.6

∴ The volume of the hemisphere can be obtained by using the formula

$$\frac{1}{2}V = \pi \int_a^b y^2 \, dx$$

∴ $V = 2\pi \int_a^b y^2 \, dx = 2\pi \int_a^b (r^2 - x^2) dx$ [where y is replaced by x from $y = f(x)$]

or $\quad V = 2\pi \int_0^r (r^2 - x^2) dx$ (Note the lower limit of the integral)

$$= 2\pi \left[r^2 x - \frac{x^3}{3} \right]_0^r = 2\pi \left[r^3 - \frac{r^3}{3} \right] = 2\pi \cdot \frac{2}{3} r^3 = \frac{4}{3}\pi r^3 \qquad \text{Ans.}$$

We can also obtain the same result when a *semicircle of radius "r" is rotated about a diameter*. *Assuming the diameter to be the x-axis*, we find that the semicircle intersects the x-axis at $x = -r$ and $x = r$. Thus, we have

$$V = \int_{-r}^r \pi y^2 \, dx = \int_{-r}^r \pi (r^2 - x^2) \, dx \text{ [where } y \text{ is replaced by } x, \text{ using Equation (4)]}$$

$$= \pi \left[r^2 x - \frac{x^3}{3} \right]_{-r}^r = \pi \left[r^2 x - \frac{x^3}{3} \right]_{x=r} - \pi \left[r^2 x - \frac{x^3}{3} \right]_{x=-r}$$

$$= \pi \left[r^3 - \frac{r^3}{3} \right] - \pi \left[-r^3 - \frac{(-r)^3}{3} \right]$$

$$= \frac{2}{3}\pi r^3 - \pi \left[-r^r + \frac{r^3}{3} \right] = \frac{2}{3}\pi r^3 - \left[-\frac{2}{3}\pi r^3 \right]$$

$$= \frac{2}{3}\pi r^3 + \frac{2}{3}\pi r^3 = \frac{4}{3}\pi r^3 \text{ cubic units} \qquad \text{Ans.}$$

Note: Now, we are in a position to find the volume of part of the sphere, between two parallel (vertical) planes. Consider two parallel planes at distances from (the origin) O, given by $OA = a$, $OB = b$ (Figure 8b.7). Then, the volume V of the part of the sphere is given by the integral,

$$V = \int_a^b \pi y^2 \, dx = \int_a^b \pi (r^2 - x^2) \, dx$$

Here, as usual, "y" has been replaced by "x", since the variable of integration is x. For this purpose, we use the given equation $x^2 + y^2 = r^2$.

Example (4): Find the volume of the solid of revolution bounded by the circle $x^2 + y^2 = 36$, and the lines $x = 0$, $x = 4$, about the x-axis

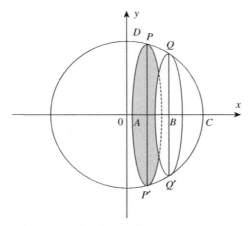

FIGURE 8b.7

Solution: We have to compute the volume of the part of sphere when the part of the circle (i.e., $x^2 + y^2 = 36$) from $x = 0$ to $x = 4$ is revolved about the x-axis. Obviously, the limits of integration in this case are from $x = 0$ to $x = 4$ (Figure 8b.8).

From the equation of the circle, we have $y^2 = 36 - x^2$
∴ Required volume

$$V = \int_0^4 \pi y^2 \, dx = \int_0^4 \pi (36 - x^2) \, dx$$

$$= \pi \left[36\, x - \frac{x^3}{3} \right]_0^4 = \pi \left[144 - \frac{64}{3} \right] = \frac{368\,\pi}{3} \text{ cubic units} \qquad Ans.$$

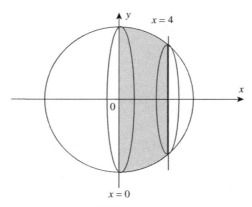

FIGURE 8b.8

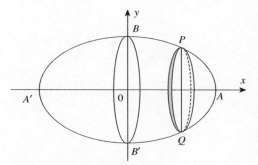

FIGURE 8b.9

Example (5): Volume of an ellipsoid of revolution

This is the solid formed by the rotation of an ellipse

(i) about its major axis,

(ii) about its minor axis.

(i) *Rotation about the Major Axis*
Let the equation of the ellipse be

$$\frac{x^2}{a^2} + \frac{y^2}{b^2} = 1 \qquad \therefore \quad b^2x^2 + a^2y^2 = a^2b^2$$

Note that, $2a = AA'$ and $2b = BB'$ Then, the center of the ellipse coincides with the origin. Thus, the rotation (of the curve) is to be about OX (i.e., x-axis).
From the above equation, we have

$$a^2y^2 = a^2b^2 - b^2x^2 = b^2(a^2 - x^2)$$
$$\therefore \quad y^2 = \frac{b^2}{a^2}(a^2 - x^2) \tag{5}$$

Let V be the volume of the ellipsoid. Consider the volume marked out by the rotation of the quadrant OAB. Clearly, the limits are from 0 to a (Figure 8b.9).
(Note that, this rotation will generate half the volume of the ellipsoid.)

$$\frac{V}{2} = \int_0^a \pi y^2 \, dx \quad \text{or} \quad V = 2\pi \int_0^a y^2 \, dx$$

$$= 2\pi \int_0^a \frac{b^2}{a^2}(a^2 - x^2)dx \qquad \text{[using Equation (5)]}$$

$$= \frac{2\pi b^2}{a^2} \int_0^a (a^2 - x^2)dx = \frac{2\pi b^2}{a^2}\left[a^3 - \frac{x^3}{3}\right]_0^a = \frac{2\pi b^2}{a^2}\left[\frac{2}{3}a^3\right]$$

$$V = \frac{4}{3}\pi ab^2 \text{ cubic units}$$

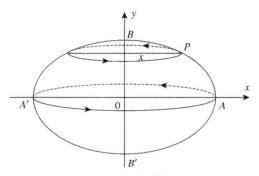

FIGURE 8b.10

Remark: Observe that if $b = a$, the ellipsoid becomes a sphere. (In Equation (5) "a", stands for the *semi-major axis*, and "b" for the *semi minor axis*, i.e., $2a = AA'$ and $2b = BB'$.)

(ii) *Rotation about the Minor Axis*

Now, it can be shown that if the rotation is considered about the *minor axis "b"*, then the *volume of the ellipsoid* will be given by (see Figure 8b.10)

$$V = \frac{4}{3}\pi a^2 b \text{ cubic units}$$

(Prove this result as an exercise.)

Note: The student should not try to memorize the formulae obtained so far, and many others similar to them—that will follow later, in this connection. The important point is that one should master the technique of setting them up.

Example (6): Paraboloid of revolution

This is the solid generated by the *rotation* of a *part of parabola, about its axis.*
(Since the parabola *is not a closed curve*, we shall consider only the *solid generated* by a *part of the curve*, so that the volume generated is a finite quantity.) There are two cases.

Case (i) When the axis of the parabola coincides with the x-axis (i.e., OX)
In this case, the general form of the equation is

$$y^2 = 4ax \tag{6}$$

In Figure 8b.11, OP represents a *part of the curve*, where P(x, y) is a point on the curve. PA is the *ordinate* of P (i.e., PA $= y$), and let, OA $= c$.
    OP rotates around OX, generating a solid, with a *circular base* PQR.
    Now, it must be clear, that the element of volume is $\pi y^2 \, dx$, and that the limits of x are 0 to c. Let V be the volume

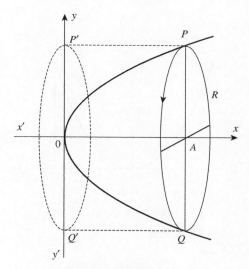

FIGURE 8b.11

$$V = \int\limits_{0}^{c} \pi y^2 \, dx$$

$$= \pi \int_0^c 4ax \, dx \quad [\text{Here, } y \text{ is replaced by } x, \text{ using Equation (6)}]$$

$$= 4\pi a \int\limits_0^c x \, dx = 4\pi a \left[\frac{x^2}{2}\right]_0^c = 4\pi a \left[\frac{c^2}{2} - 0\right]$$

$$= 2\pi a c^2 \text{ cubic units} \qquad \text{Ans.}$$

Note: Now consider the *cylinder* indicated by the *dotted lines* in the above figure. This is the *circumscribing cylinder of the paraboloid.*

The volume of this cylinder $= 4\pi y^2 \cdot OA, \quad (y \text{ at } x = 0)$

$$= \pi(4ax) \cdot OA \quad (\text{since } y^2 = 4ax \quad \text{at } x = c)$$

$$= \pi(4ac) \cdot c \quad (\text{since } OA = c)$$

$$= 4\pi a c^2 \text{ cubic units} \qquad \text{Ans.}$$

Remark: The *volume of the paraboloid* (i.e., $2\pi a \cdot c^2$) *equals half* that of the *circumscribing cylinder*.

Case (ii) When the axis of the parabola coincides with the y-axis (i.e., OY)

In Figure 8b.12, we consider a *part of a parabola*, whose axis is the y-axis.

The *general equation* of such a parabola is

$$y = ax^2 \tag{7}$$

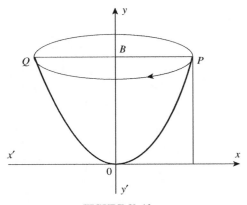

FIGURE 8b.12

Let P(x, y) be any point on the curve. Let PB be its *abscissa* (i.e., x coordinate), such that PB = x, and let OB = b. (The length OB *suggests the height of the part* of parabola that generates the paraboloid.)

Obviously, *the element of volume*, in this case is,

$$\pi x^2 \, dy$$

The limits of y are 0 to b. Therefore, the desired volume V is given by

$$V = \int_0^b \pi x^2 \, dy$$

$$= \pi \int_0^b \frac{y}{a} \, dy \qquad \text{[where, } x \text{ is replaced by ``}y\text{'', using Equation (7)]}$$

$$= \frac{\pi}{a} \int_0^c y \, dy$$

$$= \frac{\pi}{a} \left[\frac{y^2}{2} \right]_0^b = \frac{\pi}{a} \left[\frac{b^2}{2} - 0 \right] = \frac{1}{2} \frac{\pi b^2}{a} \quad \text{cubic units} \qquad \text{Ans.}$$

Example (7): Calculate the volume of a *paraboloid of revolution*, when a "parabolic triangle" *bounded by the upper half* of a parabola

$$y^2 = cx$$

is rotated about the x-axis, and bounded up to the straight line x = a.

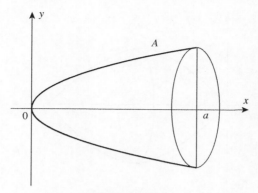

FIGURE 8b.13

Solution: The equation of the parabola is

$$y^2 = cx \tag{8}$$

When the *upper half* of this parabola is revolved about the x-axis, the paraboloid is generated. We have to find the volume of the solid (in question) bounded up to the straight line $x = a$ (Figure 8b.13).

Let the volume in question is denoted by V.

Thus we write, $V = \displaystyle\int_0^a \pi y^2 \, dx$

$\qquad\qquad\quad = \displaystyle\int_0^a \pi(cx) \, dx \qquad [\because \quad y^2 = cx \text{ by Equation (8)}]$

$\qquad\qquad\quad = \pi c \displaystyle\int_0^a x \, dx$

$\qquad\qquad\quad = \pi c \left[\dfrac{x^2}{2}\right]_0^a = \pi c \left[\dfrac{a^2}{2} - 0\right] = \dfrac{\pi c a^2}{2} \text{ cubic units} \qquad \text{Ans.}$

Example (8): Find the volume of a solid resulting from the rotation about the x-axis, *of a figure bounded* by the x-axis and *half-wave of the sine curve* $y = \sin x$ (Figure 8b.14)

Solution: The equation of the curve is

$$y = \sin x \tag{9}$$

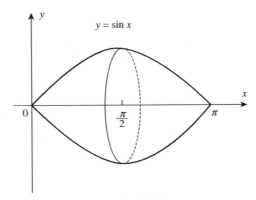

FIGURE 8b.14

∴ The desired volume is given by

$$\therefore \quad V = \int_0^\pi \pi y^2 \, dx = \pi \int_0^\pi \sin^2 x \, dx \quad [\because \quad y = \sin x, \text{ by Equation (9)}]$$

$$= \pi \int_0^\pi \left[\frac{1 - \cos 2x}{2}\right] dx$$

$$= \pi \int_0^\pi \left[\frac{1}{2} - \frac{\cos 2x}{2}\right] dx$$

$$= \pi \left[\frac{x}{2} - \frac{\sin 2x}{4}\right]_{x=\pi} - \left[\frac{x}{2} - \frac{\sin 2x}{4}\right]_{x=0}$$

$$= \pi \left[\frac{\pi}{2} - \frac{\sin 2\pi}{4}\right] - [0 - 0]$$

$$= \pi \left[\frac{\pi}{2} - 0\right] = \frac{\pi^2}{2} \text{ cubic units} \qquad \text{Ans.}$$

Example (9): Volume of a cone

A cone is generated by the rotation of *right-angled triangle*, whose axis of rotation is *one of the sides containing the right angle.*

Solution: As shown in Figure 8b.15, let the *radius of the base* of the cone be *r* (= AM) and *the height of the cone* be *h* (= OM).

Let OA be the *straight line* y = mx, A being any point on the line.

(Our interest lies in computing the volume of a cone *when the line segment* OA *is rotated* around OX.)

Let θ be the angle made by the line OA, with OX (i.e., *x*-axis).

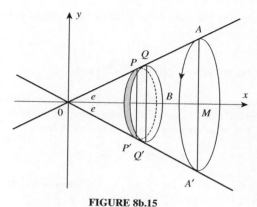

FIGURE 8b.15

Then, $\tan\theta = m = \frac{AM}{OM} = \frac{r}{h}$

Let V be the volume of the cone of which "O" is the *vertex*, the *circle* ABA', the base, and the height "h" ($=$ OM). (The *small element* PQ on rotating, describes a *small slice of the cone*, of which the ends are the circles, described by P and Q.)

Thus, the element of volume is $\pi y^2\, dx$.

$$\therefore\quad V = \int_0^h \pi y^2\, dx = \pi \int_0^h (mx)^2\, dx \quad (\text{where } y = mx)$$

$$= \pi m^2 \int_0^h x^2\, dx = \pi m^2 \left[\frac{x^3}{3}\right]_0^h = \frac{\pi m^2 h^3}{3} \text{ cubic units} \qquad \textbf{Ans.}$$

In the above equation $(V = (1/3)\pi m^2 h^3)$, "m" is not defined in terms of known quantities. Hence, we replace it by the ratio r/h, in which both r and h are known.

Therefore, $v = \frac{1}{3}\pi\frac{r^2}{h^2}h^3 = \frac{1}{3}\pi r^2 h$. Note that the quantity "πr^2" represents *the area of the base* (of cone). Hence, we can write, V (i.e., volume of the cone) $= 1/3$ (area of base)·(height).

Remark: Volume of a right circular cone is one-third that of a cylinder whose base area and height are same.

It is important that we have a method for calculating the length(s) of arc(s) of regular curve(s). [5]

---

[5] Recall that, earlier, we did not have any method for calculating the length of (even) a circular arc. However, based on the fact that the circumference of a circle is proportional to its radius "r", the radian measure of an angle suggests that the ratio of the length of circumference to its radius is a constant, denoted by 2π. This idea permits us to say that the length of the circumference of a circle is $2\pi r$. Of course, the exact value of π is not known but its approximate value (based on practical methods) is found to be 3.14159..., and used for finding the length of the circumference of a circle.

In other words, the length of circular arcs could be defined in terms of the length of a straight line segment (i.e., radius "r"). In principle, no part of a curve, however small, can be superimposed on any portion of a straight line. On the other hand, integration supplies a method of determining the length of any regular curve. In the entire process, the most important role is played by the concept of limit. Thus, it is important to learn this useful concept thoroughly.

Now we may proceed to calculate the areas of curved surfaces of regular solids. Let us discuss.

8b.6   AREA(S) OF SURFACE(S) OF REVOLUTION

When the curve $y = f(x)$ is revolved about the x-axis, *a surface is generated* (Figure 8b.16). To find the area of this surface, we consider *the area generated by an element* of the arc "ds". This area is roughly that of a cylinder of radius y, and we write

$$dS = 2\pi y \, ds$$

where "dS" stands for the small element of surface area generated by rotation of small element "ds" of the curve about the x-axis.

Note: The important point to be remembered is that we can compute *the surface area of revolution* of a curve of *finite length* only.
Summing all such *elements of surface area*, we get

$$S = \int_a^b 2\pi y \, ds \qquad \text{or} \qquad S = 2\pi \int_a^b y \, ds^{(6)}$$

Thus, the surface area in question is given by

$$S = 2\pi \int_a^b y \cdot \sqrt{1 + \left(\frac{dy}{dx}\right)^2} \, dx$$

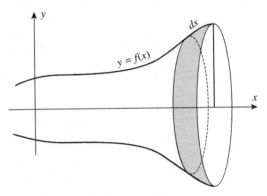

FIGURE 8b.16

---

[6] Recall that we have already shown earlier that an element of arc length of a curve in cartesian coordinates is given by $ds = \sqrt{1 + \left(\frac{dy}{dx}\right)^2} \, dx$.

Note: Appropriate modifications of this formula will be necessary, if the curve is *revolved about some other line* or if *polar coordinates* are used, and so on.

Notation: We denote the (finite) length of a curve by "*s*", whereas the surface area generated by a revolving the curve (of finite length) is denoted by "*S*".

Example (10):   Find the surface area of a sphere of radius "*r*"

Solution: This area can be generated by revolving the *upper half* of the circle $x^2 + y^2 = r^2$, about the *x*-axis. Here, $y = \sqrt{r^2 - x^2}$, $a = -r$, $b = r$ (Figure 8b.17).

Here, we have $dS = 2\pi y\, ds$

Therefore, the surface area (of the sphere) in question is given by

$$S = \int_a^b 2\pi y\, ds = \int_a^b 2\pi y \sqrt{1 + \left(\frac{dy}{dx}\right)^2}\, dx$$

$$\therefore \quad S = 2\pi \int_{-r}^{r} y \sqrt{1 + \left(\frac{dy}{dx}\right)^2}\, dx \tag{10}$$

Differentiating the equation $x^2 + y^2 = r^2$, we get

$$2x + 2y \cdot \frac{dy}{dx} = 0 \quad \text{or} \quad \frac{dy}{dx} = -\frac{x}{y}$$

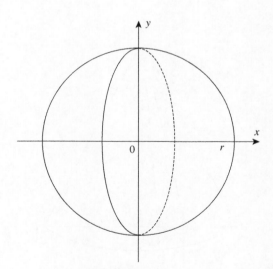

FIGURE 8b.17

Using the result in Equation (10), we get

$$S = 2\pi \int_{-r}^{r} y \cdot \sqrt{1 + \frac{x^2}{y^2}}\,dx = 2\pi \int_{-r}^{r} y\sqrt{\frac{x^2 + y^2}{y^2}}\,dx$$

$$= 2\pi \int_{-r}^{r} r\,dx \text{ (Note that "}r\text{" is a constant)}$$

$$\therefore \quad S = 2\pi r \int_{-r}^{r} dx = 2\pi r[x]_{-r}^{r} = 2\pi r[r - (-r)]$$

$$= 4\pi r \text{ square units} \qquad \text{Ans.}$$

Example (11): Surface area of a cone

The surface area of a right circular cone can be obtained by two methods:

(a) By using geometry, and the concept of radian measure of an angle.
(b) By using Calculus.

Method (a): Using geometry
Consider a right circular cone with vertex "A", the slant height ℓ and the radius of the circular base "r". If the vertex "A" is regarded as remaining at a fixed point on the table, and the base of the cone is rolled across the table so that the base completes one full turn. Then, the base will trace out the sector of a circle with radius ℓ ($= AB$), as shown in Figure 8b.18. The length of the

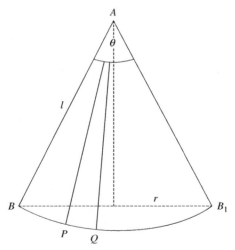

FIGURE 8b.18

arc BB_1 (traced out by the base of the cone) $= 2\pi r$. Obviously, the surface area of the cone is the circular triangle $A\,BB_1$, where BB_1 is the *arc of the circle*, whose center is at "A" and radius ℓ. Thus, the calculation of the (curved) surface area of the cone is reduced to the calculation of *the area of a sector* of a circle, whose radius is $\ell\,(= AB)$. We can calculate the *area of the sector* described by the rolling of the cone, as follows. We know that the length of the arc $BB_1 = 2\pi r$ (where "r" is radius of the base of the cone).

Let the angle BAB_1 be θ *radians*. Then by *the rules of circular measure*, we have,

$$\frac{\text{Area of the circular sector } ABB_1}{\text{Area of the (whole) circle with radius } \ell} = \frac{\text{Angle subtended by the arc } BB_1 \text{ (at the center)}}{\text{Angle subtended by the whole circle (at the center)}}$$

$$\text{i.e.,} \quad \frac{\text{(curved) surface area of the cone}}{\pi \ell^2} = \frac{\theta}{2\pi}$$

$$\therefore \quad \text{(curved) surface area of the cone} = \frac{\theta \pi \ell^2}{2\pi} = \frac{1}{2}\ell^2 \theta$$

In the above equation, the angle "θ" can be replaced in terms of known quantities. Note that, by *the rules of circular measure*, we can write

$$\theta = \frac{\text{length of the arc } BB_1 \text{ (generated by rolling one full turn of the cone, about A)}}{\text{radius "}\ell\text{" of the circle (whose center is A)}}$$

$$\therefore \quad \theta = \frac{2\pi r}{\ell}$$

Therefore, (curved) surface area of the cone $= \frac{1}{2}\ell^2\left(\frac{2\pi r}{\ell}\right) = \pi r \ell$

Thus, the formula for the (curved) surface of the cone is $s = \pi r \ell$.

Note that, *the above method does not use Calculus at all*. The same result can however be obtained by Calculus as follows.

Method (b): Using Calculus

In Figure 8b.18, PQ is a *small part of the circumference* of the base of a cone. Since PQ is a *very small part of the circumference*, it will be *very nearly straight*. As PQ is made smaller, the *sector APQ approaches the form of a triangle*.[7]

This line of thinking (in terms of limit concept) suggests to us *that we obtain the approximate* area of the sector by computing the area of the triangle, using the formula

Area of the triangle $= \frac{1}{2}$(height)·(base)

Now, the height of the triangle can be obtained by drawing a perpendicular from A to the small (arc) segment PQ. We may denote it by AS (not shown in the diagram), and obviously $AS \approx \ell$.

The area of the triangle is therefore $\frac{1}{2}\ell \times PQ$

---

[7] Note carefully that we are applying the concept of limit. Though a sector will *never become a triangle*, yet, by reducing the length of PQ, we can bring its area "*in the limit*", closer and closer to that of a triangle.

Taking the above expression, A is the limiting value of "*the element of area*", we may compute *the area of surface of the cone* by summing up all the small areas like this.

$$\therefore \quad \text{surface area of cone}$$

$$= \frac{1}{2}\ell \times \text{sum of all arcs like PQ}$$

$$= \frac{1}{2}\ell \times \text{circumference of base}$$

$$= \frac{1}{2}\ell \times 2\pi r = \pi r \ell.$$

Exercise

Q. (1) Find the length of the arc of the parabola $y^2 = 4x$ from $x=0$ to $x=4$.

Ans. $2\sqrt{5} + \log(2+\sqrt{5})$.

Q. (2) Find the length of the arc of the curve $y^2 = x^3$ from $x=0$ to $x=5$.

Ans. $\dfrac{335}{27}$ units.

Q. (3) Find the length of the arc of the circumference of the circle whose equation is $r = 2a\cos\theta$.

Ans. $2\pi a$.

Q. (4) Find the volume of solid generated by rotating the area bounded by $x^2 + y^2 = 36$ and the lines $x=0$, $x=3$, about the x-axis.

Ans. 99π cubic units.

Q. (5) Find the volume of the sphere with radius 3 units.

Ans. 36π cubic units.

Q. (6) If the region enclosed between the parabola $y=x^2+1$ and the line $y=2x+1$ is revolved about the x-axis, find the volume of the solid of revolution.

Ans. $\dfrac{104\pi}{15}$ cubic units.

Q. (7) The region bounded by $y^2=4x$, $y=0$, and $x=4$ is rotated about the x-axis. Find the volume generated.

Ans. 32π cubic units.

Q. (8) The region bounded by the lines $x-2y+6=0$, $y=0$, and $x=2$ is rotated about the x-axis. Find the volume generated.

Ans. $42\dfrac{2}{3}\pi$ cubic units.

Q. (9) Find the area of the surface of the solid generated by the rotation of the straight line $y = \frac{3}{4}x$, around the x-axis, between the values $x = 0$ and $x = 3$.

Ans. $\dfrac{135}{16}\pi$ square units.

Q. (10) The part of the curve of $x^2 = 4y$ that is intercepted between the origin and the line $y = 8$ is rotated around y-axis. Find the area of the surface of the solid that is generated.

Ans. $\dfrac{208}{3}\pi$ square units.

Q. (11) Find the area of the zone cut off a sphere of radius "r" by two parallel planes, the distance between which is h.

Ans. $2\pi rh$ square units.

9a Differential Equations: Related Concepts and Terminology

Nature's voice is mathematics; its language is differential equations.

9a.1 INTRODUCTION

Algebraic equations describe *relations among varying quantities. Differential equations* go one step further. They describe, in addition to relations among changing quantities, *the rates at which they change.*

Definition: An equation involving *derivatives* or *differentials* is called a differential equation. In other words, a differential equation is an equation connecting the independent variable x, the unknown function "y", and its derivatives or differentials.

The topic of differential equations is so vast that it is identified as a separate subject.

9a.1.1 Ordinary and Partial Differential Equations

A differential equation is said to be *ordinary* if the unknown function y depends solely on *one independent variable.* Such a function may be expressed in the form $y = f(x)$. Some examples of *ordinary differential equations* are:

$$\frac{dy}{dx} = -32x \tag{1}$$

$$\frac{dy}{dx} + 5y = 12e^{7x} \tag{2}$$

$$\frac{d^2y}{dx^2} + 8\frac{dy}{dx} + 16y = 0 \tag{3}$$

$$\frac{d^2y}{dx^2} + \sin y = 0 \tag{4}$$

$$dy = \cos x \, dx \tag{5}$$

Differential equations and their solutions 9a-Basic concepts and terminology. Formation of differential equation by eliminating the arbitrary constant(s) from the given equation(s). Solution of differential equations and the types of solutions. The simplest type of differential equation and the method of solving it.

Introduction to Integral Calculus: Systematic Studies with Engineering Applications for Beginners, First Edition. Ulrich L. Rohde, G. C. Jain, Ajay K. Poddar, and A. K. Ghosh.
© 2012 John Wiley & Sons, Inc. Published 2012 by John Wiley & Sons, Inc.

$$\frac{d^2y}{dx^2} = 0 \tag{6}$$

$$\frac{d^3y}{dx^3} + x^2 \left(\frac{d^2y}{dx^2}\right)^3 = 0 \tag{7}$$

$$\frac{d^2y}{dx^2} = \sqrt{1 + \frac{dy}{dx}} \tag{8}$$

$$y = x\frac{dy}{dx} + \frac{a}{(dy/dx)} \tag{9}$$

$$r = \frac{\left[1 + (dy/dx)^2\right]^{\frac{3}{2}}}{(d^2y/dx^2)} \tag{10}$$

Later on, it will be seen that *an ordinary differential equation* can describe many phenomena in physics and other sciences. For instance,

- the motion of a falling body, as in Equation (1),
- change in the size of population, as in Equation (2).
- flow of current in an electric circuit, as in Equation (3), and
- the motion of a pendulum, as in Equation (4), and so on.

A *partial differential equation* is one in which the unknown function y depends on more than one *independent variable*, such that the *derivatives occurring in it are partial derivatives*.
   For example, if $z = f(x,y)$ is a function of x and y, then the equation

$$y^2\frac{\partial z}{\partial x} + xy\frac{\partial z}{\partial y} = nxz \text{ is a } partial \ differential \ equation.$$

Similarly, if $w = f(x,y,z,t)$ is a function of *time "t"* and *the rectangular coordinates (x, y, z) of a point in space*, then the equation:

$$a^2 \cdot \left(\frac{\partial^2 w}{\partial x^2} + \frac{\partial^2 w}{\partial y^2} + \frac{\partial^2 w}{\partial z^2}\right) = \frac{\partial w}{\partial t} \text{ is a } partial \ differential \ equation.$$

In general, *partial differential equations* arise in the physics of *continuous media*—in problems involving *electric fields, fluid dynamics, diffusion*, and *wave motion*. Their theory is *very different* from that of *ordinary differential equations*, and is *much more difficult* in almost every respect.
   In this book, we shall confine our attention exclusively to very simple ordinary differential equations. Accordingly, the term differential equation will always stand for *an ordinary differential equation*.

9a.1.2   The Basic Concepts, Ideas, and Terminology

(a) *The order of a differential equation is the order of the highest derivative* appearing in the equation.
   In Equations (1), (2), (5), and (9) *the order is one*; in Equations (3), (4), (6), (8), and (10) *the order is two*, and in Equation (7) *the order is three*.

Note: There are situations of academic interest which demand extra care while identifying the *order of a differential equation.*

A situation which might create confusion in identifying the order of a differential equation:

Example: Consider a differential equation

$$\frac{d^2y}{dx^2} + 8\frac{dy}{dx} + \int y\,dx = \frac{x^2}{6}$$

One may be tempted to say that the order of the above equation is "2", as it appears. However, this is not correct. To find its order, we have to eliminate the term $\int y\,dx$. In this case, it is quite simple to eliminate $\int y\,dx$, since y is a function of x. By differentiating both sides of the given equation, we get $\frac{d^3y}{dx^3} + 8\frac{d^2y}{dx^2} + y = \frac{x}{3}$. Obviously, the order of the given differential equation is "3".

Recall from Chapter 6a, wherein, we have seen that the operations of differentiation and integration are the inverse processes of each other. Using this fact, we have been able to remove the term $\int y\,dx$ and obtained the (new) differential equation, free from the term $\int y\,dx$.

Thus, if there is any term of the form $\int y\,dx$ in the given differential equation, then it must be made free from such a term before deciding the order of the differential equation.

(Of course, we shall not be dealing with such differential equations in our study.)

9a.2 IMPORTANT FORMAL APPLICATIONS OF DIFFERENTIALS (dy AND dx)

Note that, the differential Equation (5) involves differentials dy and dx. We have discussed at length (in Chapter 16 of Part I) that Equation (5) can also be expressed in the form dy/dx = cos x, which is a *differential equation of order one.* Here, the symbol dy/dx represents a limit (which is a single symbol) and we call it the derivative of y with respect to x. However, it can also be looked upon as a ratio of the differential dy to the differential dx (where dy is the differential of the dependent variable "y", and dx stands for the differential of the independent variable "x"). In fact, such a ratio (of differentials) can be denoted conveniently in the form dy/dx. However, since both the forms mean the same thing, there is no confusion in expressing the differential equation dy = cos x dx in the form dy/dx = cos x. However, the question is: Can we interpret the equation dy = cos x dx in the same way as we have interpreted the other form of the equation. First, it must be emphasized that this flexibility in expressing dy/dx (in two ways) is *very useful in many formal transformations.* In fact, we have already seen such transformations in computing antiderivatives (or indefinite integrals), involving *the method of substitution.*

For a given differential equation, in the form of dy = $f(x)$ dx, our interest will always be *to find the "unknown" function y,* which must satisfy the given differential equation. This process is known as *finding a solution* of the given differential equation. To see how the formal transformation(s) play their role, let us consider the differential equation

$$\frac{dy}{dx} = f(x) \tag{11}$$

The above equation says that y is a function of x, whose derivative is $f(x)$.

Hence, our problem reduces to finding an antiderivative of $f(x)$. The method of substitution helps in converting many complicated functions to some standard form.

In an integral $y = \int f(x)dx$ if the substitution is $x = \phi(t)$ then, $dx = (dx/dt)dt$, and *to find y, we have to simplify the function f* $[\phi(t)](dx/dt)$, and express it in a standard form, so that its integral can be written. *(The expression f* $[\phi(t)](dx/dt)$ *appears to be a complicated function of* "*t*," *but the process of expressing it in the standard form is quite simple, as we have seen in the* method of integration by substitution). Now, suppose we have to obtain a solution of the differential equation $dy = f(x)dx$, then, instead of writing it in the form of Equation (11), we can also express it in the following form:

$$\frac{dy}{dx}dx = f(x)dx \qquad (12)$$

(Here, *the differential dx on the left-hand side is treated like an algebraic quantity.*)

Now, the above Equation (12) clearly says that $f(x)$ *represents the derivative of some function* "*y*". (How?) Hence, to find the function $y [= F(x)$, say], we have to evaluate $\int f(x)dx$. *If the concept of differential is clearly understood by the student, then there should not be any confusion in writing the equation* $dy = f(x)dx$ *in the form* $\frac{dy}{dx}dx = f(x)dx$. (With this manipulation, the student should be able to appreciate better, the beauty of the subject.)

(b) *The degree of a differential equation is the (algebraic) degree of the highest derivative* (appearing in the differential equation) *when the differential coefficients are free from radicals and fractions.*

(In other words, to determine the degree of a differential equation, the derivative should not be in denominator or under radical sign.)

In the (differential) Equations (1)–(7), the degree is one. In the differential Equation (8), the derivative is under the *radical sign*, which can be removed *by squaring both sides of Equation* (8). This gives us,

$$\left(\frac{d^2y}{dx^2}\right)^2 = 1 + \frac{dy}{dx}.$$

Thus, the degree of differential Equation (8) is two. Similarly, *the degree of the Equation* (9) which is,

$$y = x\frac{dy}{dx} + \frac{a}{dy/dx}$$

is obtained by making the equation *free from fraction.*

We get $y\frac{dy}{dx} = x\left(\frac{dy}{dx}\right)^2 + a$. Thus, the *degree of Equation* (9) *is two.*

Note: To find *the degree of a differential equation*, the important requirement is that, *it must be expressed in the form of a polynomial equation* in derivatives. Once this is done, *the highest positive integral index of the highest order derivative involved* in the given differential equation represents *the degree of the differential equation.*

Example: Consider the differential Equation (10), i.e.,

$$r = \frac{\left[1 + (dy/dx)^2\right]^{\frac{3}{2}}}{(d^2y/dx^2)} \quad \text{or} \quad \frac{d^2y}{dx^2} = (1/r)\left[1 + \left(\frac{dy}{dx}\right)^2\right]^{\frac{3}{2}}$$

Though we have freed the above equation from the fraction, a *fractional index* (called *radical*) still remains on the right-hand side. Hence, squaring both sides of the above equation, we get

$$r^2 \cdot \left(\frac{d^2y}{dx^2}\right)^2 = \left[1 + \left(\frac{dy}{dx}\right)^2\right]^3$$

Thus, *the degree* of the above equation is two.

9a.2.1   A Situation When the Degree of a Differential Equation is not Defined

If a differential equation cannot be expressed in the form of a polynomial equation in derivatives, then the degree of such a differential equation cannot be defined.

Example (1): Consider the differential equation

$$\frac{dy}{dx} + \sin\left(\frac{dy}{dx}\right) = 0$$

Note that this equation is not a polynomial in dy/dx. Hence, it is not possible to define the degree of this differential equation.

Remark: The order of a differential equation and its degree (if it is defined) both are always positive integers.

9a.2.2   Formation of a Differential Equation

Consider a relation in x and y involving "n" arbitrary constants. Thus, we consider a relation of the type,

$$f(x, y, c_1, c_2, \ldots c_n) = 0 \tag{13}$$

We can obtain a differential equation from Equation (13), as follows:

Differentiate Equation (13) with respect to x, successively "n" times, to get "n" more equations as follows:

$$\begin{aligned}
f_1(x, y, y', c_1, c_2, \ldots c_n) &= 0 \\
f_2(x, y, y', y'', c_1, c_2, \ldots c_n) &= 0 \\
f_3(x, y, y', y'', y''', c_1, c_2, \ldots c_n) &= 0 \\
f_n(x, y, y', y'', y''', \ldots y^{(n)}, c_1, c_2, \ldots c_n) &= 0
\end{aligned} \tag{14}$$

Thus, we have in all, $(n + 1)$ equations, as clear from Equations (13) and (14) given above.

Eliminating "n" arbitrary constants from these $(n + 1)$ equations, we get

$$F(x, y, y', y'', y''', \ldots y^{(n)}) = 0$$

$$\text{or} \quad F\left(x, y, \frac{dy}{dx}, \frac{d^2y}{dx^2}, \frac{d^3y}{dx^3}, \ldots \frac{d^ny}{dx^n}\right) = 0 \tag{15}$$

Equation (15) is the desired differential equation obtained from the Equation (13). We also say that Equation (13), containing n arbitrary constants, is the general solution of the differential equation (15). (We shall have a detailed discussion on these matters shortly.)

Now, we propose to discuss, *through solved examples*, the method of obtaining differential equation(s), *by eliminating arbitrary constants* involved in the given equation. However, it is useful to first get a feel of *the solution of a differential equation and the types of solutions.*

Definition (1): A solution (or an integral) of a differential equation is *a relation between the variables, by means of which the derivatives obtained therefrom,* the given differential equation is satisfied.

Definition (2): Any relation which reduces a differential equation to an identity, when substituted for the dependent variable (and its derivatives), is called a solution (or an integral) of the given differential equation.

Remark: Solution of a differential equation is a relation *between variables, not a number,* that satisfy the differential equation.

9a.2.3 Types of Solutions

(I) The general solution (or the general integral) of a differential equation is a solution in which *the number of arbitrary constants* is equal to *the order* of the differential equation. Thus, the general solution of the *first-order* differential equation contains *only one* arbitrary constant, a *second-order* differential equation contains *only two* arbitrary constants, and so on.

(II) *A particular solution* of a differential equation, is that obtained from the general solution by *giving particular values to the arbitrary constant(s).*

The values of arbitrary constants are obtained from the given initial conditions (of the argument and the function). The above definitions and related concepts will become clearer with the following examples.

Example (2): Consider the differential equation

$$\frac{dy}{dx} = 4x^3 \tag{16}$$

(This is a differential equation of *order one* and *degree one*.) It is easy to see that $y = x^4$ is a solution of Equation (16). Further, by actual substitution, we see that $y = x^4 + c$ is also a solution of Equation (16), where c is an arbitrary constant.

- The solution $y = x^4 + c$, *involving an arbitrary constant* is called a *general solution* of Equation (16).
- The solution $y = x^4$ which *does not involve any arbitrary constant* is called a *particular solution* of the above differential equation. (Here, we have chosen $c = 0$, but we can also give any other value to c.)

Note: Any solution that does not involve an arbitrary constant is called a *particular solution*.

Remark: *A particular solution* of a differential equation is *just one solution*, whereas, the *general solution* of a differential equation is *a set of infinite number of solutions* corresponding to the infinite number of arbitrary values which can be assigned to "*c*".

Example (3):   Consider the differential equation

$$\frac{d^2y}{dx^2} + y = 0 \tag{17}$$

(It is of order two and degree one.)
  We can easily show that the relation

$$y = \cos x \tag{18}$$

is a solution of differential equation (17).[1]
From Equation (18), we get

$$\frac{dy}{dx} = -\sin x, \quad \text{and} \quad \frac{d^2y}{dx^2} = -\cos x$$

$$\therefore \quad \frac{d^2y}{dx^2} + y = 0$$

Note that the substitution $y = \cos x$ implies that $\frac{d^2y}{dx^2} = -\cos x$, as indicated above. Thus, Equation (18), together with the derivatives obtained from it, satisfy the differential equation (17). In other words, the substitution $y = \cos x$ in Equation (17) turns the equation to an identity $0 = 0$. Therefore, Equation (18) is *a solution* (or *an integral*) of the differential equation (17). Similarly, it can be verified that the relation,

$$y = \sin x \tag{19}$$

is also *a solution* of the differential equation (17). [Note that both Equations (18) and (19) do not contain any arbitrary constants.]
Again, by considering the relation $y = a \cos x$, we get

$$\frac{dy}{dx} = -a \sin x \quad \text{and} \quad \frac{d^2y}{dx^2} = -a \cos x$$

$$\therefore \frac{d^2y}{dx^2} + y = (-a \cos x) + (a \cos x) = 0.$$

---

[1] Here, one should not bother about how the solution $y = \cos x$ was reached at. We shall be learning the method(s) of solving differential equations in the chapters to follow. Here, we make use of the available information, for our discussion.

It follows, that the relation

$$y = a \cos x \tag{20}$$

where a is *an arbitrary constant* is a solution of the differential equation (17). Also, note that the solution $y = a \cos x$ includes the solution $y = \cos x$. (This becomes clear by assigning "a", the particular value of unity.) Similarly, it can be verified that

$$y = b \sin x, \tag{21}$$

is a solution of the differential equation (17) and that it includes the solution $y = \sin x$.

In fact, "a" and "b" can be assigned any real values. From this point of view, it is logical to say that the solutions at Equations (20) and (21) are more general than the solutions at Equations (18) and (19). Further, it can be shown that the relation

$$y = a \cos x + b \sin x \tag{22}$$

where a and b are arbitrary constants is also a solution of the differential equation (17).

It is useful to prove this.

From the Equation (22), we get, on differentiation,

$$\frac{dy}{dx} = -a \sin x + b \cos x \tag{23}$$

$$\therefore \frac{d^2 y}{dx^2} = -a \cos x - b \sin x \tag{24}$$

Adding Equations (24) and (22), we get,

$\frac{d^2 y}{dx^2} + y = 0$, which is the differential equation (17). Thus, Equation (22) is a yet "*more general solution*", from which all the preceding solutions of Equation (17) are obtained by giving particular values to "a" and "b".

Observations:

(i) The solutions at Equations (18) and (19) do not contain arbitrary constant(s).

(ii) The solutions at Equations (20) and (21), contain *one* arbitrary constant each. *These solutions are more general than those at* Equations (18) and (19).

(iii) The solution at Equation (22) is a *yet more general solution* of the differential equation, than those at Equation (20) and (21).

From the above observations, one is tempted to ask the question: How many arbitrary constants must the most general solution of a differential equation contain?

The answer to this question is obtained *from the consideration of the formation of a differential equation*, from a relation of the type.

$$f(x, y, c_1, c_2, c_n) = 0 \tag{25}$$

by eliminating the "n" arbitrary constants. Using the procedure discussed earlier in Section 9a.2, we know that when all the arbitrary constants (c_1 to c_n) are eliminated from Equation (25), we get a differential equation of order "n".

The concepts developed in the above examples are expressed in the following two definitions.

Definition (1): The solution of a differential equation of order "n", which contains exactly "n" arbitrary constants is said to be *the most general solution* of the given differential equation.

(Shortly, we will show that all *such* arbitrary constants must be *independent*, which will mean that their number cannot be reduced.)

Definition (2): A solution of a differential equation that can be obtained from its general solution, by giving particular values to the arbitrary constants in it, is called a *particular solution*.

Note (1): A differential equation can have a solution, which is *neither the general solution nor a particular solution*. Such solutions are called *singular solutions*. In this book, we will not discuss differential equation having *singular solutions*.

Note (2): Our interest will be to find *either* the most general solution or a *particular solution* of the given differential equation.

Note (3): To avoid confusion in terminology, we shall use the term *general solution* to mean *the most general solution* of the differential equation. First, let us consider some more examples.

Example (4): Consider the differential equation

$$\frac{dy}{dx} = \sec^2 x \tag{26}$$

(It is of order one or degree one)

Integrating both the sides of Equation (26) with respect to x, we get

$$\int dy = \int \sec 2x dx$$

$$\text{or} \quad \int \left(\frac{dy}{dx}\right) dx = \int (\sec^2 x) dx + c \tag{27}$$

where "c" is the *constant of integration*.

We can write the solution Equation (27) in the (simplified) form,

$$y = \tan x + c \tag{28}$$

which is the *(most) general solution* of the differential equation (26), in the sense discussed above.

Remarks:

- In writing Equation (28) [from Equation (27)], we have made use of the *antiderivative* (i.e., *indefinite integral*) of the function involved. Thus, any solution of the given *differential equation* is also called *an integral* of the differential equation.
- Observe that, the differential Equation (26) is of *order one* and its solution Equation (28) contains *one arbitrary* constant. Hence, the Equation (28) is the *general solution* of the differential Equation (26).
- The general solution Equation (28) includes all solutions of the differential equation (26), which can be obtained by assigning particular values to "*c*". Accordingly, for $c = 0$, 5, and $(-\pi)$, we can write *the particular solutions of Equation* (26) as, $y = \tan x$, $y = \tan x + 5$, and $y = \tan x - \pi$, respectively.
- In Example (3), we have seen that the *general solution* of the differential equation $\frac{d^2y}{dx^2} + y = 0$, is $y = a\cos x + b\sin x$, where a and b are two arbitrary constants.

Accordingly, all other relations obtained by assigning particular values to a and b, (e.g., $y = 2\cos x - 5\sin x$, etc.) are called the particular solution(s) of the above differential equation. It is useful to verify this fact. This is really very simple.

9a.2.4   An Initial Condition and a Particular Solution

Any differential equation has *an infinite number of solutions*. A *particular solution* becomes important, whenever we are interested in not all the solutions, but in one of the solutions, which satisfies a particular condition.

Example (5):   Consider the differential equation

$$\frac{dy}{dx} = 3 \tag{29}$$

Now, suppose we wish to find the *particular solution* satisfying the condition $y = 1$ when $x = 4$. Here, the *general solution* is given by

$$\int \left(\frac{dy}{dx}\right) dx = \int (3)dx + c$$

or
$$y = 3x + c \tag{30}$$

where c is an *arbitrary constant*.

Note that, the Equation (30) represents a *family of parallel lines* corresponding to different values of c, each having *the same slope* 3. To find *the particular solution* satisfying the given condition, we must choose "*c*" suitably, as follows.

Put $x = 4$ and $y = 1$ in *the general solution* given by Equation (30), we get,

$$1 = (3) \cdot (4) + c \quad \therefore \quad c = -11.$$

Hence, *the required particular solution* is

$$y = 3x - 11 \quad \text{Ans.}$$

9a.3 INDEPENDENT ARBITRARY CONSTANTS (OR ESSENTIAL ARBITRARY CONSTANTS)

(This concept is useful with reference to the solution of a differential equation.)

Example (6): Consider the relation

$$y = mx + a + b \tag{31}$$

where m, a, and b are *arbitrary constants*.

By assigning particular values to m, a, and b, we get an equation in x and y that defines y *as a function of* x.

Now, if we write $a + b = c$ in Equation (31), we get the relation

$$y = mx + c \tag{32}$$

Observe that *every function which can be obtained* by assigning particular values to the arbitrary constants in Equation (31), *can also be obtained from Equation* (32) by assigning particular values to the arbitrary constants in it, and vice versa.

Note that, the number of *arbitrary constants* in Equation (31) appear to be *three* whereas in Equation (32) the number of arbitrary constants is reduced by one so that there are only two. Observe that, both the relations *still* include the same functions. In other words, there is *no loss of generality*, even when the number of arbitrary constants is reduced. Therefore, we say, that arbitrary constants m, a, and b [in Equation (31)] are *dependent*, not independent.

Now the question is: *Can we reduce the number of arbitrary constants in Equation* (32), *without loss of generality*? Let us discuss.

One may be tempted to write $c = 0$ in the Equation (32), Then, we get the relation,

$$y = mx \tag{33}$$

which contains only *one* arbitrary constant.

Now observe that *Equation* (32) *includes the function* $y = 2x + 3$, but Equation (33) does not include it. Thus, Equation (32) is *more general* than the Equation (33). In other words, *there is a loss of generality* when the number of arbitrary constants is reduced in Equation (32). The conclusion is that in Equation (32) *we cannot reduce the number of arbitrary constants without loss of generality*. Hence, the arbitrary constants m and c in Equation (32) are said to be *independent*.

Example (7): Consider the relations, defined by the equation

$$cy = a \cos x + b \sin x \tag{34}$$

where a, b, and c are *arbitrary constants*. [Note that Equation (34) is meaningful only when $c \neq 0$]. Now, since $c \neq 0$, dividing throughout by c, Equation (34) can be written as:

$$y = \frac{a}{c} \cos x + \frac{b}{c} \sin x$$

or
$$y = A \cos x + B \sin x, \tag{35}$$

where $A = (a/c)$, and $B = (b/c)$.

Further, observe that *every function obtained from Equation (34), by giving particular values to a, b, and c, can also be obtained from Equation (35) by giving particular values to A and B*. Thus, we can obtain Equation (35) from Equation (34), by reducing the number of arbitrary constants, *without loss of generality*. Hence, the arbitrary constants a, b, c in Equation (34) are said to be *dependant*. On the other hand, the arbitrary constants in Equation (35) *cannot be reduced without loss of generality*. Hence, the arbitrary constants A and B in Equation (35) are said to be *independent*. (The terms *independent arbitrary constants* and *essential arbitrary constants* have the same meaning. Besides, the word *parameter(s)* will be frequently used to stand for these terms.)

9a.4 DEFINITION: INTEGRAL CURVE

The graph of particular solution of a differential equation is called an *integral curve* of the differential equation.

To the general solution of a differential equation, there corresponds a family of (integral) curves.

9a.4.1 Family of Curves

Let an *equation of a curve* be represented by a relation,

$$f(x, y, c) = 0 \tag{36}$$

where c has some fixed value.

Then, by assigning different values to c, we get *different curves of a similar nature*. Thus, the Equation (36) represents a *family of curves* with *one parameter*. Similarly, the relation

$$f(x, y, a, b) = 0 \tag{37}$$

represents a *family of curves* with *two parameters*, a and b.

Now we can give a *more refined definition* of the general solution of a differential equation.

Definition: The general solution (or the general integral) of a differential equation is a solution in which the number of essential arbitrary constants is equal to the order of the differential equation.

9a.5 FORMATION OF A DIFFERENTIAL EQUATION FROM A GIVEN RELATION, INVOLVING VARIABLES AND THE ESSENTIAL ARBITRARY CONSTANTS (OR PARAMETERS)

Now we shall discuss through solved examples, the methods of obtaining differential equations, by eliminating the *independent arbitrary constants* [or parameter(s)], involved in the given relation.[2]

Example (8): Consider the equation

$$y = mx \tag{38}$$

Here, m is a parameter and for each value of m, we get a straight line with slope "m", with each line passing through the origin.

Here the slope of each line is different, but all the lines satisfy a common property that they pass through the origin (see Figure 9a.1).

Differentiating both the sides of Equation (38), w.r.t. x, we get

$$\frac{dy}{dx} = m$$

by substituting the values of m in Equation (38), we get

$$y = \frac{dy}{dx} x$$

or $$x\frac{dy}{dx} - y = 0 \tag{39}$$

which is free from the parameter "m". This is the *required differential equation*. Note that Equation (38) which contains one arbitrary constant "m", is the *general solution* of differential Equation (39), which is of order one.

Example (9): Consider the equation

$$y = mx + c \tag{40}$$

Here m and c are two parameters (i.e., arbitrary constants).
By giving different values to the parameters m and c, we get different members of the family, for example,

$$y = 2x \quad (m = 2, c = 0)$$

$$y = 2x + 8 \quad (m = 2, c = 8)$$

---

[2] The purpose of examples to be discussed here is to get acquainted with the types of situations, which may be faced in the process of obtaining differential equations. This will also make one understand: What is a differential equation about? In the next chapter, we shall introduce some methods of solving differential equations of order one and degree one. However, the reader will be able to appreciate the difficulties that may be faced in solving differential equations. Later on (in higher classes), it will be discovered that only certain types of differential equations can be solved, and not all.

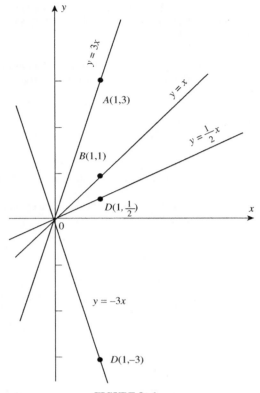

FIGURE 9a.1

$$y = x + 1 \qquad (m = 1, c = 1)$$

$$y = -x + 1 \qquad (m = -1, c = 1)$$

$$y = \sqrt{5}x - 7 \quad (m = \sqrt{5}, c = -7)$$

$$y = x \qquad\qquad (m = 1, c = 0)$$

$$y = -x \qquad\quad\ (m = -1, c = 0)$$

Thus, Equation (40) represents a *family of straight lines*, where m and c are parameters. The slope of each line is given by the value of m and the number "c" gives the y-coordinate of the point at which the line intersects the y-axis. Now, our interest is to form a differential equation, which should be satisfied by each member of the family having Equation (40). Besides, the differential equation must be free from both the parameters. (Because they are different for different members of the family.) To obtain the desired differential equation, we differentiate Equation (40) with respect to x twice, *successively*. We get

$$\frac{dy}{dx} = m$$

$$\frac{d^2y}{dx^2} = 0 \tag{41}$$

and

The Equation (41), is the *desired differential equation*. Again, remember that the general solution of the Equation (41), is represented by the Equation (40).

Note: We may also say that the *differential equation* (41) represents the *family of straight lines*, given by Equation (40).

Remark: In Example (8), the Equation (38) contained *only one parameter*, namely, "*m*". To eliminate the (single) parameter, *one differentiation* was found sufficient. In Example (9), we considered Equation (40), containing *two parameters* namely "*m*" and "*c*". To eliminate these parameters, it was necessary to apply *the process of differentiation twice successively*, and we obtained the differential Equation (41), which is of the *order two*. Also, recall that *the general solution* of the differential equation,

$$\frac{d^2y}{dx^2} + y = 0$$

was found to be $y = a\cos x + b\sin x$, where *a* and *b* are *two arbitrary constants*.

Example (10):   Consider the relation

$$y = 2x + c \tag{42}$$

This relation obviously represents a *family of lines*. The common property of all the lines in the family is that *they are mutually parallel to each other*, with slope 2. Again, corresponding to each value of the parameter "*c*", there is a line that intersects the *y*-axis at the point $(0, c)$. The desired differential equation is obviously, $dy/dx = 2$, which represents the slope of each straight line given by Equation (42).

Remark: It will be seen that, whereas in some cases the equations may represent a known family of curves, in others, the family of curves may not be known. The difficulties involved in eliminating the parameter(s) from the given equations will generally depend on the nature of the relations.

Example (11):   Obtain the differential equation by eliminating the arbitrary constants *a* and *b* from the relation,

$$y = ae^x + be^{-x}$$

Solution: The given relation is

$$y = ae^x + be^{-x} \tag{43}$$

It contains *two arbitrary constants*. (This suggests that we have to differentiate the given relation *twice*.)

Differentiating both sides of Equation (43) w.r.t. x, we get

$$\frac{dy}{dx} = ae^x - be^{-x} \left[\because \frac{d}{dx}(e^{-x}) = -e^{-x} \right]$$

Again differentiating both sides of the above equation, we get

$$\frac{d^2y}{dx^2} = ae^x + be^{-x} \tag{44}$$

Using Equation (43) in Equation (44), we get

$$\frac{d^2y}{dx^2} = y \quad \text{or} \quad \frac{d^2y}{dx^2} - y = 0 \tag{45}$$

This is the *required differential* equation of order two. Ans.

Example (12): Find the differential equation of the family of curves,

$$y = Ae^{3x} + Be^{5x}$$

where A and B are parameters.

Solution: The given equation is

$$y = Ae^{3x} + Be^{5x} \tag{46}$$

Differentiating w.r.t. x, both sides, we get

$$\frac{dy}{dx} = 3Ae^{3x} + 5Be^{5x} \tag{47}$$

Again, differentiating w.r.t. x, we get

$$\frac{d^2y}{dx^2} = 9Ae^{3x} + 25Be^{5x} \tag{48}$$

Multiplying Equation (47) by 5, we get

$$5\frac{dy}{dx} = 15Ae^{3x} + 25Be^{5x} \tag{49}$$

But from Equation (48)

The difference, Equation (49) - Equation (48) gives us,

$$5\frac{dy}{dx} - \frac{d^2y}{dx^2} = 6Ae^{3x}$$

$$\therefore \quad Ae^{3x} = \frac{5}{6} \cdot \frac{dy}{dx} - \frac{1}{6} \cdot \frac{d^2y}{dx^2} \tag{50}^{(3)}$$

Now, multiplying Equation (47) by 3, we get

$$3\frac{dy}{dx} = 9Ae^{3x} + 15Be^{5x} \tag{51}$$

Again, the difference of Equation (48) - Equation (51) gives us the result as,

$$\frac{d^2y}{dx^2} - 3\frac{dy}{dx} = 10Be^{5x}$$

$$\therefore \quad Be^{5x} = \frac{1}{10} \cdot \frac{d^2y}{dx^2} - \frac{3}{10} \cdot \frac{dy}{dx} \tag{52}$$

Using Equations (50) and (52) in Equation (46), we get

$$y = \frac{5}{6} \cdot \frac{dy}{dx} - \frac{1}{6} \cdot \frac{d^2y}{dx^2} + \frac{1}{10} \cdot \frac{d^2y}{dx^2} - \frac{3}{10} \cdot \frac{dy}{dx}$$

$$y = \frac{8}{15} \cdot \frac{dy}{dx} - \frac{1}{15} \cdot \frac{d^2y}{dx^2}$$

$$\therefore \quad \frac{d^2y}{dx^2} - 8\frac{dy}{dx} + 15y = 0$$

which is the required differential equation. Ans.

Note: In the Examples (11) and (12) above, we have applied the "*the method of elimination*", which is fairly simple and straight-forward. It is frequently used for solving *simultaneous linear equations* in two or three unknowns. However, *if the number of unknowns exceeds three, the process (of elimination) becomes extremely tedious and time consuming*.[4]

In this book, we shall consider the relations involving *at most two parameters*, and obtain differential equations from them, by applying *the usual method of elimination* (of parameters) or by applying the *more general procedure*, using determinants, discussed below.

---

[3] Similarly, we can obtain the value of Be^{5x}. Finally, using these values in Equation (46), we can obtain the desired differential equation.

[4] Fortunately, it is possible to solve simultaneous equations with more number of unknowns by a systematic general procedure, which involves the concept of a determinant and Cramer's Rule for evaluating it. Here, we shall apply this general procedure, just to get a feel that the general procedure is really simpler, in handling certain complicated relations as the one given in Example (12).

9a.6   GENERAL PROCEDURE FOR ELIMINATING *"TWO"* INDEPENDENT ARBITRARY CONSTANTS (USING THE CONCEPT OF DETERMINANT)

In Class XI, we have studied about *determinants*, their *properties*, and the method of computing their *value(s)*. Here, we shall recall some important points, pertaining to determinants, which will be useful in the process of obtaining the desired differential equation using the concept.[5]

(i) Suppose we are given *three equations in the form*:

$$a_1p + b_1q + c_1 = 0 \tag{53}$$

$$a_2p + b_2q + c_2 = 0 \tag{54}$$

$$a_3p + b_3q + c_3 = 0 \tag{55}$$

in which p and q are *independent arbitrary constants* (i.e., parameters). Then, finding the values of p and q by solving these equations, we can substitute their values in any of the equations, and obtain a relation, which will be free from these parameters (p and q).

Something like this has to be done in the process of forming a differential equation by eliminating parameters. For this purpose, we shall differentiate the given relation involving two parameters and then form three equations. (Of course, two of these equations will contain derivatives.) Finally, we will eliminate the two parameters using the properties of determinants.

(ii) The symbol,

$$\begin{vmatrix} a_1 & b_1 & c_1 \\ a_2 & b_2 & c_2 \\ a_3 & b_3 & c_3 \end{vmatrix} \tag{56}$$

where $a_1b_1c_1 \ldots$ and so on may be real *numbers* or *functions* (including derivatives) is called a *determinant* of *order* 3. Note that the determinant in Equation (56) is *free from the parameters p and q*.

(iii) The *arrangement* or *the symbol*, $\begin{vmatrix} a_1 & b_1 \\ a_2 & b_2 \end{vmatrix}$ is called a *determinant of order* 2. *It stands for the number* or *the quantity* $[a_1b_2 - a_2b_1]$ and it is called the value (or the result) of the determinant. The diagonal from the *upper left* to the *lower right* is called the *principal diagonal*.[6]

(iv) The determinant in Equation (56) consists of *three rows* and *three columns*. *To find the value* (or the result) of a *determinant of order three*, the determinant must be expanded along the elements of *first row* or the *first column*.[7]

The determinant in Equation (56) (repeated below for convenience)

---

[5] This is just for understanding the general procedure. Of course, in simple problems [as in the case of Example (11)], the usual method of elimination may be found more convenient.

[6] A determinant whose elements are real numbers and ultimately gets reduced to real numbers.

[7] In fact, a determinant can be expanded along the elements of any row or any column. Also, any row (or column) can be shifted to the first position. However, in any such shift, the sign of the determinant changes. Further, any number (or function) common in any row (or column) can be taken out as a common factor of a determinant, without making any change in the sign of the determinant. (All three facts are covered under the properties of determinants.)

$$\begin{vmatrix} a_1 & b_1 & c_1 \\ a_2 & b_2 & c_2 \\ a_3 & b_3 & c_3 \end{vmatrix}$$

when *expanded along the first row,* will appear as follows:

$$a_1 \begin{vmatrix} b_2 & c_2 \\ b_3 & c_3 \end{vmatrix} - b_1 \begin{vmatrix} a_2 & c_2 \\ a_3 & c_3 \end{vmatrix} + c_1 \begin{vmatrix} a_2 & b_2 \\ a_3 & b_3 \end{vmatrix} \tag{57}$$

Note: It is important to remember that we attach a *positive sign* to the *first term*, a *negative sign* to the *second term*, and, finally a *positive sign* to the *third term*. The sum of all the three terms gives the expansion of the above determinant, as indicated in Equation (57).

(v) *The rule for expansion of a determinant is as follows:* Starting from a_1, *delete the row and column passing through* a_1, thus getting the 2×2 determinant $\begin{vmatrix} b_2 & c_2 \\ b_3 & c_3 \end{vmatrix}$.

Then, the product $a_1 \begin{vmatrix} b_2 & c_2 \\ b_3 & c_3 \end{vmatrix}$ is the *first term* in the expansion in Equation (57).

Similarly, we obtain the second and the third terms, with b_1 and c_1, respectively.

Remark: Irrespective of whether we expand a determinant *along a row* (or a *column*), we get the same value (or result) of the determinant. In other words, a *determinant represents a definite value* (or result).

Now, we proceed to obtain the required differential equation from the given relation, involving *two parameters.*

Example (13):   Find the differential equation of the family of the curves

$$y = ae^{3x} + be^{2x} \tag{58)[1]}$$

Differentiating Equation (56) twice successively w.r.t. x, we get

$$\frac{dy}{dx} = 3ae^{3x} + 2be^{2x} \tag{59}$$

and

$$\frac{d^2y}{dx^2} = 9ae^{3x} + 4be^{2x} \tag{60}$$

Eliminating a and b from the above, the above three equations, we get

$$\begin{vmatrix} -y & e^{3x} & e^{2x} \\ \dfrac{-dy}{dx} & 3e^{3x} & 2e^{2x} \\ \dfrac{-d^2y}{dx^2} & 9e^{3x} & 4e^{2x} \end{vmatrix} = 0$$

[8] Now if we transfer the dependent variable "y" and its derivatives, respectively in the following Equations (58), (59) and (60) to the right-hand side of the corresponding equation, then each equation equals zero. Also, the transferred terms will have negative sign.

$$\therefore \quad -e^{3x} \cdot e^{23x} \begin{vmatrix} y & 1 & 1 \\ \dfrac{dy}{dx} & 3 & 2 \\ \dfrac{d^2y}{dx^2} & 9 & 4 \end{vmatrix} = 0$$

[Here we have taken out (-1) from first column, e^{3x} from second column, and e^{2x} from the third column.]

Again, note that $-e^{3x} \cdot e^{2x}$ (i.e., $-e^{5x}$) $\neq 0$. It follows that the determinant $= 0$.

Now, it is convenient to expand the above determinant along the first row, since it will involve simpler calculations. We have,

$$y(12 - 18) - (1)\left(4\frac{dy}{dx} - 2\frac{d^2y}{dx^2}\right) + \left(9\frac{dy}{dx} - 3\frac{d^2y}{dx^2}\right)$$

$$\therefore \quad -6y - 4\frac{dy}{dx} + 2\frac{d^2y}{dx^2} + 9\frac{dy}{dx} - 3\frac{d^2y}{dx^2} = 0$$

or
$$\frac{d^2y}{dx^2} - 5\frac{dy}{dx} + 6y = 0,$$

which is the required differential equation. Ans.

Note: Observe that the above procedure for obtaining the differential equation is comparatively simpler than the method of elimination used in Example (12).

Example (14): Obtain the differential equation whose general solution $Ax^2 + By^2 = 1$, where A and B are parameters.

Solution: Given relation (or the solution) is

$$\begin{aligned} Ax^2 + By^2 &= 1 \\ \text{or} \quad Ax^2 + By^2 - 1 &= 0 \end{aligned} \tag{61}$$

Differentiating both the sides of Equation (61) w.r.t. x, we get

$$2Ax + 2By\frac{dy}{dx} + 0 = 0 \tag{62}$$

Again, differentiating both the sides of Equation (62) w.r.t. x, we get,

$$A + B\left[y\frac{d^2y}{dx^2} + \left(\frac{dy}{dx}\right)\left(\frac{dy}{dx}\right)\right] = 0$$

$$A + B\left[y\frac{d^2y}{dx^2} + \left(\frac{dy}{dx}\right)^2\right] + 0 = 0 \tag{63}$$

Eliminating A and B from Equations (61), (62) and (63), we get

$$\begin{vmatrix} x^2 & y^2 & -1 \\ x & y\dfrac{dy}{dx} & 0 \\ 1 & \left\{\left(\dfrac{dy}{dx}\right)^2 + y\left(\dfrac{d^2y}{dx^2}\right)^2\right\} & 0 \end{vmatrix} = 0$$

Now, we expand the above determinant along the elements of the third column. We get

$$(-1)\left\{x\left[\left(\frac{dy}{dx}\right)^2 + y\left(\frac{d^2y}{dx^2}\right)^2\right] - (1)\left(y\left(\frac{dy}{dx}\right)\right)\right\} = 0$$

Since, the number $(-1) \neq 0$, hence the required differential equation is,

$$xy\left(\frac{d^2y}{dx^2}\right) + x\left(\frac{dy}{dx}\right)^2 - y\frac{dy}{dx} = 0 \quad \text{Ans.}$$

Example (15): Find the differential equation whose general solution is

$$y = c^2 + \frac{c}{x}$$

where c is an arbitrary constant.

Solution: The given solution is

$$y = c^2 + \frac{c}{x} \tag{64}$$

which contains *only one arbitrary constant*. (This suggests that we will have to differentiate only once.)

Differentiating both sides of Equation (64), w.r.t. x, we get

$$\frac{dy}{dx} = 0 - \frac{c}{x^2}$$

$$\therefore \quad c = -x^2\frac{dy}{dx} \tag{65}$$

Using Equation (65) in Equation (64), we get

$$y = \left[-x^2 \frac{dy}{dx}\right]^2 + \frac{1}{x}\left[-x^2 \frac{dy}{dx}\right]$$

$$\text{or} \qquad y = x^4 \left(\frac{dy}{dx}\right)^2 - x\frac{dy}{dx} \tag{66}$$

This is the required differential equation. Ans.

Remark: Observe that whereas the *order of the differential Equation* (66) is *one*, its *degree is two*. Thus, depending on *the nature of the given equation*, we get the required differential equation.

Example (16): Obtain the differential equation whose solution is

$$xy = ae^x + be^{-x}$$

Solution: The given relation is

$$xy = ae^x + be^{-x} \tag{67}$$

Differentiating both sides of Equation (67) w.r.t. x, we get

$$x\frac{dy}{dx} + y = ae^x - be^{-x} \tag{68}$$

Again, differentiating Equation (68) w.r.t. x, we get

$$\left(x\frac{d^2y}{dx^2} + \frac{dy}{dx}\right) + \frac{dy}{dx} = ae^x + be^{-x}$$

$$= xy \text{ [using Equation (67)]}$$

$$\therefore \quad x\frac{d^2y}{dx^2} + 2\frac{dy}{dx} = xy$$

which is the required differential equation. Ans.

Observation: Compare the Equation (67) in this example, with Equation (43) given in the earlier Example (11). *What useful conclusion can be drawn from this observation?*

Example (17): Obtain the differential equation by eliminating the arbitrary constants, from the relation

$$y = a\sin(wt + c)$$

Solution: The given relation is

$$y = a\sin(wt + c) \tag{69}$$

that contains two arbitrary constants, a and c.
Differentiating Equation (69) w.r.t. x, we get

$$\frac{dy}{dx} = a[\cos(wt + c)] \cdot w$$

or
$$\frac{dy}{dx} = a \cdot w \cdot \cos(wt + c) \tag{70}$$

Again, differentiating Equation (70) w.r.t. x, we get

$$\frac{d^2y}{dx^2} = a \cdot w \cdot [-\sin(wt + c)] \cdot w$$

or
$$\frac{d^2y}{dx^2} = -a \cdot w^2 \sin(wt + c)$$

$$= -w^2[a\sin(wt + c)]$$

$$= -w^2 \cdot y \quad \text{[using Equation (69)]}$$

or
$$\frac{d^2y}{dx^2} + w^2 \cdot y = 0,$$

which is the required differential equation. Ans.

Example (18): Prove that the relation

$$y = x^3 + ax^2 + bx + c$$

is a solution of the differential equation $\dfrac{d^3y}{dx^3} = 6$

Solution: Given,

$$y = x^3 + ax^2 + bx + c \tag{71}$$

This relation contains *three arbitrary constants a, b,* and *c,* hence *it will be differentiated thrice.*
Differentiating Equation (71) w.r.t. x, we get

$$\frac{dy}{dx} = 3x^2 + 2ax + b \tag{72}$$

Differentiating again, w.r.t. x, we get

$$\frac{d^2y}{dx^2} = 6x + 2a \tag{73}$$

Differentiating again w.r.t. x, we get

$$\frac{d^3y}{dx^3} = 6 \tag{74}$$

It follows, that Equation (71) *is the solution of the differential Equation* (74). It must also be clear that it is the *(most) general solution*. (Why?) Ans.

Example (19): Obtain the differential equation whose general solution is,

$$y = a\cos x + b\sin x$$

where a and b are *arbitrary constants*.

Solution: The given relation is,

$$y = a\cos x + b\sin x \tag{75}$$

Note that the given relation contains two arbitrary constants. Hence, we have to differentiate it twice.

Differentiating Equation (75) w.r.t. x, we get

$$\frac{dy}{dx} = -a\sin x + b\cos x \tag{76}$$

Now, differentiating Equation (76) w.r.t. x, we get

$$\begin{aligned}
\frac{d^2y}{dx^2} &= -a\cos x - b\sin x \\
&= -(a\cos x + b\sin x) \\
&= -y[\text{using (1)}]
\end{aligned}$$

$$\therefore \frac{d^2y}{dx^2} + y = 0.$$

This is the required differential equation. Ans.

Note: In this example, *the arbitrary constants are easily eliminated*. Similarly, the next example is very simple. However, it is important to remember that in a particular case, a special method is more convenient, as we have seen in Examples (13) and (14). Also, it will be observed that some extra care is needed in Example (21), to follow.

Example (20): Find the differential equation of the family of curves:

$$y = e^x(a\cos x + b\sin x)$$

Solution: The given differential equation is,

$$y = e^x(a \cos x + b \sin x) \tag{77}$$

[In this case, we will have to differentiate twice. (Why?)]
Differentiating Equation (77) w.r.t. x, we get

$$\frac{dy}{dx} = e^x[-a \sin x + b \cos x] + [a \cos x + b \sin x]e^x$$

$$= e^x[b \cos x - a \sin x] + y, \quad [\text{using Equation (77)}]$$

$$\therefore \frac{dy}{dx} - y = e^x[b \cos x - a \sin x] \tag{78}$$

Now, differentiating Equation (78) w.r.t. x, we get

$$\frac{d^2y}{dx^2} - \frac{dy}{dx} = e^x[-b \sin x - a \cos x] + [b \cos x - a \sin x]e^x$$

$$= -e^x[b \sin x + a \cos x] + \left[\frac{dy}{dx} - y\right] \quad [\text{using Equation (78)}]$$

$$= -e^x[a \cos x + b \sin x]$$

$$\therefore \quad \frac{d^2y}{dx^2} - \frac{dy}{dx} - \frac{dy}{dx} + y = -y \, [\text{using Equation (77)}]$$

$$\therefore \quad \frac{d^2y}{dx^2} - 2\frac{dy}{dx} + 2y = 0,$$

which is the required differential equation.   Ans.

Note: The above example gives a complicated look due to the presence of e^x, but elimination of the arbitrary constants (a and b) is quite simple. However, the above solution suggests that one has to be more careful, for obtaining the required differential equation.

Example (21): Form the differential equation from the relation,

$$(x - a)^2 + (y - b)^2 = 16$$

where a, b are constants[9]

Solution: The given relation is,

$$(x - a)^2 + (y - b)^2 = 16 \tag{79}$$

---

[9] The given relation represents a family of circles having a radius of 4 units. Also, any point (a, b) in the xy-plane, can be the center of the circle with Equation (79). The only requirement is that its radius must be of 4 units. Obviously, a, b are arbitrary constants in the given relation.

Differentiating Equation (79) w.r.t. x, we get

$$2(x - a) + 2(y - b)\frac{dy}{dx} = 0$$

$$(x - a) + (y - b)\frac{dy}{dx} = 0 \tag{80}$$

Again, differentiating Equation (80) w.r.t. x, we get

$$1 + \left[(y - b)\frac{d^2y}{dx^2} + \frac{dy}{dx}\left(\frac{dy}{dx}\right)\right] = 0$$

$$\therefore 1 + (y - b)\frac{d^2y}{dx^2} + \left(\frac{dy}{dx}\right)^2 = 0 \tag{81}$$

$$\therefore (y - b) = \frac{-\left(\frac{dy}{dx}\right)^2 - 1}{\left(\frac{d^2y}{dx^2}\right)} = -\frac{\left(\frac{dy}{dx}\right)^2 + 1}{\left(\frac{d^2y}{dx^2}\right)} \tag{82}$$

Using Equation (82) in Equation (80), we get

$$(x - a) - \left[\frac{\left(\frac{dy}{dx}\right)^2 + 1}{\left(\frac{d^2y}{dx^2}\right)}\right] \cdot \frac{dy}{dx} = 0$$

$$\therefore (x - a) = \left[\frac{\left(\frac{dy}{dx}\right)^2 + 1}{\left(\frac{d^2y}{dx^2}\right)}\right] \cdot \frac{dy}{dx} \tag{83}$$

Using Equations (82) and (83) in Equation (79), we get

$$\therefore \left[\left(\frac{\left(\frac{dy}{dx}\right)^2 + 1}{\left(\frac{d^2y}{dx^2}\right)}\right) \cdot \frac{dy}{dx}\right]^2 + \left[(-1)\frac{\left(\frac{dy}{dx}\right)^2 + 1}{\left(\frac{d^2y}{dx^2}\right)}\right]^2 = 16$$

$$\therefore \left[\left(\frac{dy}{dx}\right)^2 + 1\right]^2 \left[\left(\frac{dy}{dx}\right)^2 + 1\right] = 16\left(\frac{d^2y}{dx^2}\right)^2$$

$$\therefore \left[\left(\frac{dy}{dx}\right)^2 + 1\right]^3 = 16\left(\frac{d^2y}{dx^2}\right)^2 \quad \text{Ans.}$$

Note: The above solution appears to be lengthy, but it is quite simple. Moreover, it is not convenient to obtain the desired differential equation using the method of determinant. Check this.

Example (22): Find the differential equation of the family of curves whose equation is $y = (c_1 + c_2 x)e^x$, where c_1 and c_2 are two parameters.

Solution: The given relation is,

$$y = (c_1 + c_2 x)e^x \qquad (84)$$

[Since, there are two parameters, hence, we must differentiate Equation (84) twice.]
Differentiating Equation (84) w.r.t. x, we get

$$\frac{dy}{dx} = (c_1 + c_2 x)e^2 + e^x(0 + c_2)$$

$$\frac{dy}{dx} = y + c_2 e^x$$

$$\therefore c_2 = \frac{((dy/dx) - y)}{e^x} \qquad (85)$$

Again differentiating Equation (85) w.r.t. to x, we get

$$0 = \frac{e^x \cdot \dfrac{d}{dx}\left(\dfrac{dy}{dx} - y\right) - \left(\dfrac{dy}{dx} - y\right)\dfrac{d}{dx}(e^x)}{(e^x)^2} = \frac{\left[\dfrac{d^2y}{dx^2} - \dfrac{dy}{dx}\right] - \left(\dfrac{dy}{dx} - y\right)}{e^x}$$

$$\therefore 0 = \frac{d^2y}{dx^2} - \frac{dy}{dx} - \frac{dy}{dx} + y$$

or $$\frac{d^2y}{dx^2} - 2\frac{dy}{dx} + y = 0 \quad \text{Ans.}$$

Exercise

Q. (1) Determine the order and degree (if defined) of differential equations given below:

(a) $y = y\left(\frac{dy}{dx}\right)^2 + 2x\frac{dy}{dx}$

(b) $\left(\frac{d^2y}{dx^2}\right) + \sin\left(\frac{d^2y}{dx^2}\right) = 0$

(c) $\frac{d^2y}{dx^2} = \sqrt{1 + \frac{dy}{dx}}$

(d) $y = x\frac{dy}{dx} + \frac{a}{(dy/dx)}$

(e) $\left(\frac{ds}{dt}\right)^3 + 3s\left(\frac{d^2s}{dt^2}\right)^2 = 0$

(f) $\left(\frac{d^3y}{dx^3}\right)^2 + \sin(y'') = 0$

(g) $\left(\frac{d^2y}{dx^2}\right)^2 + \cos\left(\frac{dy}{dx}\right) = 0$

(h) $(y''')^2 + (y'')^3 + (y)^4 + y^5 = 0$

ANSWERS:

(a) Order one, degree two.

(b) Order two, degree not defined.

(c) Order two, degree two.

(d) Order one, degree two.

(e) Order two, degree two.

(f) Order three, degree not defined.

(g) Order two, degree not defined.

(h) Order three, degree two.

Q. (2) Find the order and degree of the differential equations

(i) $\sqrt{\frac{d^2y}{dx^2}} + \frac{dy}{dx} + xy^2 = 0$

(ii) $\left(\frac{d^4y}{dx^4}\right) = \left[1 + \left(\frac{dy}{dx}\right)^2\right]^{3/2}$

(iii) $e^{\frac{dy}{dx}} + \frac{dy}{dx} = x$

(iv) $\sqrt{1 + \frac{1}{(dy/dx)^2}} = \left(\frac{d^2y}{dx^2}\right)^{3/2}$

ANSWERS:

(i) Order two, degree one.

(ii) Order four, degree two.

(iii) Order one, degree not defined.

(iv) Order two, degree three.

[Hint: R.H.S. $= \sqrt{\left(\frac{d^2y}{dx^2}\right)^3}$. On squaring both sides, the result follows.]

Q. (3) Obtain the differential equation by eliminating a and b from the relation

$$y = a\cos 4x + b\sin 4x$$

$$\left(\text{Ans.} \quad \frac{d^2y}{dx^2} + 16y = 0\right)$$

Q. (4) Obtain the differential equation by eliminating c from the relation

$$e^x + c \cdot e^y = 1$$

$$\left[\text{Ans.} \quad \frac{dy}{dx} = \left(\frac{dy}{dx} - 1\right)e^x\right]$$

Q. (5) Obtain the differential equation whose general solution is

$$y = ax^2 + b \cdot x \text{ (Try both the methods)}$$

$$\left(\text{Ans.} \quad x^2\frac{d^2y}{dx^2} - 2x \cdot \frac{dy}{dx} + 2y = 0\right)$$

Q. (6) Obtain the differential equation by eliminating m from the relation

$$y = mx + \frac{4}{m}$$

$$\left[\text{Ans.} \quad y\frac{dy}{dx} = x\left(\frac{dy}{dx}\right)^2 + 4 \right]$$

Q. (7) Form the differential equation for

$$y = Ae^{2x} + Be^x + C$$

$$\left(\text{Ans.} \quad \frac{d^3y}{dx^3} - 3\frac{d^2y}{dx^2} + 2\frac{dy}{dx} = 0\right)$$

Q. (8) Obtain the differential equation of the family of curves, whose equation is $y = (c_1 + c_2 x)e^{3x}$, where c_1 and c_2 are parameters.

$$\left(\text{Ans.} \quad \frac{d^2y}{dx^2} - 6\frac{dy}{dx} + 9y = 0\right)$$

Q. (9) If $y = Ce^{\sin^{-1}x}$ then form the corresponding differential equation.

$$\left(\text{Ans.} \quad \sqrt{1 - x^2}\frac{dy}{dx} - y = 0\right)$$

Q. (10) Choose the correct answer

(i) The degree of the differential equation

$$\left(\frac{d^2y}{dx^2}\right)^4 + \left(\frac{dy}{dx}\right)^2 + \cos\left(\frac{dy}{dx}\right) + 1 = 0, \text{ is}$$

  (a) 4.
  (b) 2
  (c) 1
  (d) Not defined.

(ii) The order of the differential equation

$$\left(\frac{d^2y}{dx^2}\right)^3 = \cos 4x + \sin 4x \text{ is,}$$

  (a) 2
  (b) 3
  (c) 4
  (d) Not defined.

(iii) The order of the differential equation

$$\left[1+\left(\frac{dy}{dx}\right)^2\right]^3 = 16\left(\frac{d^2y}{dx^2}\right)^2 \text{ is } \underline{\hspace{2cm}}.$$

(iv) The order of the differential equation

$$\left(\frac{dy}{dx}\right)^3 + 2a\frac{d^2y}{dx^2} = 0 \text{ is } \underline{\hspace{3cm}} \text{and its degree is } \underline{\hspace{3cm}}.$$

(v) The number of arbitrary constants in the general solution of a differential equation of order three must be

(a) 1

(b) 2

(c) 3

(d) Arbitrary.

(vi) The number of arbitrary constants in the particular solution of a differential equation of order three must be

(a) 3

(b) 2

(c) 1

(d) 0.

(vii) The number of arbitrary constants in the equation $y = a + \log b \, x$ is $\underline{\hspace{3cm}}$.

(viii) The number of arbitrary constants in the equation $y = ae^{x^2-b}x$ is $\underline{\hspace{3cm}}$.

(ix) Which of the following differential equations has $y = c_1e^x + c_2e^{-x}x$ as the general solution.

(a) $\frac{d^2y}{dx^2} + y = 0$

(b) (b) $\frac{d^2y}{dx^2} - y = 0$

(c) $\frac{d^2y}{dx^2} + 1 = 0$

(d) $\frac{d^2y}{dx^2} - 1 = 0$

[Hint: The desired equation should not have any constant term. Further, it is easily seen from the given relation that $\frac{d^2y}{dx^2} = c_1e^x + c_2e^{-x} = y$.]

(x) Which of the following differential equation has $y = x$ as the particular solution

(a) $\frac{d^2y}{dx^2} - x^2\frac{dy}{dx} + xy = x$

(b) $\frac{d^2y}{dx^2} + x\frac{dy}{dx} + xy = x$

(c) $\frac{d^2y}{dx^2} - x^2\frac{dy}{dx} + xy = 0$

(d) $\frac{d^2y}{dx^2} + x\frac{dy}{dx} + xy = 0$

[Hint: The differential equation should be satisfied by the condition $dy/dx = 1$.]

ANSWERS:

(i) (d), (ii) (a), (iii) order is two, degree is two, (iv) order is two, degree is one, (v) (c), (vi) (d), (vii) one, (viii) one [Hint: $y = a + \log b + \log x = (a + \log b) + \log x$], (ix) (b), (x) (c).

9a.6.1 Forming the Differential Equation Representing a Family of Curves

Sometimes, equations of the family of the curves (i.e., an equation *with arbitrary constants*) are not given directly. *From the given conditions, the equation representing the family of curves, has to be formed first*, and then it is to be differentiated in order to get the *required differential equation*. Given below are some examples.

Example (23): Find the differential equation of the family of concentric circles with the origin at (0,0), and radius r units.

Solution: Equation of the circle with center at the origin and radius "r" is given by

$$x^2 + y^2 = r^2 \tag{86}$$

By giving different values to r, we get different members of the family, e.g., $x^2 + y^2 = 1$, $x^2 + y^2 = 4$, $x^2 + y^2 = 9$, and so on (see Figure 9a.2).
In Equation (86), "r" is an *arbitrary constant*.
Differentiating Equation (86) w.r.t. x, we get

$$2x + 2y\frac{dy}{dx} = 0$$

$$\therefore x + y\frac{dy}{dx} = 0 \tag{87}$$

which represents the family of concentric circles, given by Equation (86). Ans.

Note: Generally, from a given differential equation, it is not possible to say what the differential equation represents, till we solve it. Here, the differential equation (87) is obtained from Equation (86), which represents the family of concentric circles with center at (0, 0). Hence we make the same statement for the differential equation (87).

Example (24): Find the differential equation of the family of circles of a fixed radius "r", with center on the x-axis.

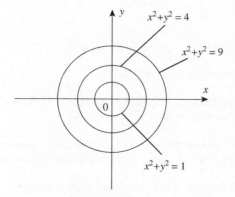

FIGURE 9a.2

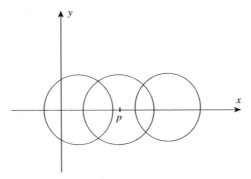

FIGURE 9a.3

Solution: We know that if the center of the circle lies on the x-axis, then its coordinates are given by ordered pair of the form $(a, 0)$, where $a \in (-\infty, \infty)$. Also, it is given that the radius "r" is fixed.

$\therefore$ The equation of family of such circles is given by,

$$(x - a)^2 + y^2 = r^2 \tag{88}$$

where "a" is *arbitrary constant*. For different values of "a," we get different circles, *which have the same radius "r."*

Differentiating Equation (88) w.r.t. x, we get

$$2(x - a) + 2y\frac{dy}{dx} = 0 \quad \therefore (x - a) + y\frac{dy}{dx} = 0$$

$$\therefore (x - a) = -y\frac{dy}{dx} \tag{89}$$

Using Equation (89) in Equation (88), we get (see Figure 9a.3)

$$\left[-y\frac{dy}{dx}\right]^2 + y^2 = r^2$$

$$\therefore y^2 \left(\frac{dy}{dx}\right)^2 + y^2 = r^2$$

which is the desired differential equation. Ans.

Example (25): Find the differential equation of the family of circles whose centers are on the x-axis and which touches the y-axis.

Solution: Since the centers are on the x-axis, the coordinates of the center may be taken as $(a, 0)$. Further, since the circles touch the y-axis, it follows that *the circles pass through the origin.* (*Why?*)

(This family of circles can touch the y-axis, *only if they pass through the origin.*) Accordingly, the radius must be "a" *units*, which is variable (see Figure 9a.4).

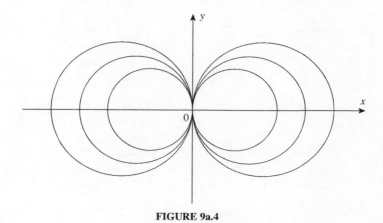

FIGURE 9a.4

The equation of the family of circles may be written as

$$(x - a)^2 + y^2 = a^2 \tag{90}$$

Note: In this problem, the first important requirement was to find the equation of the given family, which is given by Equation (90). Since, there is only one parameter, only one differentiation is needed to eliminate "*a*". Also, it is easy to guess that the desired differential equation will be of order one.

Equation (90) can be written in the form

$$x^2 + a^2 - 2ax + y^2 = a^2$$

$$\therefore x^2 + y^2 = 2ax \tag{91}$$

Differentiating Equation (91) w.r.t. *x*, we get

$$2x + 2y\frac{dy}{dx} = 2a \quad \text{or} \quad x + y\frac{dy}{dx} = a$$

Using this value of "*a*" in Equation (91), we get

$$x^2 + y^2 = 2\left[x + y\frac{dy}{dx}\right]x$$

or
$$x^2 + y^2 = 2x^2 + 2xy\frac{dy}{dx}$$

or
$$x^2 - y^2 + 2xy\frac{dy}{dx} = 0$$

which is the desired differential equation. Ans.

Example (26): Find the differential equation of parabolas with vertex at the origin and foci on *x*-axis.

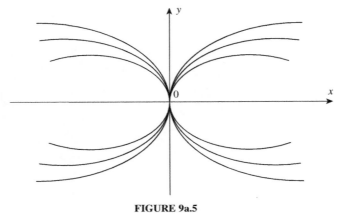

FIGURE 9a.5

Solution: The equation of the family of parabolas with *vertex at the origin and foci on x-axis* is

$$y^2 = 4ax \tag{92}$$

where "*a*" is a parameter (Figure 9a.5).
Differentiating Equation (92) w.r.t. x, we get

$$2y \frac{dy}{dx} = 4a$$

Substituting in Equation (92), we get

$$y^2 = \left[2y \frac{dy}{dx}\right] x$$

$$y^2 - 2xy \frac{dy}{dx} = 0 \tag{93}$$

which is the required differential equation. Ans.

Remark: Even if we consider the parabolas only to the positive direction of x-axis, *the differential Equation (93) will remain unchanged.*

Example (27): Form the differential equation of the family of ellipses having foci on the x-axis and center at the origin.

Solution: The equation of the family of ellipses, in question is,

$$\frac{x^2}{a^2} + \frac{y^2}{b^2} = 1 \tag{94}$$

where a and b are parameters (see Figure 9a.6).

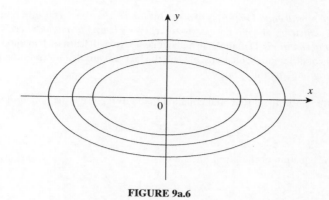

FIGURE 9a.6

For convenience, Equation (94) may be written as

$$b^2x^2 + a^2y^2 = a^2b^2$$

Differentiating both sides w.r.t. x, we get

$$2b^2x + 2a^2y\frac{dy}{dx} = 0$$

or

$$b^2x + a^2y\frac{dy}{dx} = 0$$

or

$$y\frac{dy}{dx} = -\frac{b^2}{a^2}x$$

or

$$\frac{y}{x}\frac{dy}{dx} = -\frac{b^2}{a^2} \tag{95}$$

Now, differentiating both sides of Equation (95) w.r.t. x, we get

$$\frac{y}{x}\frac{d^2y}{dx^2} + \frac{dy}{dx}\frac{d}{dx}\left[\frac{y}{x}\right] = 0$$

or

$$\frac{y}{x}\frac{d^2y}{dx^2} + \frac{dy}{dx}\left[\frac{x\cdot\dfrac{dy}{dx} - y(1)}{x^2}\right] = 0$$

Multiplying both sides by x^2, we get

$$xy\frac{d^2y}{dx^2} + \frac{dy}{dx}\left[x\frac{dy}{dx} - y\right] = 0$$

$$\therefore xy\frac{d^2y}{dx^2} + x\left(\frac{dy}{dx}\right)^2 - y\frac{dy}{dx} = 0 \tag{96}$$

This is the required differential equation. Ans.

Note: *The differential equation (96) is of order two and degree* one. This differential equation has *variable coefficients*, and it is not easy to solve, *till we learn the methods needed for solving such differential equations.* Definitely, in higher classes, we will learn such methods for solving different types of differential equations, but still the fact remains that *we can solve only certain types of differential equations, not all.*

Example (28): Form the differential equation that will represent the *pair of lines.*

$$y = 2x + 3 \quad \text{and} \quad y = 4x + 5.$$

Solution: The equation of the given lines, are

$$y = 2x + 3 \tag{97}$$

and
$$y = 4x + 5 \tag{98}$$

Differentiating Equations (97) and (98) w.r.t. x, we get respectively

$$\frac{dy}{dx} = 2 \quad \text{or} \quad \frac{dy}{dx} - 2 = 0 \tag{99}$$

and
$$\frac{dy}{dx} = 4 \quad \text{or} \quad \frac{dy}{dx} - 4 = 0 \tag{100}$$

∴ The differential equation which will represent the lines having Equations (97) and (98) is given by the product,

$$\left(\frac{dy}{dx} - 2\right) \cdot \left(\frac{dy}{dx} - 4\right) = 0 \tag{101}$$

Equation (101) is the required differential equation. Ans.

Exercise

Q. (1) Obtain the differential equation of the family of circles touching the x-axis at the origin.[10]

$$\left(\text{Ans.} \quad \frac{dy}{dx} = \frac{2xy}{x^2 - y^2}\right)$$

Q. (2) Obtain the differential equation of all circles having their centers on y-axis and passing through the origin.

$$\left[\text{Ans.} \quad (x^2 - y^2)\frac{dy}{dx} - 2xy = 0\right]$$

---

[10] Note that the centers of all such circles must lie on the y-axis.

Q. (3) Obtain the differential equation of the family of rectangular hyperbolas.[11]

(i) $x^2 - y^2 = a^2$
(ii) $xy = c^2$

Ans.   (i) $x - y\dfrac{dy}{dx} = 0$

Ans.   (ii) $x\dfrac{dy}{dx} + y = 0$

9a.7   THE SIMPLEST TYPE OF DIFFERENTIAL EQUATIONS

We know that a differential equation is an equation connecting the *unknown function "y"*, its *derivatives* (or differentials) and *the independent variable "x"*. [To be clearer, a differential equation connects the *values of a (unknown) function*, its *derivatives* (or differentials), and its *argument* (i.e., the independent variable).]

Remark: In differential equations, the *"part of the unknown"* is played by *a function that satisfies the (given) differential equation*. We ask the question: *Why is solving the differential equation more difficult than solving any other equation?* The reason is as follows:

We can regard *the problem of integration of the function f(x) as a problem of solving a differential equation*$(dy/dx) = f(x)$, where *"y"* is the required function.

Note that, the above equation is the *simplest differential equation*. It is an *ordinary* differential equation of the *order one* and *degree one*.

In the simplest case, the solution of a differential equation *reduces to integration*. However, this is *not always possible*, since differential equations of *higher orders* and *higher degrees* cannot be reduced to the above form by applying basic algebraic operations. Therefore, solving a differential equation proves to be more complicated than integration.

The most general form of an ordinary differential equation of the first order and first degree is,

$$\frac{dy}{dx} = f(x, y) \tag{102}$$

or $M(x, y)dx + N(x, y)dx = 0$

$$\tag{103}$$

where M and N are functions of x and y

We have seen that a differential equation of the Equation form (102) can be represented by geometric means. It means that the slope of the graph of the required function *"y"* (i.e., the integral curve), which passes through the point $P(x_0, y_0)$ is known beforehand, and that it is $f(x_0, y_0)$.

Thus, the Equation (102) defines the so-called *direction field*, i.e., the direction of the integral curves. In other words, Equation (102) tells the *direction of the tangent lines to the integral curves*, at all points of the domain of the function $f(x, y)$. In the next chapter, we shall discuss different methods of solving differential equations of the first order and of first degree.

---

[11] Note that in this problem the equations of the curves are given. Thus, the name of the curve should not confuse the student.

Extending the Definition of Definite Integrals: Improper Integrals—Integrals with Infinite Limits

Up till now, when speaking of definite integrals we assumed that the interval of integration was finite and closed and that the integrand was continuous.

However, it often becomes necessary to extend the definition of definite integrals of functions defined over unbounded domains. These are the infinite semi-intervals of the forms $[a, +\infty]$ or infinite intervals of the form $(-\infty, \infty)$.

Definition: When one (or both) of the limits of integration is infinite or the integrand itself becomes infinite at (or between) the limits of integration, the integral is called an improper integral (see Figure 9a.1).

First, we consider the case of *infinite limits of integration.*

The question arises as to what *meaning is to be attached to a definite integral* of the form:

$$\int_a^{-\infty} f(x)dx, \quad \int_{-\infty}^{-b} f(x)dx, \quad \int_{-\infty}^{-\infty} f(x)dx,^{(12)}$$

Let function $f(x)$ be defined for all $x \geq a$ and continuous on every finite closed interval $a \leq x \leq b$, where a is a given number and b ($b \geq a$) is any arbitrary number. Then, $f(x)$ is *integrable* on $[a, b]$, and other similar interval.[13]

To make the notion of the definite integral applicable to unbounded intervals of integration, we consider the function,

$$I(b) = \int_a^b f(x)dx \text{ of the variable } b(b \geq a).$$

The integral varies with b is continuous function of b. Let us consider the behavior of the integral when $b \rightarrow +\infty$.

Definition: If there exists a finite limit as $b \rightarrow +\infty$, then this limit is called the improper integral of the function $f(x)$ on the interval $[a, +\infty)$ and is denoted by the symbol

$$\int_a^{-\infty} f(x)dx$$

Thus, by definition, we have,

$$\int_a^b f(x)dx = \lim_{x \to a} \int_a^b f(x)dx$$

In this case, we say that the improper integral $\int_a^{+\infty} f(x)dx$ converges (or exists). [if $\lim_{x \to a} \int_a^b f(x)dx$ is not a finite number, one says that $\int_a^{+\infty} f(x)dx$ does not exist (or that it)

---

[12] We encounter such integrals when computing the potential of gravitational or electrostatic forces.

[13] This is true due to the existence of theorem for the definite integral, which we restate here for convenience. If the function $f(x)$ is continuous in the finite-closed interval $[a, b]$, its integral sum tends to a definite limit as the length of the greatest subinterval tends to be zero (see Chapter 5).

diverges. In this case no numerical value is assigned to this interval]. We similarly define the improper integral of other infinite intervals:

$$\int_{-\infty}^{b} f(x)dx = \lim_{x \to a} \int_{a}^{b} f(x)dx \tag{104}$$

$$\int_{-\infty}^{+\infty} f(x)dx = \int_{-\infty}^{c} f(x)dx + \int_{c}^{+\infty} f(x)dx \tag{105}$$

Equation (105) should be understood as follows: if each of the improper integrals on the right exists, *then by definition the integral on the left also exists (or converges).*

Example (29): Evaluate the integral $\displaystyle\int_{0}^{+\infty} \frac{dx}{1 + x^2}$

$$\int_{0}^{+\infty} \frac{dx}{1 + x^2} = \lim_{b \to +\infty} \int_{0}^{b} \frac{dx}{1 + x^2}$$

$$= \lim_{b \to +\infty} [\tan^{-1} x]_{0}^{b}$$

$$= \lim_{b \to +\infty} [\tan^{-1} b - \tan^{-1} 0]$$

$$= \lim_{b \to +\infty} [\tan^{-1} b]$$

$$= \frac{\pi}{2} \qquad\qquad \text{Ans.}$$

9b Methods of Solving Ordinary Differential Equations of the First Order and of the First Degree

9b.1 INTRODUCTION

A differential equation is said to be *ordinary* if the *unknown function* depends *solely* on *one independent variable*.

The *simplest type* of differential equation is an *ordinary* differential equation of the *first order* and of the first *degree*. [A differential equation of the *first order* is an equation containing *derivatives* (or differentials) of *not higher than the first order*.]

The *most general form* of an ordinary differential equation of the *first order* and of *first degree* is

$$\frac{dy}{dx} = f(x, y) \text{ or } M(x, y)dx + N(x, y)dy = 0 \tag{1}$$

where M and N are functions of x and y.

9b.1.1 Solving a Differential Equation

Solving a differential equation means finding the *general solution* of the given differential equation.

Note: One might expect that solving differential equations of the form (1), should be quite simple. *Of course*, in some cases, it is so. However, *all differential equations of the first order and first degree cannot be solved.*

There is *no single method* of solving a differential equation of *any order* and *any degree*. Differential equations are *classified into certain standard forms*, and methods of solving them (in such standard forms) have been evolved.

Only those differential equations that belong to (or which can be reduced to) these standard forms can be solved by standard methods.

Differential equations and their solutions 9b-Ordinary differential equations of the first order and of first degree. (Methods of solving them and their applications)

Introduction to Integral Calculus: Systematic Studies with Engineering Applications for Beginners, First Edition. Ulrich L. Rohde, G. C. Jain, Ajay K. Poddar, and A. K. Ghosh. © 2012 John Wiley & Sons, Inc. Published 2012 by John Wiley & Sons, Inc.

9b.2   METHODS OF SOLVING DIFFERENTIAL EQUATIONS

The various *methods of solving differential equations* of the form (1) are classified as

(1) *Type (I)*: *Variable separable* or *reducible to variable separable* by *substitution*.

(2) *Type (II)*: *Linear* or *reducible* to the *linear form*.

(3) *Type (III)*: *Exact* (differential) *equation or equations reducible to the exact form* by *an integrating factor* (IF).

9b.2.1   Type (I): Variable Separable Form

In a differential equation of the form $dy/dx = f(x, y)$, if the function $f(x, y)$ *can be expressed as a product* $g(x) \cdot h(y)$, where, $g(x)$ is a function of x, and $h(y)$ is a function of y, then the given differential equation is said to be of *variable separable type*. Accordingly, we can write the given differential equation in the form:

$$\frac{dy}{dx} = h(y) \cdot g(x) \tag{2}$$

To solve this equation, it is first necessary to separate the variables. Accordingly, *rearranging terms*, we express Equation (2) in the form

$$\frac{1}{h(y)} \cdot dy = g(x) \cdot dx \tag{3}$$

At this stage, it is important to check that the function $h(y)$ does not assume the value zero [so that $(1/h(y))dy$ is defined]. Then, integrating both sides of Equation (3), we get

$$\int \frac{1}{h(y)} \cdot dy = \int g(x)dx \tag{4}$$

Thus, we get the solutions of the given differential equation in the form $H(y) = G(x) + c$, where $H(y)$ and $G(x)$ are the *antiderivatives* of $(1/h(y))$ and $g(x)$, respectively, and c is *an arbitrary constant*.

Further, note that if the differential Equation (1) is in the form

$$M(x, y)dx + N(x, y)dy = 0$$

Then, we have $N(x, y)dy = -M(x, y)dx$, therefore (neglecting the sign) we may write

$$\frac{dy}{dx} = \frac{M(x, y)}{N(x, y)}$$

For the above equation, *to be variable separable type*, it is necessary that both the functions $M(x, y)$ and $N(x, y)$ be capable of being expressed in the form of products $g_1(x) \cdot h_1(y)$ and $g_2(x) \cdot h_2(y)$, respectively, where $g_1(x)$, $g_2(x)$ are the functions of x only, and $h_1(y)$, $h_2(y)$ are the functions of y only.

In other words, if the given differential equation can be put in the form such that dx *and all the terms containing x are at one place*, and dy *along with all the terms containing y are on the other place*, then the variables are said to be separable. Once the given equation is expressed in the form $f(x)dx + g(y)dy = 0$, we can always write its solution by integrating the above equation. We write

$$\int f(x)dx + \int g(y)dy = c$$

[Note that, the problem now reduces to finding antiderivatives of $f(x)$ and $g(y)$.]

This is the *general solution* of the given differential equation, wherein c is an *arbitrary constant*.

Note: We know that the equation, $\dfrac{dy}{dx} = f(x)$ can always be expressed in the form $dy = f(x)\,dx$, and vice versa.

The fact, that we can deal with the symbols dy and dx like separate entities (exactly as if they were ordinary numbers) permits us to use them in many calculations and formal transformations, involving Calculus.

For instance, to solve the (differential) equation $(dy/dx) = f(x)$, we have to find a function $y[= F(x)]$ whose derivative is $f(x)$. *This is equivalent to saying that we have to find an antiderivative of $f(x)$.*

Recall that, if we consider the above equation in the form $dy = f(x)dx$, then the definition of differential dy also suggests the same thing (Chapter 16 of Part I).

Further, this fact can also be visualized if we express the equation, $dy = f(x)dx$, in the form of

$$dy\frac{dx}{dx} = f(x)dx \quad \text{or} \quad \frac{dy}{dx}dx = f(x)dx \quad \text{or} \quad \frac{dy}{dx} = f(x)$$

(by canceling "dx" from both sides).

Now, let us solve some examples of variable separable form.

Example (1): Find the general solution of the differential equation, $\dfrac{dy}{dx} = \dfrac{1+y^2}{1+x^2}$

Solution: Note that $1 + y^2 \neq 0$, and $1 + x^2 \neq 0$, therefore, separating the variables, we have

$$\frac{dy}{1+y^2} = \frac{dx}{1+x^2} \tag{5}$$

Integrating both sides of Equation (5), we get

$$\int \frac{dy}{1+y^2} = \int \frac{dx}{1+x^2}$$

or $\qquad\qquad\qquad \tan^{-1} y = \tan^{-1} x + c \quad$ **Ans.**

This is the general solution of the Equation (5).

Example (2): Find the general solution of the differential equation, $\dfrac{dy}{dx} = \dfrac{x+1}{2-y}$, $\quad (y \neq 2)$

Solution: We have

$$\frac{dy}{dx} = \frac{x+1}{2-y} \tag{6}$$

Separating the variables, we get

$$(2 - y)dy = (x + 1)dx \qquad (7)$$

Integrating both sides of Equation (7), we get

$$\int (2 - y)dy = \int (x + 1)dx$$

or $\quad 2y - \dfrac{y^2}{2} = \dfrac{x^2}{2} + x + k \quad$ or $\quad \dfrac{x^2}{2} + \dfrac{y^2}{2} + x - 2y + k = 0$

or $\quad x^2 + y^2 + 2x - 4y + 2k = 0$

or $\quad x^2 + y^2 + 2x - 4y + c = 0, \quad$ where $2k = c$

This is the *general solution* of Equation (6)   Ans.

Note: Recall that $\dfrac{f'(x)}{\sqrt{f(x)}} dx = 2\sqrt{f(x)}$. We shall make use of this result in the following solved examples.

Example (3): Find the general solution of the equation $x(1 + y^2)dx = y\, dy$

Solution: We have

$$x(1 + y^2)dx = y\, dy \qquad (8)$$

separating the variables, we get

$$x\, dx = \dfrac{y\, dy}{1 + y^2} \qquad (9)$$

Integrating both sides of Equation (9), we get

$$\int x\, dx = \int \dfrac{y\, dy}{1 + y^2}$$

or $\quad \dfrac{x^2}{2} = \dfrac{1}{2}\log_e(1 + y^2) + c_1$

where c_1 is an arbitrary constant

or $\quad \dfrac{x^2}{2} = \dfrac{1}{2}\log_e(1 + y^2) + \dfrac{1}{2}\log_e c \qquad$ Ans.

Example (4): Solve $\dfrac{y}{x} \cdot \dfrac{dy}{dx} = \sqrt{1 + x^2 + y^2 + x^2 y^2}$

Solution: We have

$$\dfrac{y}{x} \cdot \dfrac{dy}{dx} = \sqrt{1 + x^2 + y^2 + x^2 y^2} \qquad (10)$$

(In such problems, one may try to simplify the expression like the one on the right-hand side as done below.)

Consider the expression $1 + x^2 + y^2 + x^2y^2$

$$= (1 + x^2) + y^2(1 + x^2) = (1 + x^2)[1 + y^2]$$

$\therefore$ Equation (10) can be written in the form

$$\frac{y}{x} \cdot \frac{dy}{dx} = \sqrt{1 + x^2} \cdot \sqrt{1 + y^2}$$

Separating the variables, we get

$$\frac{y}{\sqrt{1 + y^2}} dy = x\sqrt{1 + x^2} \cdot dx$$

Integrating, we get

$$\therefore \quad \frac{1}{2}\int \frac{2y}{\sqrt{1 + y^2}} dy = \frac{1}{2}\int 2x\sqrt{1 + x^2}\, dx + c^{(1)}$$

$$\therefore \quad \frac{1}{2}\left[2\sqrt{1 + y^2}\right] = \frac{1}{2}\left[\frac{2}{3}(1 + x^2)^{3/2}\right] + c$$

$$\therefore \quad \sqrt{1 + y^2} = \frac{1}{3}\ (1 + x^2)^{3/2} + c$$

This is the *general solution*.   Ans.

Note: Since the *arbitrary constant* c_1 can take *any numerical value*, we have preferred to choose it as $(1/2)\log_e c$, instead of c_1, for convenience of further *transformation*. Now, we can write the general solution of the given equation, in the form $x^2 = \log_e[c(1 + y^2)]$, where "c" is an *arbitrary constant*.

Example (5):   Solve $\dfrac{dy}{dx} = \dfrac{x(2 \log x + 1)}{\sin y + y \cos y}$

Solution: The given equation is $\dfrac{dy}{dx} = \dfrac{x(2 \log x + 1)}{\sin y + y \cos y}$

Separating the variables, we get

$$(\sin y + y \cos y)dy = x(2 \log_e x + 1)dx \qquad (11)$$

Integrating both sides of Equation (11), we get

$$\int (\sin y + y \cos y)dy = \int x(2 \log_e x + 1)dx + c$$

---

[1] At this stage, we may use the result indicated in the note above. It must also be clear that the above result is obtained by applying the method of substitution, which could also be applied directly in solving such problems.

$$\therefore \quad -\cos y + y \sin y + \cos y = 2\left[(\log_e x)\cdot \frac{x^2}{2} - \int \frac{1}{x^2}\cdot \frac{x^2}{2}\cdot dx\right] + \frac{x^2}{2} + c^{(2)}$$

$$\therefore \quad y \sin y = x^2 \log_e x - \frac{x^2}{2} + \frac{x^2}{2} + c$$

or $$\qquad\qquad y \sin y = x^2 \log_e x + c$$

This is the general solution.   Ans.

Example (6): Find the particular solution of the equation $s \tan t\, dt + ds = 0$ satisfying the initial conditions $s = 4$ for $t = \dfrac{\pi}{3}$

Solution: The given differential equation is $s \tan t\, dt + ds = 0$
Separating the variables, we get

$$\tan t \quad dt + \frac{ds}{s} = 0 \tag{12}$$

Integrating both sides of Equation (12), we get

$$\int \tan t\, dt + \int \frac{ds}{s} = \log_e c \quad \text{or} \quad -\log_e \cos t + \log_e s = \log_e c$$

$$\text{or}\quad \log_e s = \log_e \cos t + \log_e c = \log_e c \cos t$$

or $$\qquad\qquad s = c \cos t \text{ (by taking antilog)}$$

where c is an *arbitrary constant.*
    This is the *general solution* of the given equation. In order that, the above solution should satisfy the given condition, we substitute the values $t = (\pi/3)$ and $s = 4$ into the general solution.We get

$$4 = c \cos\left(\frac{\pi}{3}\right) \quad \text{or} \quad 4 = c\cdot\frac{1}{2} \quad \therefore \quad c = 8$$

Consequently, the desired particular solution satisfying the given conditions has the form

$$s = 8 \cos t \qquad \text{Ans.}$$

9b.2.2   Equations Reducible to Variable Separable Form

9b.2.2.1   Method of Substitution   Some equations that are not in a variable separable form can be reduced to that form by using proper substitution. We consider some simple examples of such differential equations. The important point is to be able to identify easily, *the differential equations that can be reduced to variable separable form.*

---

[2] To achieve this result, we have applied the method of integration by parts. In practice, it is useful to remember the rule of integration by parts, and be able to apply it for the simple products of functions.

Example (7):   Solve $\dfrac{dy}{dx} = \cos(x + y)$

Solution: The given differential equation is

$$\frac{dy}{dx} = \cos(x + y) \tag{13}$$

Note that the differential Equation (13) is *not in variable separable form*. Also, observe that in the term $\cos(x + y)$, the expression $(x + y)$ is *linear*. In such cases, we put the *linear expression*,

$$x + y = v \tag{14}$$

$$\therefore \quad 1 + \frac{dy}{dx} = \frac{dv}{dx}$$

$$\therefore \quad \frac{dy}{dx} = \left(\frac{dv}{dx} - 1\right) \tag{15}$$

Using Equations (14) and (15) in the given Equation (13),
   We get

$$\frac{dv}{dx} - 1 = \cos v \quad \therefore \quad \frac{dv}{dx} = \cos v + 1$$

(Now, this equation is in the variable separable form.)
   We can simplify it further. We write,

$$\frac{dv}{dx} = \left(2\cos^2\frac{v}{2} - 1\right) + 1 = 2\cos^2\frac{v}{2}$$

$$\therefore \quad \frac{dv}{2\cos^2(v/2)} = dx \quad \text{or} \quad \frac{1}{2}\sec^2\frac{v}{2} \cdot dv = dx$$

Integrating, $\displaystyle\int \frac{1}{2}\sec^2\frac{v}{2} \cdot dv = \int dx$

$\therefore \quad \tan\dfrac{v}{2} = x + c$, where c is an *arbitrary constant*.

   This is the required *general solution*.   Ans.

Note: It is sufficient to put the constant of integration, at the end.

Example (8):   Solve $(2x - 2y + 3)dx - (x - y + 1)dy = 0$, given that $y = 1$ where $x = 0$

Solution: The given equation is $(2x - 2y + 3)dx - (x - y + 1)dy = 0$
(Observe that this equation *cannot be expressed in variable separable form*.)
   For convenience, we write this equation in the form,

$$(x - y + 1)dy = (2x - 2y + 3)dx$$

$$\therefore \quad \frac{dy}{dx} = \frac{2x - 2y + 3}{x - y + 1} = \frac{2(x - y) + 3}{(x - y) + 1} \tag{16}$$

[Observe that the expression $(x - y)$ in both, the numerator and the denominator, is a *linear expression*.]

We put

$$x - y = v \tag{17}$$

$$\therefore \quad 1 - \frac{dy}{dx} = \frac{dv}{dx}$$

$$\therefore \quad \frac{dy}{dx} = 1 - \frac{dv}{dx} \tag{18}$$

Using Equations (17) and (18), in Equation (16), we get

$$1 - \frac{dv}{dx} = \frac{2v + 3}{v + 1}$$

$$\therefore \quad \frac{dv}{dx} = 1 - \frac{2v + 3}{v + 1} = \frac{v + 1 - 2v - 3}{v + 1} = \frac{-v - 2}{v + 1}$$

$$\therefore \quad \frac{dv}{dx} = -\frac{v + 2}{v + 1}$$

(This is in the variable separable form)

$$\therefore \quad \frac{v + 1}{v + 2} dv = -dx$$

Integrating, both sides, we get

$$\int \frac{v + 1}{v + 2} dv = -\int dx$$

$$\therefore \quad \int \left[1 - \frac{1}{v + 2} \right] dv = -\int dx$$

$$\therefore \quad v - \log_e(v + 2) = -x + c$$

$$\therefore \quad x - y - \log_e(x - y + 2) + x = c$$

or
$$\therefore \quad 2x - y - \log_e(x - y + 2) = c$$

This is the *general solution* of Equation (16). We *have to find the particular solution that satisfies the given condition.* Thus, to determine the *particular solution*, we put $y = 1$ and $x = 0$ in the general solution. We get

$$\therefore \quad -1 - \log_e(0 - 1 + 2) = c \quad \therefore \quad -1 = c$$

Therefore, the *required particular solution is*

$$\therefore \quad 2x - y - \log_e(x - y + 2) + 1 = 0 \qquad \text{Ans.}$$

(Once the method of solving a differential equation is learnt, we can easily drop many steps from the above solution.)

Example (9): Solve $(4x + y)^2 = \dfrac{dy}{dx}$

Solution: The given equation is

$$\frac{dy}{dx} = (4x + y)^2 \tag{19}$$

The given Equation (19) *cannot be expressed* in variable separable form.

Also, observe that the term on the right-hand side involves an expression $(4x + y)$, which is *linear*. We put

$$4x + y = v \tag{20}$$

$$\therefore \quad 4 + \frac{dy}{dx} = \frac{dv}{dx}$$

$$\therefore \quad \frac{dy}{dx} = \frac{dv}{dx} - 4 \tag{21}$$

Using Equations (20) and (21) in Equation (19), we get

$$\frac{dv}{dx} - 4 = v^2$$

or

$$\frac{dv}{dx} = v^2 + 4$$

(Now, this equation is in the variable separable form.) We write,

$$\frac{dv}{v^2 + 2^2} = dx \quad \text{or} \quad dx = \frac{dv}{v^2 + 2^2}$$

On integrating, we get

$$x = \frac{1}{2}\tan^{-1}\left(\frac{v}{2}\right) + k \quad \text{or} \quad \therefore \quad 2x = \tan^{-1}\left(\frac{4x + y}{2}\right) = c,$$

where c is an *arbitrary constant*.

Note: In Example (8) above, we have solved the differential equation of the form $\therefore \quad \dfrac{dy}{dx} = \dfrac{a_1 x + b_1 y + c_1}{a_2 x + b_2 y + c_2}$, where $\dfrac{a_1}{a_2} = \dfrac{b_1}{b_2}$ using substitution. A natural question arises: *Can we solve similar looking differential equations of the form*:

$$\therefore \quad \frac{dy}{dx} = \frac{a_1 x + b_1 y + c_1}{a_2 x + b_2 y + c_2} \quad \text{in which} \quad \frac{a_1}{a_2} \neq \frac{b_1}{b_2}?$$

The answer is *yes*. (We shall discuss the method of solving such equations shortly, under the heading: *Equations reducible to homogenous form*. Again, such equations are finally reduced to the *variable separable form* as we will see.)

9b.2.2.2 Homogeneous Differential Equations in x and y

First, it is useful to understand clearly *the meaning of a homogeneous function of degree "n" in x and y.*

Definition: A function $f(x, y)$ is called a *homogeneous* function of degree "*n*" in *x*, *y*, if *each term* in $f(x, y)$ is of the *same degree "n"*.

Definition: A differential equation of the form $\frac{dy}{dx} = \frac{f(x,y)}{g(x,y)}$, where $f(x, y)$ and $g(x, y)$ are

homogeneous functions of the *same degree*, is called a *homogeneous differential equation* in *x* and *y*. Remember that the degree of a homogeneous function is a whole number "*n*" (i.e., $n = 0$, 1, 2, 3, ...).

Note: In a homogeneous function $F(x, y)$ of degree "*n*", if we replace *x* by *kx*, and *y* by *ky*, where *k* is a *nonzero constant*, then we get

$$F(kx, ky) = k^n F(x, y) \quad \text{for any } n \in w$$

This observation suggests a method of defining a homogeneous function of degree "*n*". Let us discuss.

Consider the following functions in *x* and *y*, and let us find their *degree*.

(1) $F_1(x, y) = x^3 + 2xy^2 + x^2 y$

$\therefore \quad F_1(kx, ky) = k^3 x^3 + 2(kx)(k^2 y^2) + (k^2 x^2)(ky)$

$\qquad\qquad = k^3 x^3 + k^3(2xy^2) + k^3(x^2 y)$

$\qquad\qquad = k^3[x^3 + 2xy^2 + x^2 y] = k^3 F_1(x, y)$

Thus, the degree of $F_1(x, y)$ is 3.

(2) $F_2(x, y) = \sin\left(\dfrac{x}{y}\right)$

$\qquad F_2(kx, ky) = \sin\left(\dfrac{kx}{ky}\right) = \sin\left(\dfrac{x}{y}\right) = 1 \sin\left(\dfrac{x}{y}\right) = k^0 F_2(x, y)$

Thus, the degree of $F_2(x, y)$ is "0".

(3) $F_3(x, y) = x^2 + 3xy$

$\therefore \quad F_3(kx, ky) = k^2 x^2 + 3kx \cdot ky$

$\qquad\qquad = k^2 x^2 + 3k^2 x \cdot y$

$\qquad\qquad = k^2[x^2 + 3xy]$

$\qquad\qquad = k^2 F_3(x, y)$

Therefore, the degree of $F_3(x, y)$ is 2.

(4) $F_4(x, y) = \sin x + \cos x$

$\therefore$ $F_4(kx, ky) = \sin(kx) + \cos(ky)$

$\neq k^n F_4(x, y)$,   for any $n \in W$

Note that here, it is *not possible* to define the degree of $F_4(x, y)$.

Accordingly, we say that $F_4(x, y)$ is *not a homogeneous function.*

[Compare $F_4(x, y)$ with $F_2(x, y)$, and get convinced why the degree of $F_4(x, y)$ cannot be defined.]

Definition: A function $F(x, y)$ is said to be homogeneous function of degree "n", if $F(kx, ky) = k^n F(x, y)$, where k is *a nonzero constant.*

Note (1): A differential equation that involves a homogeneous function (*of any degree*) is said to be a *homogeneous differential equation.*

A differential equation in the form $\frac{dy}{dx} = g\left(\frac{y}{x}\right)$ involves a *homogeneous function of degree* "0", Hence, it is a *homogeneous differential equation* of degree zero. Now, it must be clear that the function,

$$F(x, y) = 2x^3 - 5xy^2 + 3x^2 y + \frac{x^4}{y} \sin \frac{y}{x}$$

is a *homogeneous function* of degree 3. [Note that in the expression on right-hand side, the degree of the *component function*$\sin(y/x)$ is *zero*].

It will be observed that the method of solving a *homogeneous differential equation* does not depend on *the degree of the homogeneous function involved.* In other words, the method of solving all homogeneous differential equations (of order 1 and degree 1) is the same.

To Illustrate the Method for Solving Homogeneous Differential Equation of Order 1 in x and y: A homogeneous differential equation (in x and y) can be written in the form:

$$\frac{dy}{dx} = \frac{f(x, y)}{g(x, y)} \tag{I}$$

where $f(x, y)$ and $g(x, y)$ are *homogeneous expressions* of the *same degree,* (say "r")

Then, $f(x, y) = x^r f_1\left(\frac{y}{x}\right)$ and $f_2\left(\frac{y}{x}\right) = x^r f_2\left(\frac{y}{x}\right)$, hence the equation (I) becomes

$$\frac{dy}{dx} = \frac{f_1(y/x)}{f_2(y/x)} = h\left(\frac{y}{x}\right) \tag{II}$$

[The homogeneous functions $f_1(y/x)$ and $f_2(y/x)$, each being of *degree zero,* can be combined and jointly viewed in the form $h(y/x)$ that is a function of *degree zero.*]

In fact, a homogeneous differential equation of the form (I) can be always expressed in the form (II).[3]

---

[3] Consider the differential equation $\dfrac{dy}{dx} = \dfrac{x + 2y}{x - y}$. Let $f_1(x, y) = x + 2y$ and $f_2(x, y) = x - y$. Then, right-hand side of the above equation is $\dfrac{f_1(x, y)}{f_2(x, y)}$. Thus, we can write $\dfrac{dy}{dx} = \dfrac{x + 2y}{x - y} = \dfrac{x(1 + 2(y/x))}{x(1 - 2(y/x))} = \dfrac{(1 + 2(y/x))}{(1 - (y/x))} = g\left(\dfrac{y}{x}\right)$ which is a homogeneous function of degree zero.

Now, if we substitute $\frac{y}{x} = v$ [i.e., $y = vx$] in equation (II), we get the right-hand side as $\frac{f_1(v)}{f_2(v)}$, and the left-hand side is obtained by differentiating $y = vx$.

We get $\frac{dy}{dx} = v + x\frac{dv}{dx}$ (since v is a function of x and y)
Thus, equation (II) becomes

$$v + x\frac{dv}{dx} = \frac{f_1(v)}{f_2(v)} \tag{III}$$

Note that right-hand side of (III) is only a function v[say$F(v)$], so that

$$v + x\frac{dv}{dx} = F(v)$$

Now, it is clear that *we can easily separate the variables.* We get

$$\frac{dv}{f(v) - v} - \frac{dx}{x} = 0 \tag{IV}$$

On integrating, we will get the solution in terms of v and x. Finally, by substituting (back) $v = (y/x)$, we get the required solution.

Thus, *we conclude* that a homogeneous differential equation, of order 1, *can be converted* to the *variable separable form*, which can be solved by integration (i.e., by finding the antiderivatives) of the functions involved.

Note: To convert a given *homogeneous differential equation* of order 1 to the *variable separable form*, a *convenient substitution* is chosen as follows:

- If the given equation is in the form $(dy/dx) = F(x, y)$, where $F(x, y)$ is a *homogeneous function* of any nonzero degree "n" ($n = 1, 2, 3, \ldots$), then we make the substitution $(y/x) = v$, that is, $y = vx$. Similarly, if the differential equation is in the form $(dy/dx) = F(x, y)$, then we make the substitution $(x/y) = v$, that is, $x = vy$.
- If the equation is in the form $(dy/dx) = F(x, y)$ or $f(x, y)dx + g(x, y)dy = 0$, wherein a term involving y/x appears, then we choose the substitution $(y/x) = v$, that is, $y = vx$. On the other hand, if the term x/y occurs then we make the substitution $(x/y) = v$, that is, $x = vy$. *These substitutions make finding the solution convenient.*

Now consider the following examples:

Example (10): Solve the differential equation

$$xy^2\,dy - (y^3 - 2x^3)dx = 0$$

Solution: From the given equation, we write

$$xy^2\,dy = (y^3 - 2x^3)dx$$

or

$$\frac{dy}{dx} = \frac{(y^3 - 2x^3)}{xy^2} \tag{22}$$

(This is a homogeneous differential equation.)

$$\text{Put } \frac{y}{x} = v \quad \text{or} \quad y = vx \tag{23}$$

Using Equation (23) in Equation (22), we get

$$\text{LHS} = v + x\frac{dv}{dx}$$

and

$$\text{RHS} = \frac{v^3 x^3 - 2x^3}{xv^2 x^2} = \frac{x^3(v^3 - 2)}{x^3(v^2)}$$

$\therefore$ Equation (22) changes to

$$v + x\frac{dv}{dx} = \frac{v^3 - 2}{v^2}$$

$$\therefore \quad x\frac{dv}{dx} = \frac{v^3 - 2}{v^2} - v = \frac{v^3 - 2 - v^3}{v^2} = \frac{-2}{v^2}$$

or

$$x\frac{dv}{dx} = \frac{-2}{v^2} \text{ (Now variables are separable)}$$

$$\therefore \quad v^2\, dv = -2\frac{dx}{x}$$

Integrating, we get

$$\int v^2\, dv = -2\int \frac{dx}{x}$$

$$\therefore \quad \frac{v^2}{3} = -2\log_e|x| + c$$

or

$$\frac{v^2}{3} + 2\log_e|x| = c$$

where "c" is an arbitrary constant. This is the required general solution.

Example (11):   Solve the differential equation $(x + y)dx - (x - y)dy = 0$

Also find the particular solution when $y(1) = 0$.
Solution: The given differential equation is

$$(x + y)dx - (x - y)dy = 0$$

or

$$\frac{dy}{dx} = \frac{x + y}{x - y} \tag{24}$$

(This is a homogeneous differential equation.)

$$\text{Put } \frac{y}{x} = v \quad \text{or} \quad y = vx \tag{25}$$

From Equation (25), we get

$$\frac{dy}{dx} = v + x\frac{dv}{dx} \tag{26}$$

Using Equations (25) and (26) in Equation (24), we get

$$v + x\frac{dv}{dx} = \frac{x + vx}{x - vx} = \frac{x(1 + v)}{x(1 - v)}$$

$$\therefore \quad v + x\frac{dv}{dx} = \frac{(1 + v)}{(1 - v)}$$

$$\therefore \quad x\frac{dv}{dx} = \frac{(1 + v)}{(1 - v)} - v = \frac{1 + v - v + v^2}{1 - v} = \frac{1 + v^2}{1 - v}$$

$$\therefore \quad x\frac{dv}{dx} = \frac{1 + v^2}{1 - v} \quad \text{(Now variables are separable)}$$

$$\frac{1 - v}{1 + v^2}dv = \frac{dx}{x}$$

$$\therefore \quad \frac{1}{1 + v^2}dv - \frac{v}{1 + v^2}dv = \frac{dx}{x}$$

Integrating both sides, we get

$$\therefore \quad \int \frac{dv}{1 + v^2} - \int \frac{v\,dv}{1 + v^2} = \int \frac{dx}{x}$$

$$\therefore \quad \tan^{-1} v - \frac{1}{2}\log_e|1 + v^2| = \log_e|x| + c$$

Substituting the value of $v\left(= \dfrac{y}{x}\right)$, we get

$$\therefore \quad \tan^{-1}\left(\frac{y}{x}\right) - \frac{1}{2}\log_e\left|\frac{x^2 + y^2}{x^2}\right| = \log_e|x| + c$$

or

$$\therefore \quad \tan^{-1}\left(\frac{y}{x}\right) - \frac{1}{2}\log_e|x^2 + y^2| - [-\log_e|x|] = \log_e|x| + c$$

$$\tag{27}$$

or

$$\therefore \quad \tan^{-1}\left(\frac{y}{x}\right) - \log_e\sqrt{x^2 + y^2} + \log_e|x| = \log_e|x| + c$$

or

$$\therefore \quad \tan^{-1}\left(\frac{y}{x}\right) - \log_e\sqrt{x^2 + y^2} = c$$

This is the required general solution. To obtain the particular solution, we use $y(1) = 0$ [i.e., $x = 1$, $y = 0$] in the Equation (27), we get

$$\tan^{-1}\left(\frac{0}{1}\right) = \log_e \sqrt{1 + 0} + c$$

$$\therefore \quad 0 = \log_e(1) + c$$

$$\therefore \quad c = 0$$

Putting this value of "c" in Equation (27), we get the particular solution as

$$\therefore \quad \tan^{-1}\left(\frac{y}{x}\right) = \log\sqrt{x^2 + y^2} \qquad \text{Ans.}$$

Example (12): Solve $\dfrac{dy}{dx} = \dfrac{y}{x - \sqrt{xy}}$

Solution: The given differential equation is

$$\frac{dy}{dx} = \frac{y}{x - \sqrt{xy}} \tag{28)$^{(4)}$}$$

$$\text{Put } \frac{y}{x} = v, \quad \text{i.e., } y = vx \tag{29}$$

$$\therefore \quad s\frac{dy}{dx} = v + x\frac{dv}{dx} \tag{30}$$

Using Equations (29) and (30) in Equation (28), we get

$$v + x\frac{dv}{dx} = \frac{vx}{x - \sqrt{x \cdot vx}} = \frac{vx}{x(1 - \sqrt{v})} = \frac{v}{(1 - \sqrt{v})}$$

$$x\frac{dv}{dx} = \frac{v}{(1 - \sqrt{v})} - v = \frac{v - v + v\sqrt{v}}{1 - \sqrt{v}}$$

$$x\frac{dv}{dx} = \frac{v\sqrt{v}}{(1 - \sqrt{v})} \quad \text{(This is variable separable form)}$$

$$\therefore \quad \frac{1 - \sqrt{v}}{(v\sqrt{v})}dv = \frac{dx}{x}$$

or $\qquad v^{-3/2}\,dv - v^{-1}\,dv = \dfrac{dx}{x}$

$^{(4)}$ The student must convince himself that Equation (28) is a homogeneous differential equation.

Integrating, we get

$$\frac{v^{-1/2}}{-1/2} - \log_e|v| = \log_e|x| + c$$

or

$$-2\sqrt{v} = \log_e|x| + \log_e|v| + c$$

$$\therefore \quad \log_e|vx| + 2\sqrt{v} + c = 0$$

$$\therefore \quad \log_e|y| + 2\sqrt{\frac{y}{x}} + c = 0 \left[\because \quad v = \frac{y}{x}\right]$$

This is the general solution.   Ans.

Example (13): $\left(x + y\cot\frac{x}{y}\right)dy - y\,dx = 0$

Solution: Presence of $\frac{x}{y}$ indicates that we use the substitution $\frac{x}{y} = v$, that is, $x = vy$. The given equation can be written in the form,

$$x + y\cot\frac{x}{y} - y\frac{dx}{dy} = 0 \tag{31}$$

$$\text{Put } x = vy \tag{32}$$

$$\therefore \quad \frac{dx}{dy} = v + y\frac{dv}{dy} \tag{33}$$

Using Equations (32) and (33) in Equation (31), we get

$$vy + y\cot v - y\left(v + y\frac{dv}{dy}\right) = 0$$

$$\therefore \quad vy - vy + y\cot v - y^2\frac{dv}{dy} = 0$$

or

$$y\cot v - y^2\frac{dv}{dy} = 0$$

or

$$\cot v = y\frac{dv}{dy} \quad \text{(This is variable separable form)}$$

$$\therefore \quad \frac{dy}{y} = \frac{dv}{\cot v} = \tan v\,dv$$

Integrating both sides, we get

$$\log_e|y| = \log_e|\sec v| + \log|c| = \log_e|\sec v + c|$$

$\therefore$ $y = c \sec v$ (Taking antilogarithms on both sides)

or $\therefore$ $y = c \sec \frac{x}{y}$, where c is an *arbitrary constant*.

This is the general solution.   Ans.

Example (14): Solve the differential equation

$$\left(1 + e^{x/y}\right)dx + \left(1 - \frac{x}{y}\right)e^{x/y}\,dy = 0 \tag{34}$$

Solution: The given equation is

$$\left(1 + e^{x/y}\right)dx + \left(1 - \frac{x}{y}\right)e^{x/y}\,dy = 0$$

Presence of the expression x/y suggests that the proper substitution will be, $x/y = v$, that is, $x = vy$, and that we must write the Equation (34) in the form,

$$\frac{dx}{dy} + \frac{(1 - (x/y))e^{x/y}}{1 + e^{x/y}} = 0 \tag{35}$$

To obtain the expression for dx/dy, we differentiate the relation, $x = vy$.
  We get

$$\frac{dx}{dy} = v + y\frac{dx}{dy}$$

Using the above result, and the substitution $((x/y) = v)$ in Equation (35), we get

$$v + y\frac{dv}{dy} + \frac{(1 - v)e^v}{1 + e^v} = 0$$

or

$$y\frac{dv}{dy} + \frac{(1 - v)e^v}{1 + e^v} + v = 0$$

or

$$y\frac{dv}{dy} + \frac{e^v - ve^v + v + ve^v}{1 + e^v} = 0$$

or

$$y\frac{dv}{dy} + \frac{e^v + v}{1 + e^v} = 0$$

or

$$\frac{e^v + v}{1 + e^v} = -y\frac{dv}{dy}$$

$$\frac{dy}{y} = -\frac{1 + e^v}{e^v + v}dv$$

$$\therefore \quad \frac{dy}{y} + \frac{1 + e^v}{e^v + v}dv = 0$$

Integrating both sides, we get

$$\log_e |y| + \log_e |e^v + v| = \log_e |c| \left[\because \frac{1 + e^v}{v + e^v} = \frac{f'(v)}{f(v)} \right]$$

where c is an arbitrary constant.

$$\therefore \quad \log_e |y(e^v + v)| = \log_e |c|$$

Taking antilog both sides, we get

$$\therefore \quad y(e^v + v) = c$$

This is the general solution of the given differential equation.   Ans.

Example (15): Solve $(1 + 2 e^{x/y}) dx + 2 e^{x/y} \left(1 - \frac{x}{y} \right) dy = 0$

Solution: The given differential equation is

$$(1 + 2 e^{x/y}) dx + 2 e^{x/y} \left(1 - \frac{x}{y} \right) dy = 0 \tag{36}$$

This equation is homogenous differential equation of the degree zero. The presence of x/y suggests that we use the substitution $x/y = v$, that is,

$$x = vy \tag{37}$$

Differentiating w.r.t. y, we get

$$\frac{dx}{dy} = v + y \frac{dx}{dy} \tag{38}$$

The given differential equation can be written in the form,

$$(1 + 2 e^{x/y}) \frac{dx}{dy} + 2 e^{x/y} \left(1 - \frac{x}{y} \right) = 0$$

or $\dfrac{dx}{dy} = \dfrac{2 e^{x/y} ((x/y) - 1)}{1 + 2 e^{x/y}}$ (Negative sign is absorbed in the numerator)

$$v + y \frac{dv}{dy} = \frac{2 e^v (v - 1)}{1 + 2 e^v}$$

$$\therefore \quad y \frac{dv}{dy} = \frac{2 e^v (v - 1)}{1 + 2 e^v} - v$$

$$= \frac{2v e^v - 2 e^v - v - 2v e^v}{1 + 2 e^v}$$

$$y\frac{dv}{dy} = \frac{-(2\,e^v + v)}{(2\,e^v + 1)} \quad \text{(This is variable separable form)}$$

$$\therefore \quad \frac{(2\,e^v + v)}{(2\,e^v + 1)} = -\frac{dy}{y}$$

[Now observe that $\frac{d}{dv}[2\,e^v + v] = (2e^v + 1)$]

Therefore, on integrating, we get

$$\log|2\,e^v + v| = -\log|v| + \log c$$

or
$$\log|2\,e^v + v| + \log|v| = \log c$$

or
$$\log|y(2\,e^v + v)| = \log c$$

Taking antilog, we get

$$y(2\,e^v + v) = c$$

or
$$y\left(2\,e^{x/y} + \frac{x}{y}\right) = c$$

or
$$2y\,e^{x/y} + x = c$$

which is the required general solution.   Ans.

Note: The above problem can also be solved using the same substitution (i.e., $x = vy$), but differentiating it with respect to x, rather than by v.

Example (16): Solve $\left(1 + 2\,e^{x/y}\right)dx + 2\,e^{x/y}\left(1 - \frac{x}{y}\right)dx = 0$

Solution: Given differential equation is

$$\left(1 + 2\,e^{x/y}\right)dx + 2\,e^{x/y}\left(1 - \frac{x}{y}\right)dx = 0$$

This equation is *homogeneous equation* of degree zero. (Presence of an expression x/y in the equation suggests that we use the substitution $x/y = v$).

$$\text{i.e.,} \quad x = vy \tag{39}$$

The important point is that we still differentiate Equation (39) w.r.t. x, and obtain from Equation (39),

$$1 = v\frac{dy}{dx} + y\frac{dv}{dx} \quad \therefore \quad v\frac{dy}{dx} = 1 - y\frac{dv}{dx}$$

$$\text{Note that} \quad \therefore \quad dx = v\,dy + y\,dv \tag{40}$$

(We have treated the differentials like algebraic quantities.)

Using Equations (39) and (40), we get

$$(1 + 2\,e^v)(v\,dy + y\,dv) + 2\,e^v(1 - v)dy = 0$$

$$\therefore \quad (1 + 2\,e^v)v\,dy + (1 + 2\,e^v)y\,dv + 2\,e^v(1 - v)dy = 0$$

$$\therefore \quad [(1 + 2\,e^v)v + 2\,e^v(1 - v)]dy + (1 + 2\,e^v)y\,dv = 0$$

$$\therefore \quad [1 + 2v\,e^v + 2\,e^v - 2v\,e^v]dy + (1 + 2\,e^v)y\,dv = 0$$

$$\therefore \quad \frac{dy}{y} = -\frac{1 + 2\,e^v}{v + 2\,e^v}dv$$

On integrating, we get

$$\log_e|y| = -\log_e|v + 2\,e^v| + c_1$$

where c_1 is the constant of integration.

(We may choose $c_1 = \log_e c$, where c is a constant.)

But, we have used the substitution $v = (x/y)$. Hence, substituting it back for v, we get

$$\log_e|y| + \log_e\left|\frac{x}{y} + 2\,e^{x/y}\right| = \log_e c$$

or

$$\log_e\left|y\left(\frac{x}{y} + 2\,e^{x/y}\right)\right| = \log_e c$$

or

$$\log_e\left|x + 2y\,e^{x/y}\right| = \log_e c$$

Taking antilog, we get

$$x + 2y\,e^{x/y} = c$$

where c is an *arbitrary constant*.

This is the required general solution.   Ans.

9b.2.2.3 Nonhomogeneous Differential Equations (of Order 1 and Degree 1) Reducible to Homogeneous Form

A nonhomogeneous linear equation (in question) in x and y can be written in the form

$$\frac{dy}{dx} = \frac{a_1 x + b_1 y + c_1}{a_2 x + b_2 y + c_2}$$

Two cases exist:

Case (1): When $\frac{a_1}{a_2} = \frac{b_1}{b_2}$

We have already solved such an equation earlier [see Example (11) in Section 9b.2.2.1]. We have seen that the method of solving such differential equations (using substitution) is quite simple.

Case (2): When $\dfrac{a_1}{a_2} \neq \dfrac{b_1}{b_2}$

In this case, we can convert the given differential equation to the *homogeneous differential equation,* by the following transformation equations:

$$\text{Put } x = X + h \quad \text{and} \quad y = Y + k$$

where h and k are *constants.*

$$\therefore \quad dx = dX \quad \text{and} \quad dy = dY$$

Accordingly, *the given equation*

i.e., $\dfrac{dy}{dx} = \dfrac{a_1 x + b_1 y + c_1}{a_2 x + b_2 y + c_2}$ changes to $\dfrac{dY}{dX} = \dfrac{a_1(X+h) + b_1(Y+k) + c_1}{a_2(X+h) + b_2(Y+k) + c_2}$

or

$$\frac{dY}{dX} = \frac{a_1 X + b_1 Y + (a_1 h + b_1 k + c_1)}{a_2 X + b_2 Y + (a_2 h + b_2 k + c_2)} \tag{41}$$

If, we now choose constants h *and* k so *that*

$$(a_1 h + b_1 k + c_1) = 0 \quad \text{and} \quad (a_2 h + b_2 k + c_2) = 0$$

Then, the given equation changes to,

$$\frac{dY}{dX} = \frac{a_1 X + b_1 Y}{a_2 X + b_2 Y} \tag{42}^{(5)}$$

Observe that, the above equation is a *homogeneous differential equation*, and hence, it can be solved by the substitution

$$\frac{Y}{X} = v, \text{ i.e., } Y = vX$$

If the solution of the Equation (42) be

$$f(X, Y) = c$$

then the *solution of the original equation* is $f(x - h, y - k) = c$.

Now, let us solve some examples.

---

(5) In practice, finding the value(s) of h and k is very simple. However, one may also use the results $h = \dfrac{b_1 c_2 - b_2 c_1}{a_1 b_2 - a_2 b_1}$, $k = \dfrac{c_1 a_2 - c_2 a_1}{a_1 b_2 - a_2 b_1}$, where $(a_1 b_2 - a_2 b) \neq 0$. These expressions for the values of h and k follow from the method of solving a pair of linear equations (in two variables), simultaneously.

Example (17): Solve $(3x + 2x + 4)dx - (4x + 6y + 5)dy = 0$

Solution: The equation is

$$(4x + 6y + 5)dy = (3x + 2x + 4)dx$$

$$\therefore \quad \frac{dy}{dx} = \frac{2x + 3y + 4}{4x + 6y + 5} = \frac{(2x + 3y) + 4}{2(2x + 3y) + 5} \tag{43}$$

Here $\frac{a_1}{a_2} = \frac{b_1}{b_2}$, hence the method for solving this equation is very simple.

Let $2x + 3y = u$

Differentiating both sides w.r.t. x, we get

$$2 + 3\frac{dy}{dx} = \frac{du}{dx}$$

$$\therefore \quad \frac{dy}{dx} = \frac{1}{3}\left(\frac{du}{dx} - 2\right)$$

$\therefore$ The given Equation (43) becomes

$$\therefore \quad \frac{1}{3}\left(\frac{du}{dx} - 2\right) = \frac{u + 4}{2u + 5}$$

$$\therefore \quad \frac{du}{dx} - 2 = \frac{3u + 12}{2u + 5}$$

$$\therefore \quad \frac{du}{dx} = \frac{3u + 12}{2u + 5} + 2 = \frac{3u + 12 + 4u + 10}{2u + 5}$$

$$\therefore \quad \frac{du}{dx} = \frac{7u + 22}{2u + 5} \text{ (This is in \textit{variable separable form})}$$

$$\therefore \quad \frac{2u + 5}{7u + 22}du = dx$$

Consider, $\dfrac{2u + 5}{7u + 22}$

$$= \frac{1}{7}\left[\frac{14u + 35}{7u + 22}\right] = \frac{1}{7}\left[\frac{14u + 44 - 9}{(7u + 22)}\right]$$

$$= \frac{1}{7}\left[\frac{2(7u + 22) - 9}{(7u + 22)}\right]$$

$$= \frac{2}{7} - \frac{9}{7} \cdot \frac{1}{7u + 22} \text{ (This is \textit{variable separable} form)}$$

$$= \left[\frac{2}{7} - \frac{9}{7} \cdot \frac{1}{7u + 22}\right]du = dx$$

Integrating both sides, we get

$$\int \frac{2}{7} du - \frac{9}{7} \cdot \int \frac{1}{7u + 22} du = \int dx$$

$$\therefore \quad \frac{2}{7} u - \frac{9}{7} \cdot \left[\frac{1}{7} \log_e |7u + 22| \right] = x$$

Now substitute the value of $u \, (= 2x + 3y)$, we get

$$\frac{2}{7} (2x + 3y) - \frac{9}{49} \cdot \log_e |7(2x + 3y) + 22| = x$$

$$\therefore \quad \frac{2}{7} (2x + 3y) - \frac{9}{49} \cdot \log_e |14x + 21y + 22| = x$$

Multiplying both sides by 49, we get

$$28x + 42y - 9 \cdot \log_e |14x + 21y + 22| - 7x = 0$$

$$\text{or} \quad 21x + 42y - 9 \cdot \log_e |14x + 21y + 22| = 0$$

$$\text{or} \quad 21(x + 2y) - 9 \cdot \log_e |14x + 21y + 22| = 0$$

Dividing both sides by 3, we get

$$7(x - 2y) - 3 \cdot \log_e |14x + 21y + 22| = c$$

This is the *required general solution* where c is an arbitrary constant.   Ans.

Example (18): Solve $(3y - 7x + 7)dx + (7y - 3x - +3)dy = 0$

Solution: The given equation is

$$(7y - 3x - +3)dy = -(3y - 7x + 7)dx$$

or
$$\frac{dy}{dx} = -\frac{3y - 7x + 7}{7y - 3x - +3}$$

or
$$\frac{dy}{dx} = -\frac{7x - 3y - 7}{-3x + 7y + 3} \tag{44}$$

Here $\frac{a_1}{a_2} \neq \frac{b_1}{b_2}$ since $\frac{7}{-3} \neq \frac{-3}{7}$

Put $x = X + h$ and $y = Y + k$, where h and k are *constants to be determined*.
From the above transformation equations, we get

$$dx = dX \quad \text{and} \quad dy = dY$$

$$\therefore \quad \frac{dy}{dx} = \frac{dY}{dX} \text{ (Recall that, this is the property of differentials of order 1)}$$

$\therefore$ Equation (44) becomes

$$\frac{dY}{dX} = \frac{7(X+h) - 3(Y+k) - 7}{-3(X+h) + 7(Y+k) + 3}$$

$$= \frac{7X - 3Y + (7h - 3k - 7)}{-3X + 7Y + (-3h + 7k + 3)}$$

We choose h and k such that

$$7h - 3k - 7 = 0$$

$$-3h + 7k + 3 = 0$$

Solving these equations simultaneously, we have

$$h = 1, k = 0^{(6)}$$

The given equation reduces to

$$\frac{dY}{dX} = \frac{7X - 3Y}{-3X + 7Y} \tag{45}$$

This is a homogeneous differential equation in X and Y.
   We put

$$\frac{Y}{X} = v, \quad \text{i.e., } Y = v \cdot X$$

Equation (45) changes to

$$v + X\frac{dv}{dX} = \frac{7X - 3vX}{-3X + 7vX} = \frac{X(7 - 3v)}{X(-3 + 7v)}$$

$$\therefore \quad X\frac{dv}{dX} = \frac{7 - 3v}{-3 + 7v} - v = \frac{7 - 3v - v(-3 + 7v)}{-3 + 7v}$$

$$= \frac{7 - 3v + 3v - 7v^2}{-3 + 7v} = \frac{7(1 - v^2)}{7v - 3} \text{ (This is in } \textit{variable separable} \text{ form)}$$

Separating the variables,

$$\therefore \quad \frac{7v - 3}{1 - v^2}dv = 7\frac{dX}{X}$$

$$\therefore \quad \frac{7v - 3}{(1 - v)(1 + v)}dv = 7\frac{dX}{X}$$

$^{(6)}$ The procedure for obtaining these values is explained in detail at the end of this problem.

By partial fractions, we have

$$\left[\frac{2}{1-v} - \frac{5}{1+v}\right] dv = 7\frac{dX}{X}$$

Integrating both sides, we get

$$-2\log(1-v) - 5\log(1+v) = 7\log X + \log_e c$$

$$\therefore \quad -2\log(1-v) - 5\log(1+v) - 7\log X = \log_e c$$

or $$2\log(1-v) + 5\log(1+v) + 7\log X = \log_e c$$

$$\log(1-v)^2 \cdot (1+v)^5 \cdot X^7 = \log_e c$$

Taking antilog, we get

$$(1-v)^2 \cdot (1+v)^5 \cdot X^7 = c$$

Now substituting for $v\left(=\dfrac{Y}{X}\right)$, we get

$$(1-v)^2 \cdot (1+v)^5 \cdot X^7 = c$$

$$\therefore \quad \frac{(X-Y)^2}{X^2} \cdot \frac{(X+Y)^5}{X^5} \cdot X^7 = c$$

$$\therefore \quad (X-Y)^2(X+Y)^5 = c \qquad (46)$$

But, $x = X + h = X + 1$ $\therefore$ $X = x - 1$
and $y = Y + k = Y + 0$ $\therefore$ $Y = y$ (since $k = 0$)
Therefore, the Equation (46) changes to

$$(x-1-y)^2 (x-1+y)^5 = c$$

or $$(x-y-1)^2 (x+y-1)^5 = c$$

This is the solution of the given differential equation.   Ans.

Exercise

Q. (1) Solve the following differential equations:

(a) $\sqrt{1+x^2}\, dy + \sqrt{1-y^2}\, dx = 0$

Ans. $\sin^{-1} y + \sin^{-1} x = c$

(b) $(\sin x + \cos x)dy + (\cos x - \sin x)dx = 0$

Ans. $y + \log|\sin x + \cos x| = c$

(c) $\frac{dy}{dx} = \sqrt{\frac{1-y^2}{1-x^2}}$ Also find the particular solution if $x = 0$ when $y = 1$.

Ans. (i) $\sin^{-1} y - \sin^{-1} x = c$ (ii) $\sin^{-1} y - \sin^{-1} x = \frac{\pi}{2}$

(d) $\frac{dy}{dx} = 2\,e^x y^3$, given that $y(0) = \dfrac{1}{2}$

Ans.   (i) $4\,e^x + \dfrac{1}{y^2} = c$ (ii) $4\,e^x + \dfrac{1}{y^2} = 8$

(e) Find the particular solution of the differential equation

$\frac{dy}{dx} - x^2 = x^2 y$, if $x = 0$ when $y = 2$

Ans. $\log \left| \frac{1+y}{3} \right| = \dfrac{x^3}{3}$

Q. (2) Solve the following differential equation using suitable substitution:

(a) $2\frac{dy}{dx} + \cos^2(x - 2y) = 1$

Ans. $\tan(x - 2y) = x + c$

(b) $\frac{dy}{dx} = (9x + y + 2)^2$

Ans. $9x + y + 2 = 2\tan(3x + c)$

(c) $x + y\frac{dy}{dx} = x^2 + y^2$

Ans. $x^2 + y^2 = c\,e^{2x}$

Q. (3) Solve the following differential equations using proper substitution:

(a) $x\,dy - y\,dx = \sqrt{x^2 + y^2}\,dx$

Ans. $y + \sqrt{x^2 + y^2} = cx^2$

(b) $\frac{dy}{dx} = \frac{x^2 - 2xy + 5y^2}{x^2 + 2xy + y^2}$

Ans. $\log|x - y| = \dfrac{2x^2}{(y-x)^2} + \dfrac{4x}{y-x} + c$

(c) $xy^2\,dy - (y^3 - 2x^3)dx = 0$

Ans. $\dfrac{y^3}{3x^3} + 2\log|x| + c$

Q. (4) Show that the following differential equations are homogeneous:

(i) $x\cos\left(\frac{y}{x}\right)\frac{dy}{dx} = y\cos\left(\frac{y}{x}\right) + x$

(ii) $(1 + 2\,e^{x/y})dx + 2\,e^{x/y}\left(1 - \frac{x}{y}\right)dy = 0$

(iii) $2y\,e^{x/y}\,dx + (y - 2x\,e^{x/y})dy = 0$

Q. (5) Solve the differential equation $2y\,e^{x/y}\,dx + (y - 2x\,e^{x/y})\,dy = 0$ and find its particular solution given that $x = 0$ when $y = 0$

Ans. $2\,e^{x/y} + \log|y| = 2$

Q. (6) Solve the differential equation $x \cos\left(\frac{y}{x}\right)\frac{dy}{dx} = y \cos\left(\frac{y}{x}\right) + x$

Ans. $\sin\left(\frac{y}{x}\right) = \log|cx|$

Q. (7) $\left(x \tan\frac{y}{x} - y \sec^2\frac{y}{x}\right)dx + x\sec^2\frac{y}{x}\,dy = 0$

Ans. $x \tan\frac{y}{x} = c$

Q. (8) $(2x - y)e^{x/y}\,dx + (y + x\,e^{x/y})dy = 0$

Ans. $y^2 + 2x^2 + e^{x/y} = c$

Q.(9) Solve the following differential equations:

(i) $(6x - 4y + 1)\frac{dy}{dx} = 3x - 2y + 1$

Ans. $4x - 8y + \log(12x - 8y + 1)$

(ii) $\frac{dy}{dx} = \frac{x + 2y + 1}{2x + y - 3}\left[\frac{a_1}{a_2} \neq \frac{b_1}{b_2}\right]$

Ans. $(x - y)^3 = c(x + y - 2)$

(iii) $(3x - 2y + 1)dx + (3x - 2y + 4)dy = 0 \left[\frac{a_1}{a_2} \neq \frac{b_1}{b_2}\right]$
Ans. $(x - y + 1)(x + y - 3)^5 = c$

(iv) $(2x + y + 1)dx + (4x + 2y - 1)dy = 0$
Ans. $x = 2y + \log(2x + y - 1) = c$

The procedure for obtaining the values of h and k from a pair of linear equations involved in the above problem is explained below for convenience.

It can be shown (and therefore important to remember) that the system of equations

$$a_1 h + b_1 k = c_1$$

$$a_2 h + b_2 k = c_2 (c_2 \neq 0)$$

has exactly one solution (i.e., one value for x and one value for y), if $\left[\frac{a_1}{a_2} \neq \frac{b_1}{b_2}\right]$, and it is given by

$$\frac{x}{b_1 c_2 - b_2 c_1} = \frac{y}{c_1 a_2 - c_2 a_1} = \frac{-1}{a_1 b_2 - a_2 b_1}$$

Now, proceed as follows:

(i) Write down the given linear equations, one below the other in the form

$$7h - 3k = 7$$

$$3h - 7k = 3$$

so that the constants on the right-hand side in both the equations are positive integers.

(ii) We write the coefficients in the pattern

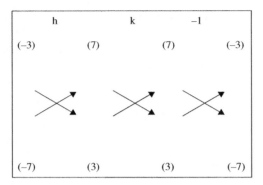

(iii) The arrows between two numbers indicate that they are to be multiplied and the second product is to be subtracted from the first.

(iv) We now write down the solution as follows:

$$\frac{h}{(-9) - (-49)} = \frac{k}{21 - 21} = \frac{-1}{(-49) - (-9)}$$

$$\therefore \quad h = \frac{-1(40)}{(-40)} = 1 \quad \text{and} \quad k = \frac{(-1)(0)}{(-40)} = 1 = 0$$

9b.3   LINEAR DIFFERENTIAL EQUATIONS

A differential equation of the form

$$\frac{dy}{dx} + Py = Q \tag{47}$$

where P and Q are *constants* or *functions of "x" only*, is known as a *first-order linear differential equation*.

Another form of the *first-order linear differential equation* is

$$\frac{dx}{dy} + P_1 y = Q_1 \tag{48}$$

where P_1 and Q_1 are *constants* or *functions of "y" only*.

Note: Equations (47) and (48) both are *standard form(s) of linear differential equation of order 1.*[7]
Observe that,

- In *both the standard forms*, the coefficients P and Q (or P_1 and Q_1) are functions *of the independent variable, or constants.*[8]
- Degree of the *dependent variable* and its *derivative* is *one.*
- The coefficient of the derivative (dy/dx) in the form (47) and that of (dx/dy) in the form (48) is one.

(These observations help us in identifying whether the differential equations are linear differential equations.)

Note: An equation of the type $R\dfrac{dy}{dx} + Sy = T$, where R, S, T are *functions of x* or *constants*, can be written in the standard form (47). We write

$$\frac{dy}{dx} + \frac{S}{R}y = \frac{T}{R}$$

Thus, the given equation represents *a linear differential equation.*

9b.3.1   The Method of Solving First-Order Linear Differential Equation

Consider the differential equation,

$$\frac{dy}{dx} + Py = Q \tag{49}^{(9)}$$

Multiply both sides of the Equation (49), by a function of "x" [say $g(x)$]. Thus, we obtain from Equation (49), the equation

$$g(x)\frac{dy}{dx} + P \cdot [g(x)] \cdot y = Q \cdot g(x) \tag{50}^{(10)}$$

Now, consider the left-hand side of Equation (50), *as if it is a derivative of some product of functions.* The first term on left-hand side of Equation (50) [i.e., $g(x)(dy/dx)$] suggests that the left-hand side *can be looked at as a derivative of the product* $g(x)y$. Thus, we choose to equate the left-hand side of (2) with the derivative of $g(x)y$. We write, $g(x)\frac{dy}{dx} + P \cdot g(x) \cdot y = g(x)\frac{dy}{dx} + y \cdot g'(x)$ [where the right-hand side is a derivative of $g(x)y$]

---

[7] A differential equation is said to be "linear" when the "dependent variable" and its "derivatives" appear only in the first degree.

In view of the above definition, the equation $\frac{d^2y}{dx^2} + P\frac{dy}{dx} + Qy = X$ is called a linear differential equation of the second order.
[8] Remember that in an expression of a derivative [i.e., $\frac{dy}{dx}$ or $\frac{dx}{dt}$ or $\frac{dv}{dt}$ or $\frac{dx}{dy}$, etc.], we always mean the derivative of "*the dependent variable*" with respect to "*the independent variable*". Thus, in the form (47), the dependent variable is y and the independent variable is x.
[9] In fact, we are going to develop a method for solving the first-order linear differential equation given at Equation (49).
[10] We have obtained Equation (50) by multiplying both sides of Equation (49) by some function of x. [Note that in the differential Equation (49), the *independent variable* is x.] Here, we have not assumed (or mentioned) any thing about the nature of $g(x)$. Accordingly, we are free to choose any suitable function $g(x)$, whenever needed.

On simplification, we get

$$Pg(x) = g'(x) \tag{51}$$

The above relation indicates the nature of the function g(x).

∴   From Equation (51), we obtain

$$P = \frac{g'(x)}{g(x)}$$

Integrating both sides w.r.t. x, we get

$$\int P\,dx = \int \frac{g'(x)}{g(x)}\,dx \quad \text{or} \quad \int P\,dx = \log_e |g(x)|$$

or

$$g(x) = e^{\int P\,dx} \tag{52}$$

Equation (52) suggests that $g(x)$ must be equal to $e^{\int P\,dx}$.

Now, it is clear that if we multiply the Equation (49) by $g(x) = e^{\int P\,dx}$, then the left-hand side becomes the *derivative of the product* $y \cdot g(x)$. Therefore, after multiplying by $e^{\int P\,dx}$ and then integrating both sides, the integral of the left-hand side will obviously be $y \cdot g(x)$. (In fact, we can write this product without any formality.) The problem then reduces to finding the integral of $Q \cdot e^{\int P\,dx}$, on the right-hand side.

The function $g(x) = e^{\int P\,dx}$ is called *the integrating factor* of the given differential equation, for obvious reason.

Substituting the value of $g(x) \left[= e^{\int P\,dx} \right]$ in Equation (50), we get

$$e^{\int P\,dx} \cdot \frac{dy}{dx} + P \cdot e^{\int P\,dx} y = Q \cdot e^{\int P\,dx} \quad \text{or} \quad \frac{d}{dx}\left(y \cdot e^{\int P\,dx} \right) = Q \cdot e^{\int P\,dx}$$

Integrating both sides w.r.t. x, we get

$$y \cdot e^{\int P\,dx} = \int \left(Q \cdot e^{\int P\,dx} \right) dx + c \tag{11}$$

which is the *general solution* of the differential Equation (49), c being an arbitrary constant. Now, we list below the steps to solve first-order linear differential equations.

9b.3.2   Steps Involved to Solve First-Order Linear Differential Equations

(I) Write the given differential equation in the (standard) form

$\frac{dy}{dx} + Py = Q$, where P and Q are *constants or functions of x only.*

(II) Find the integrating factor $= e^{\int P\,dx}$

---

[11] Gott fried Wilhelm Leibniz (1646–1716) appears to have been the first who obtained this solution. Recall that Leibniz is known to have invented differential Calculus independently of Newton. [*Introductory Course in Differential Equations* by Danial A. Murray, Longmans Green and Co.]

(III) Write the solution of the given differential equation as

$$y \cdot \text{IF} = \int (Q \times \text{IF}) dx$$

Note: In case, the first-order linear differential equation is in the form $\frac{dx}{dy} + P_1 x = Q_1$, where P_1, Q_1 are *constants*, or *functions of* y *only*, then $\text{IF} = e^{\int P_1 \, dy}$ and the *solution of the differential equation* is given by

$$x \cdot (\text{IF}) = \int (Q \times \text{IF}) dy + cx$$

Remark: A linear differential equation in the standard form

$$\frac{dy}{dx} + P \cdot y = Q \tag{53}$$

where P and Q are functions of "x", is (in fact) an equation in the form

$$\frac{dy}{dx} + F(x) \cdot y = H(x) \tag{I}$$

so that $F(x) = P$ and $H(x) = Q$, provided the coefficient of the derivative dy/dx is unity. Then, we evaluate $\int F(x) \, dx \left[= \int P \, dx \right]$ and obtain the integrating factor $e^{\int F(x) dx}$. [The important point to be remembered is that $F(x)$ is the coefficient of the dependent variable "y" in (I)]. Similarly, the second standard form of a linear differential equation

$$\frac{dx}{dy} + P_1 x = Q_1 \tag{54}$$

where P_1 and Q_1 are functions of "y" is an equation in the form

$$\frac{dx}{dy} + f(y)x = h(y) \tag{II}$$

so that $f(y) \, (= P_1)$ and $h(y) \, (= Q_1)$, provided the coefficient of the derivative is unity.

Then, to find the integrating factor, we evaluate $\int f(y) dy [= P_1 \, dy]$ and obtain the integrating factor $e^{\int f(y) dy}$. [Again, it must be remembered that $f(y)$ is the coefficient of the dependent variable "x" in (II).]

Now, we are in a position to write down the important steps for solving problems on linear differential equations.

(A) Make the coefficient of $\frac{dy}{dx}$ or $\left(\frac{dx}{dy} \right)$ as unity.

(B) Find P (or P_1) and hence the integrating factor (IF).

(C) Write the *general solution* according to the formula.

The following solved examples will make the situation clear.

Example (19): Find the general solution of the differential equation

$$x\frac{dy}{dx} + 2y = x^2 \quad (x \neq 0)$$

Solution: The given differential equation is

$$x\frac{dy}{dx} + 2y = x^2. \tag{55}$$

Dividing both sides of Equation (55) by x, we get

$$\frac{dy}{dx} + \frac{2}{x}y = x. \tag{56}$$

This is the linear differential equation of the (standard) form: $\frac{dy}{dx} + py = Q$
where $p = \frac{2}{x}$ and $Q = x$.
Therefore, IF $= e^{\int \frac{2}{x}dx}$
Now $\int \frac{2}{x}dx = 2\log_e x = \log x^2$

$$\therefore \quad \text{IF} = e^{\log_e x^2} = x^2 \left[\because e^{\log_e f(x)} = f(x) \right]$$

Therefore, solution of the given equation is given by

$$y \cdot x^2 = \int x \cdot x^2 \, dx = \int x^3 \, dx + c$$

or
$$y \cdot x^2 = \frac{x^4}{4} + c$$

This is the general solution of the given differential equation.   Ans.

Note: The above solution may also be written as

$$4y \cdot x^2 = x^4 + 4c_1 = x^4 + c$$

or we may write it (by dividing both sided by x^2) as

$$y = \frac{x^2}{4} + cx^{-2}$$

Example (20): Find the general solution of the differential equation

$$\frac{dy}{dx} - y = \cos x$$

Solution: The given differential equation is

$$\frac{dy}{dx} - y = \cos x \tag{57}$$

It is of the (standard) form $\frac{dy}{dx} + Py = Q$, where $P = -1$ and $Q = \cos x$.

$$\therefore \quad IF = e^{\int(-1)dx}$$

But
$$\int -dx = -x$$

$$\therefore \quad IF = e^{-x}$$

Multiplying both sides of Equation (57) by IF, we get

$$e^{-x}\frac{dy}{dx} - e^{-x}y = e^{-x}\cos x$$

or
$$\frac{d}{dx}(e^{-x}y) = e^{-x}\cos x$$

Integrating both sides w.r.t. x, we get

$$e^{-x}y = \int e^{-x}\cos x\, dx + c \tag{58}$$

(Now we have to evaluate the integral $\int e^{-x}\cos x\, dx$, by the method of parts.)
    Let $I = \int e^{-x}\cos x\, dx$

$$\left[\int e^{-x} = \left(\frac{e^{-x}}{-1}\right) = -e^{-x}, \quad \frac{d}{dx}(\cos x) = (-\sin x)\right]$$

or
$$I = \int(\cos x)(e^{-x})dx = \cos x\left(\frac{e^{-x}}{-1}\right) - \int(-\sin x)\cdot(-e^{-x})dx$$

$$= -\cos x\cdot e^{-x} - \int(\sin x)\cdot(e^{-x})dx$$

$$= -\cos x\cdot e^{-x} - \left[\sin x\cdot(-e^{-x}) - \int\cos x(-e^{-x})dx\right]$$

$$\therefore \quad I = -\cos x\cdot e^{-x} + \sin x\cdot e^{-x} - \int\cos x\cdot e^{-x}\, dx$$

or
$$I = -\cos x\cdot e^{-x} + \sin x\cdot e^{-x} - I$$

or
$$2I = e^{-x}(\sin x - \cos x)$$

$$\therefore \quad I = \left(\frac{\sin x - \cos x}{2}\right)e^{-x}$$

Substituting the value of I in Equation (58), we get

$$y\cdot e^{-x} = \left(\frac{\sin x - \cos x}{2}\right)e^{-x} + c$$

$$\text{or} \quad y = \left(\frac{\sin x - \cos x}{2}\right) + c \cdot e^x$$

where c is an arbitrary constant. This is the general solution of the given differential equation.

Example (21): Solve the equation $\cos x \frac{dy}{dx} + y \sin x = 1$

Solution: The given equation is

$$\cos x \frac{dy}{dx} + y \sin x = 1$$

Dividing by $\cos x$, we get

$$\frac{dy}{dx} + y \tan x = \sec x$$

This is a linear differential equation of the type

$$\frac{dy}{dx} + py = Q$$

$$\therefore \quad P = \tan x$$

$$\int P \, dx = \int \tan x \, dx = \log|\sec x|$$

$$\text{IF} = e^{\int P \, dx} = e^{\log|\sec x|} = \sec x$$

$\therefore$ The solution is given by

$$y \cdot \sec x = \int (\sec x)(\sec x) dx$$

$$= \int \sec^2 x \, dx = \tan x + c$$

$\therefore$ The solution is
$$y \sec x = \tan x + c$$

or $y = \tan x \cos x + c \cos x$, where c is an arbitrary constant

Example (22): Solve $\frac{dy}{dx} + 2y \cot x = 3x^2 \csc^2 x$

Solution: Here the coefficient of y is $2 \cot x$. Hence the integrating factor is

$$e^{\int 2 \cot x \, dx} = e^{2 \log|\sin x|} = e^{\log(\sin x)^2} = e^{\log \sin^2 x} = \sin^2 x \cdot [e^{\log f(x)} = f(x)]$$

$\therefore$ Solution of the given differential equation is

$$y \cdot \sin 2x = \int (3x^2 \cosec^2 x)\sin^2 x \, dx + c$$

$$= \int (3x^2 \, dx + c$$

$$= 3 \cdot \frac{x^3}{3} + c$$

$$= x^3 + c \qquad \text{Ans.}$$

Example (23):   Solve $(1 + x^3)\dfrac{dy}{dx} + 6x^2 y = 1 + x^2$

Solution: The given differential equation is

$$(1 + x^3)\frac{dy}{dx} + 6x^2 y = 1 + x^2$$

To make the coefficient of $\frac{dy}{dx}$ unity, we divide both sides by $(1 + x^3)$. We get

$$\frac{dy}{dx} + \frac{6x^2}{1 + x^3} y = \frac{1 + x^2}{1 + x^3}$$

Here $P = \dfrac{6x^2}{1 + x^3}$

Let $1 + x^3 = t$

$$\therefore \quad 3x^2 \, dx = dt$$

$$\therefore \quad 6x^2 \, dx = 2dt$$

$$\therefore \quad I = \int \frac{2 \, dt}{t} = 2 \log t = 2 \log(1 + x^3)$$

$\therefore$ Integrating factor is

$$e^{\int P \, dx} = e^{2 \log(1+x^3)} = e^{\log(1+x^3)^2} = (1 + x^3)^2$$

$\therefore$ The solution is given by

$$y \cdot (1 + x^3)^2 = \int \frac{(1 + x^2)}{(1 + x^3)}(1 + x^3)^2 dx = \int (1 + x^2)(1 + x^3)dx$$

$$\therefore \quad y \cdot (1 + x^3)^2 = \int (1 + x^2)(1 + x^3)dx + c$$

$$= \int (1 + x^2 + x^3 + x^5)dx + c$$

$$= x + \frac{x^3}{3} + \frac{x^4}{4} + \frac{x^6}{6} + c \qquad \text{Ans.}$$

This is the general solution.

Example (24): Solve $y\,e^y\,dx = (y^3 + 2x\,e^y)dy$

Solution: The given equation can be written as

$$\frac{dx}{dy} = \frac{dx}{dy} + P_1 x = Q_1$$

$$\frac{dx}{dy} = y^2\,e^{-y}\,\frac{2x}{y}$$

$$\frac{dx}{dy} - \frac{2}{y}\cdot x = y^2\,e - y$$

This is a linear differential equation of the type

$$\frac{dx}{dy} + P_1 x = Q_1$$

where $P_1 = -\frac{2}{y}$ and $Q_1 = y^2\,e^{-y}$

$$\int P_1\,dy = \int \frac{-2}{y}dy = -2\log y = \log\frac{1}{y^2}$$

$\therefore$ Integrating factor is $e^{\log(1/y^2)} = \dfrac{1}{y^2}$

$\therefore$ The solution is

$$x\cdot\frac{1}{y^2} = \int y^2\,e^{-y}\frac{1}{y^2}\,dy + c = e^{-y}\,dy + c = \frac{e^{-y}}{-1} + c = -e - y + c$$

or
$$x = -y^2\,e^{-y} + cy^2 \qquad \text{Ans.}$$

This is the general solution of the given differential equation.

Exercise

Solve the following differential equations:

Q. (1) $(x^2 + 1)^3\frac{dy}{dx} + 4x\,(x^2 + 1)^2 y = 1$

Ans. $y(x^2 + 1)^2 = \tan^{-1}x + c$

Q. (2) $x^2\frac{dy}{dx} = 3x^2 = y + 1$

Ans. $y = \frac{c}{x^2} + x + \frac{1}{x}$

Q. (3) $y\,e^y\,dx = y^3 + (2x\,e^y)dy$

Ans. $x = -y^2\,e^{-y} + cy^2$

Q. (4) $x\frac{dy}{dx} + 2y = x^2\ (x \neq 0)$

Ans. $y = \frac{x^2}{4} + cx^{-2}$

Q. (5) $y\,dx - (x + 2y^2)dy = 0$

Ans. $x = 2y^2 + cy$

9b.4 TYPE III: EXACT DIFFERENTIAL EQUATIONS

Definition: An *exact differential* equation of the *first order* is that equation which is *obtained from its general solution* by mere *differentiation* and *without any additional process of elimination* or *reduction*.

Example (25): Consider the equation,

$$x^3 y^4 = c \tag{59}$$

which is the *general solution* of some differential equation.

On differentiating Equation (59), both sides, we get

$$3x^2 y^4 dx + 4x^3 y^3\,dy = 0 \tag{60}$$

Equation (60) *in this form* is *an exact differential equation* whose solution is at Equation (59).
There is *another way* in which we can understand an exact differential equation.
From Equation (59), we may write $x^3 y^4 - c = 0$. Obviously, the left-hand side is a function of two variables x and y. Let us denote this function by u. Then, we have

$$x^3 y^4 - c = u$$

Differentiating the above equation (w.r.t. to x and y), we get

$$3x^2 y^4\,dx + 4x^3 y^3 = du \tag{61}$$

The expression $3x^2 y^4\,dx + 4x^3 y^3\,dy$ is called an *exact differential* of $x^3 y^4$[12]
Comparing the expressions on left-hand side of Equations (60) and (61) we note that an equation in the form,

$$M\,dx + N\,dy = 0$$

(where M and N are functions of x and y) will be an *exact differential equation* if there be some function u (of x and y), such that

$$M\,dx + N\,dy = du$$

---

[12] Recall that, if $y = f(x)$, then $dy = f'(x)dx$. Similarly, if $u = f(x, y)$, then $du = f'(x, y)dx + f'(x, y)dy$. The expression $f'(x, y)dx$ is called the partial differential of u with respect to x, and similarly $f'(x, y)dy$ is called the partial differential of u with respect to y. [Symbolically, we write $f'(x, y)dx$ as $\frac{\partial u}{\partial x}dx$ and $f'(x, y)dy$ as $\frac{\partial u}{\partial y}dy$.]

Note: The differential Equation (60) can be simplified to

$$3y\,dx + 4y\,dy = 0 \tag{62}$$

However, Equation (62) is *not an exact differential equation*, by definition. Methods of finding solution(s) of exact differential equations are quite interesting. The first requirement is to check whether the given equation is an exact differential equation or whether it can be converted to that form. It can be shown that the *condition of exactness* for an equation, in the form

$$M\,dx + N\,dy = 0$$

is that $\frac{\partial M}{\partial y}$ must be equal to $\frac{\partial N}{\partial x}$, that is, each should be equal to $\frac{\partial^2 u}{\partial x \cdot \partial y}$.

At this point, we put to an end, the discussion about exact differential equations and the methods of their solution(s). At most, it may be mentioned that depending on the given equation, there are *Rules* for finding the integrating factor(s) that help in finding the solution(s) of exact differential equations.

9b.5 APPLICATIONS OF DIFFERENTIAL EQUATIONS

Differential equations find many applications in Engineering (particularly in mechanics) and other sciences. We have already discussed some important applications of differential equations of first order and first degree in Chapter 13a of Part I.

INDEX

Introduction to Integral Calculus: Systematic Studies with Engineering Applications for Beginners, First Edition.
Ulrich L. Rohde, G. C. Jain, Ajay K. Poddar, and A. K. Ghosh.
© 2012 John Wiley & Sons, Inc. Published 2012 by John Wiley & Sons, Inc.

Printed and bound by CPI Group (UK) Ltd, Croydon, CR0 4YY
01/09/2022
03145090-0002